SWARM INTELLIGENCE

SWARM INTELLIGENCE

From Natural to Artificial Systems

Eric Bonabeau
Sante Fe Institute
Sante Fe, New Mexico USA

Marco Dorigo
FNRS
Belgium

Guy Theraulaz
CNRS
France

Sante Fe Institute
Studies in the Sciences of Complexity

New York Oxford
Oxford University Press
1999

Oxford University Press

Oxford New York
Athens Auckland Bangkok Bogotá Buenos Aires Calcutta
Cape Town Chennai Dar es Salaam Delhi Florence Hong Kong Istanbul
Karachi Kuala Lumpur Madrid Melbourne Mexico City Mumbai
Nairobi Paris São Paulo Singapore Taipei Tokyo Toronto Warsaw

and associated companies in
Berlin Ibadan

Published by Oxford University Press, Inc.
198 Madison Avenue, New York, New York 10016

Library of Congress Cataloging-in-Publication Data
Bonabeau, Eric.
Swarm intelligence : from natural to artificial intelligence / by
Eric Bonabeau, Marco Dorigo, Guy Theraulaz.
p. cm.
Includes bibliography references and index.
ISBN 0-19-513158-4 (cloth); ISBN 0-19-513159-2 (pbk.)
1. Artificial intelligence. 2. Insect societies. 3. Insects—Psychology.
4. Biological models. I. Theraulaz, Guy.
II. Dorigo, Margo. III. Title.
Q335.B59 1999
006.3—dc21 98-49821

9 8
Printed in the United States of America
on acid-free paper

About the Santa Fe Institute

The *Santa Fe Institute* (SFI) is a private, independent, multidisciplinary research and education center, founded in 1984. Since its founding, SFI has devoted itself to creating a new kind of scientific research community, pursuing emerging science. Operating as a small, visiting institution, SFI seeks to catalyze new collaborative, multidisciplinary projects that break down the barriers between the traditional disciplines, to spread its ideas and methodologies to other individuals, and to encourage the practical applications of its results.

All titles from the *Santa Fe Institute Studies in the Sciences of Complexity* series will carry this imprint which is based on a Mimbres pottery design (circa A.D. 950–1150), drawn by Betsy Jones. The design was selected because the radiating feathers are evocative of the outreach of the Santa Fe Institute Program to many disciplines and institutions.

Dedications

To Croque.

—Eric Bonabeau

To Laura and Luca for sharing with me their lives.

—Marco Dorigo

To Sylvie.

—Guy Theraulaz

Contents

Preface

The social insect metaphor for solving problems has become a hot topic in the last five years. This approach emphasizes distributedness, direct or indirect interactions among relatively simple agents, flexibility, and robustness. The number of its successful applications is exponentially growing in combinatorial optimization, communications networks, and robotics. More and more researchers are interested in this new exciting way of achieving a form of artificial intelligence, swarm intelligence—the emergent collective intelligence of groups of simple agents. Researchers have good reasons to find swarm intelligence appealing: at a time when the world is becoming so complex that no single human being can understand it, when information (and not the lack of it) is threatening our lives, when software systems become so intractable that they can no longer be controlled, swarm intelligence offers an alternative way of designing "intelligent" systems, in which autonomy, emergence, and distributed functioning replace control, preprogramming, and centralization. But the field of swarm intelligence relies to a large extent on fragmented knowledge and no attempt had been made at putting all the pieces together.

Therefore, the timing seemed just perfect to summarize for the first time the advances made in, and provide an organized survey of, this new field. Not only do we now have enough material to justify the publication of a book on swarm intelligence, but the growing interest of many researchers and the potential usefulness of explicitly connecting the functioning principles of social insect colonies to the design principles of artificial systems were also important factors in our decision to undertake the writing of this book.

Each chapter is organized around a biological example (foraging, division of labor, corpse clustering, larval sorting, nest building, and cooperative transport): the example is described, modeled (when a model exists), and used as a metaphor to design an algorithm, a multiagent system, or a group of robots. Although the resulting book is rather unusual since biological modeling and engineering are equally important, hopefully the reader will find our work enjoyable and useful.

Of course, this book would not exist without the help and care of many people. We wish to thank Ronald C. Arkin, Hajime Asama, John S. Bay, Scott Camazine, Jean-Louis Deneubourg, Pablo Funes, Sylvain Guérin, Florian Hénaux, Alain Hertz, C. Ronald Kube, Pascale Kuntz, Alcherio Martinoli, Jordan Pollack, Andrew Russell, Dominique Snyers, and Hong Zhang for providing us with valuable articles, sharing their results prior to publication, or communicating their enthusiasm. Special thanks are due to Bernd Bullnheimer, Daniele Costa, Gianni Di Caro, Thomas Stützle, and Marco Wiering for reading and commenting draft versions of the book; to Erica Jen and the Santa Fe Institute for their support; to all the researchers of the IRIDIA laboratory for their friendship; to Philippe Smets and Hugues Bersini, for supporting Marco Dorigo's research; to Greg Chirikjian, Nigel R. Franks, Pablo Funes, Owen Holland, Kazuo Hosokawa, C. Ronald Kube, Alcherio Martinoli, Jordan Pollack, and Eiichi Yoshida for help with figures and much more; and, last but certainly not least, to Jean-Louis Deneubourg for being our source of inspiration. Marco Dorigo wishes to thank Alberto Colorni and Vittorio Maniezzo, who, while he was a graduate student at Politecnico di Milano, started with him the work on ant colony optimization, and Luca Gambardella and Gianni Di Caro, who have had the faith to follow his adventure and greatly contributed with their fantasy and programming skills to the success of part of the research presented in this book. Marco Dorigo acknowledges the support received from all the members of the Artificial Intelligence and Robotics Project of Politecnico di Milano at the time he started his research on ant colony optimization, particularly from Marco Somalvico, Andrea Bonarini, and colleague and friend Marco Colombetti.

In addition, Eric Bonabeau is grateful to the Santa Fe Institute for supporting his work through the Interval Research Fellowship. Eric Bonabeau and Guy Theraulaz have benefited from a grant from the GIS (Groupement d'Intérêt Scientifique) Sciences de la Cognition. Guy Theraulaz is a Chargé de Recherche at CNRS (Centre National pour la Recherche Scientifique). Marco Dorigo is a Chercheur Qualifié at the FNRS, the Belgian Fund for Scientific Research.

Finally, this book would not be worth anything without the work, help, and care (or gentle pressure?) of Ronda Butler-Villa, Marylee McInnes, and Della Ulibarri.

—Eric Bonabeau, Marco Dorigo, and Guy Theraulaz

CHAPTER 1
Introduction

1.1 SOCIAL INSECTS

Insects that live in colonies, ants, bees, wasps,[1] and termites, have fascinated naturalists as well as poets for many years. "What is it that governs here? What is it that issues orders, foresees the future, elaborates plans, and preserves equilibrium?," wrote Maeterlinck [230]. These, indeed, are puzzling questions. Every single insect in a social insect colony seems to have its own agenda, and yet an insect colony looks so organized. The seamless integration of all individual activities does not seem to require any supervisor.

For example, Leafcutter ants (*Atta*) cut leaves from plants and trees to grow fungi (Figure 1.1). Workers forage for leaves hundreds of meters away from the nest, literally organizing highways to and from their foraging sites [174].

Weaver ant (*Oecophylla*) workers form chains of their own bodies, allowing them to cross wide gaps and pull stiff leaf edges together to form a nest (Figure 1.2). Several chains can join to form a bigger one over which workers run back and forth. Such chains create enough force to pull leaf edges together. When the leaves are in place, the ants connect both edges with a continuous thread of silk emitted by a mature larva held by a worker (Figure 1.3) [172, 174].

[1] Only a fraction of all species of bees and wasps are social: most are solitary.

FIGURE 1.1 Leafcutter ants (*Atta*) bringing back cut leaves to the nest.

FIGURE 1.2 Chains of *Oecophylla longinoda*.

In their moving phase, army ants (such as *Eciton*) organize impressive hunting raids, involving up to 200,000 workers, during which they collect thousands of prey (see chapter 2, section 2.2.3) [52, 269, 282].

In a social insect colony, a worker usually does not perform all tasks, but rather specializes in a set of tasks, according to its morphology, age, or chance. This division of labor among nestmates, whereby different activities are performed simultaneously by groups of specialized individuals, is believed to be more efficient than if tasks were performed sequentially by unspecialized individuals [188, 272].

In polymorphic species of ants, two (or more) physically different types of workers coexist. For example, in *Pheidole* species, minor workers are smaller and morphologically distinct from major workers. Minors and majors tend to perform differ-

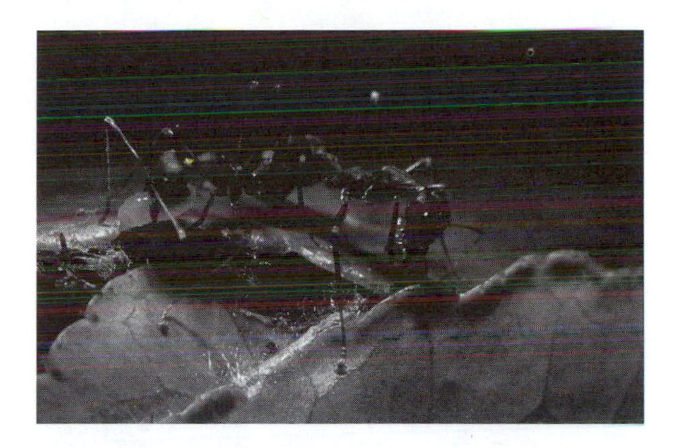

FIGURE 1.3 Two workers holding a larva in their mandibles.

ent tasks: whereas majors cut large prey with their large mandibles or defend the nest, minors feed the brood or clean the nest. Removal of minor workers stimulates major workers into performing tasks usually carried out by minors (see chapter 3, section 3.2) [330]. This replacement takes place within two hours of minor removal. More generally, it has been observed in many species of insects that removal of a class of workers is quickly compensated for by other workers: division of labor exhibits a high degree of plasticity.

Honey bees (*Apis mellifica*) build series of parallel combs by forming chains that induce a local increase in temperature. The wax combs can be more easily shaped thanks to this temperature increase [82]. With the combined forces of individuals in the chains, wax combs can be untwisted and be made parallel to one another. Each comb is organized in concentric rings of brood, pollen, and honey. Food sources are exploited according to their quality and distance from the hive (see section 1.2.2). At certain times, a honey bee colony divides: the queen and approximately half of the workers leave the hive in a swarm and first form a cluster on the branch of a nearby tree (Figure 1.4). Potential nesting sites are carefully explored by scouts. The selection of the nesting site can take up to several days, during which the swarm precisely regulates its temperature [166].

Nest construction in the wasp *Polybia occidentalis* involves three groups of workers, pulp foragers, water foragers, and builders. The size of each group is regulated according to colony needs through some flow of information among them (Figure 1.5) [189].

Tropical wasps (for example, *Parachartergus, Epipona*) build complex nests, comprised of a series of horizontal combs protected by an external envelope and connected to each other by a peripheral or central entrance hole (Figure 1.6) [187].

Some species of termites (*Macrotermes*) build even more complex nests (Figure 1.7), comprised of roughly cone-shaped outer walls that often have conspicuous

FIGURE 1.4 A swarm of honey bees *Apis mellifica.*

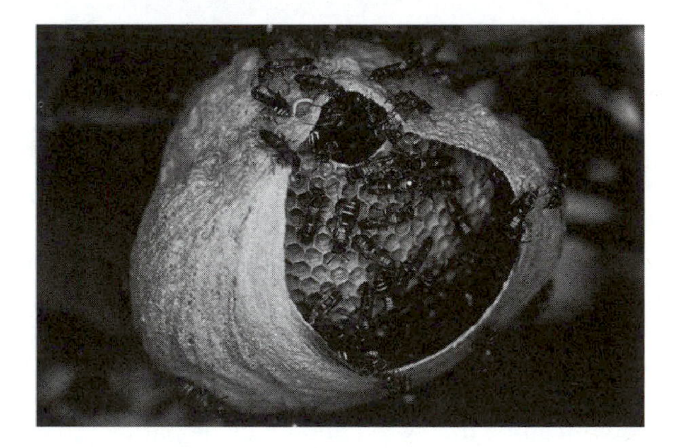

FIGURE 1.5 Nest building in the neotropical wasp *Polybia occidentalis.*

ribs containing ventilation ducts which run from the base of the mound toward its summit, brood chambers within the central "hive" area, which consists of thin horizontal lamellae supported by pillars, a base plate with spiral cooling vents, a royal chamber, which is a thick-walled protective bunker with a few minute holes in its walls through which workers can pass, fungus gardens, draped around the hive and consisting of special galleries or combs that lie between the inner hive and the outer walls, and, finally, peripheral galleries constructed both above and below ground which connect the mound to its foraging sites [227, 228].

And there are many more examples of the impressive capabilities of social insects.[2] If no one is in charge, how can one explain the complexity and sophistication

[2]Many more examples are discussed in greater detail in Camazine et al. [60].

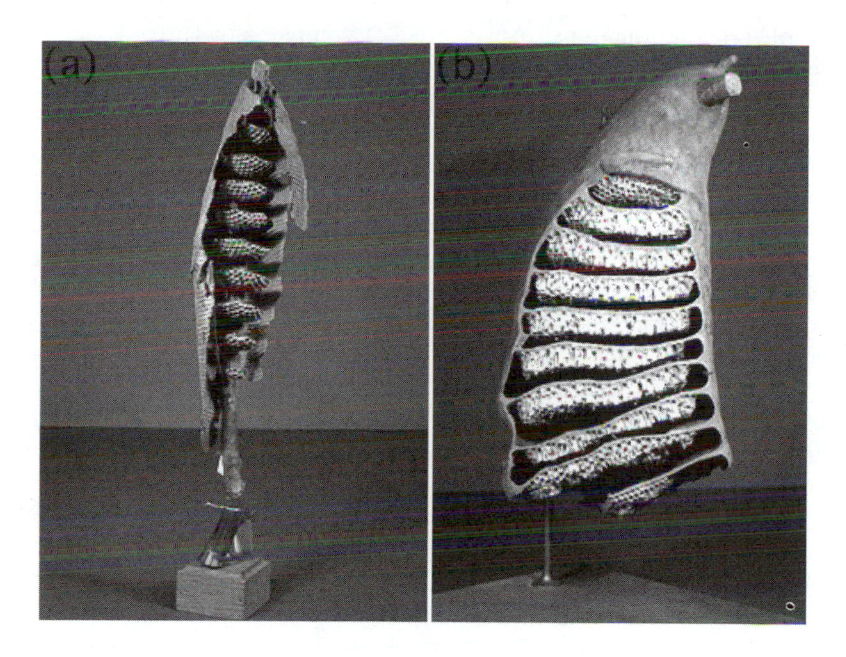

FIGURE 1.6 (a) Nest of a *Parachartergus* wasp species. (b) Nest of the wasp *Epipona tatua*.

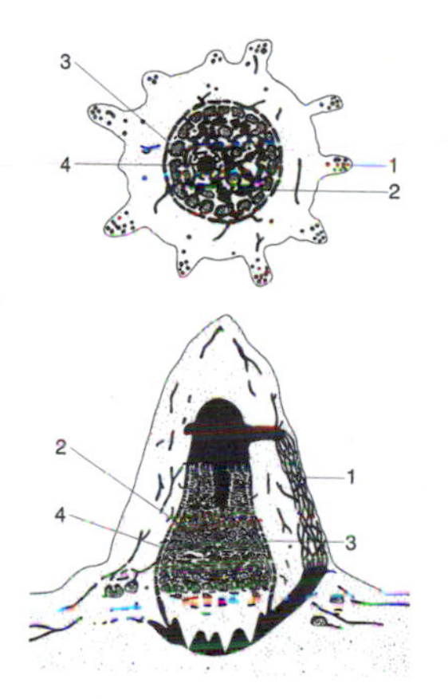

FIGURE 1.7 Cross section of a *Macrotermes* mound. (1) Walls containing ventilation ducts. (2) Brood chambers. (3) Base plate. (4) Royal chamber. After Lüscher [228]. Reprinted by permission © John L. Howard.

of their collective productions? An insect is a complex creature: it can process a lot of sensory inputs, modulate its behavior according to many stimuli, including interactions with nestmates, and make decisions on the basis of a large amount of information. Yet, the complexity of an individual insect is still not sufficient to explain the complexity of what social insect colonies can do. Perhaps the most difficult question is how to connect individual behavior with collective performance? In other words, how does cooperation arise?

Some of the mechanisms underlying cooperation are genetically determined: for instance, anatomical differences between individuals, such as the differences between minors and majors in polymorphic species of ants, can organize the division of labor. But many aspects of the collective activities of social insects are *self-organized.* Theories of self-organization (SO) [164, 248], originally developed in the context of physics and chemistry to describe the emergence of macroscopic patterns out of processes and interactions defined at the microscopic level, can be extended to social insects to show that complex collective behavior may emerge from interactions among individuals that exhibit simple behavior: in these cases, there is no need to invoke individual complexity to explain complex collective behavior. Recent research shows that SO is indeed a major component of a wide range of collective phenomena in social insects [86].

Models based on SO do not preclude individual complexity: they show that *at some level of description* it is possible to explain complex collective behavior by assuming that insects are relatively simple interacting entities. For example, flying from a food source back to the nest involves a set of complex sensorimotor mechanisms; but when one is interested in collective food source selection in honey bees (see section 1.2.2), the primitives that one may use in a model need not go into the detail of the implementation of flight, and flying back to the nest will be considered a simple behavior, although at the neurobiological level of description, flying back to the nest is certainly not simple. Nor does one need to use a model based on elementary particles to describe the aerodynamics of airplanes. Moreover, models based on SO are logically sound: they assume that it might be possible to explain something apparently complex in terms of simple interacting processes. If such an explanation is possible, then why does it have to be more complex? It is only when such an explanation fails that more complex assumptions will be put into the model.

The discovery that SO may be at work in social insects not only has consequences on the study of social insects, but also provides us with powerful tools to transfer knowledge about social insects to the field of *intelligent system design.* In effect, a social insect colony is undoubtedly a decentralized problem-solving system, comprised of many relatively simple interacting entities. The daily problems solved by a colony include finding food, building or extending a nest, efficiently dividing labor among individuals, efficiently feeding the brood, responding to external challenges, spreading alarm, etc. Many of these problems have counterparts in engineering and computer science. One of the most important features of social insects is that they can solve these problems in a very flexible and robust way: flexibility allows adaptation to changing environments, while robustness endows the colony with the ability to function even though some individuals may fail to perform their tasks. Finally, social insects have limited cognitive abilities: it is, therefore, simple to design agents, including robotic agents, that mimic their behavior at some level of description.

In short, the modeling of social insects by means of SO can help design artificial distributed problem-solving devices that self-organize to solve problems—swarm-intelligent systems. It is, however, fair to say that very few applications of swarm intelligence have been developed. One of the main reasons for this relative lack of success resides in the fact that swarm-intelligent systems are hard to "program," because the paths to problem solving are not predefined but emergent in these systems and result from interactions among individuals and between individuals and their environment as much as from the behaviors of the individuals themselves. Therefore, using a swarm-intelligent system to solve a problem requires a thorough knowledge not only of what individual behaviors must be implemented but also of what interactions are needed to produce such or such global behavior.

One possible idea is to develop a catalog of all the collective behaviors that can be generated with simple interacting agents: such a catalog would be extremely useful in establishing clear connections between artificial swarms and what they can achieve, but this could be a boring and endless undertaking. Another, somewhat more reasonable, path consists of studying how social insects collectively perform some specific tasks, modeling their behavior, and using the model as a basis upon which artificial variations can be developed, either by tuning the model parameters beyond the biologically relevant range or by adding nonbiological features to the model. The aim of this book is to present a few models of natural swarm intelligence and how they can be transformed into useful artificial swarm-intelligent devices.

This approach is similar to other approaches that imitate the way "nature" (that is, physical or biological systems) solves problems [37, 121]. Another possible, and apparently very promising, pathway is to use economic or financial metaphors to solve problems [182, 185, 196].

The expression "swarm intelligence" was first used by Beni, Hackwood, and Wang [13, 14, 15, 16, 161, 162] in the context of cellular robotic systems, where many simple agents occupy one- or two-dimensional environments to generate patterns and self-organize through nearest-neighbor interactions. Using the expression "swarm intelligence" to describe only this work seems unnecessarily restrictive: that is why we extend its definition to include any attempt to design algorithms or distributed problem-solving devices inspired by the collective behavior of social insect colonies and other animal societies. And, strangely enough, this definition only marginally covers work on cellular robotic systems, which does not borrow a lot from social insect behavior.

From an historical perspective [2], the idea of using collections of simple agents or automata to solve problems of optimization and control on graphs, lattices, and networks was already present in the works of Butrimenko [53], Tsetlin [318], Stefanyuk [292], and Rabin [79, 265]. The latter introduced moving automata that solve problems on graphs and lattices by interacting with the consequences of their previous actions [2]. Tsetlin [318] identified the important characteristics of biologically-inspired automata that make the swarm-based approach potentially powerful, randomness, decentralization, indirect interactions among agents, and self-organization. Butrimenko [53] applied these ideas to the control of telecommunications networks, and Stefanyuk [292] to cooperation between radio stations.

1.2 MODELING COLLECTIVE BEHAVIOR IN SOCIAL INSECTS

1.2.1 MODELING AND DESIGNING

The emphasis of this book is how to design adaptive, decentralized, flexible, and robust artificial systems, capable of solving problems, inspired by social insects. But one necessary first step toward this goal is understanding the mechanisms that generate collective behavior in insects. This is where modeling plays a role. Modeling is very different from designing an artificial system, because in modeling one tries to uncover what actually happens in the natural system—here an insect colony. Not only should a model reproduce some features of the natural system it is supposed to describe, but its formulation should also be consistent with what is known about the considered natural system: parameters cannot take arbitrary values, and the mechanisms and structures of the model must have some biological plausibility. Furthermore, the model should make testable predictions, and ideally all variables and parameters should be accessible to experiment.

Without going too deeply into epistemology, it is clear that an engineer with a problem to solve does not have to be concerned with biological plausibility: efficiency, flexibility, robustness, and cost are possible criteria that an engineer could use. Although natural selection may have picked those biological organizations that are most "efficient" (in performing all the tasks necessary to survival and reproduction), "flexible," or "robust," the constraints of evolution are not those of an engineer: if evolution can be seen as essentially a tinkering process, whereby new structures arise from, and are constrained by, older ones, an engineer can and must resort to whatever available techniques are appropriate. However, the remarkable success of social insects (they have been colonizing a large portion of the world for several million years) can serve as a starting point for new metaphors in engineering and computer science.

The same type of approach has been used to design artificial neural networks that solve problems, or in the development of genetic algorithms for optimization: if the brain and evolution, respectively, served as starting metaphors, most examples of neural networks and genetic algorithms in the context of engineering are strongly decoupled from their underlying metaphors. In these examples, some basic principles of brain function or of evolutionary processes are still present and are most important, but, again, ultimately a good problem-solving device does not have to be biologically relevant.

1.2.2 SELF-ORGANIZATION IN SOCIAL INSECTS

Self-organization is a set of dynamical mechanisms whereby structures appear at the global level of a system from interactions among its lower-level components. The rules specifying the interactions among the system's constituent units are executed on the basis of purely local information, without reference to the global pattern, which is an emergent property of the system rather than a property imposed upon the system by an external ordering influence. For example, the emerging structures in the case of foraging in ants include spatiotemporally organized networks of pheromone trails. Self-organization relies on four basic ingredients:

1. Positive feedback (amplification) often constitutes the basis of morphogenesis in the context of this book: they are simple behavioral "rules of thumb" that promote the creation of structures. Examples of positive feedback include recruitment and reinforcement. For instance, recruitment to a food source is a positive feedback that relies on trail laying and trail following in some ant species, or dances in bees.
   When a bee finds a nectar source, she goes back to the hive and relinquishes her nectar to a hive bee. Then she can either start to dance to indicate to other bees the direction and the distance to the food source, or continue to forage at the food source without recruiting nestmates, or she can abandon her food source and become an uncommitted follower herself. If the colony is offered two identical food sources at the same distance from the nest, the bees exploit the two sources symmetrically. However, if one source is better than the other, the bees are able to exploit the better source, or to switch to this better source even if discovered later. Let us consider the following experiment (Figure 1.8). Two food sources are presented to the colony at 8:00 a.m. at the same distance from the hive: source A is characterized by a sugar concentration of 1.00 mol/l and source B by a concentration of 2.5 mol/l. Between 8:00 and noon, source A has been visited 12 times and source B 91 times. At noon, the sources are modified: source A is now characterized by a sugar concentration of 2.5 mol/l and source B by a concentration of 0.75 mol/l. Between noon and 4:00 p.m., source A has been visited 121 times and source B only 10 times.
   It has been shown experimentally that a bee has a relatively high probability of dancing for a good food source and abandoning a poor food source. These simple behavioral rules allow the colony to select the better quality source. Seeley et al. [287] and Camazine and Sneyd [59] have confirmed with a simple mathematical model based on these observations that foragers can home in on the best food source through a positive feedback created by differential rates of dancing and abandonment based upon nectar source quality. Figure 1.9 shows a schematic representation of foraging activity; decision points (C1: become a follower?) and (C2: become a dancer?) are indicated by black diamonds. Figure 1.10(a) shows the number of different individuals that visited each feeder during the previous half hour in the experiments. Figure 1.10(b) shows the forager group size (here,

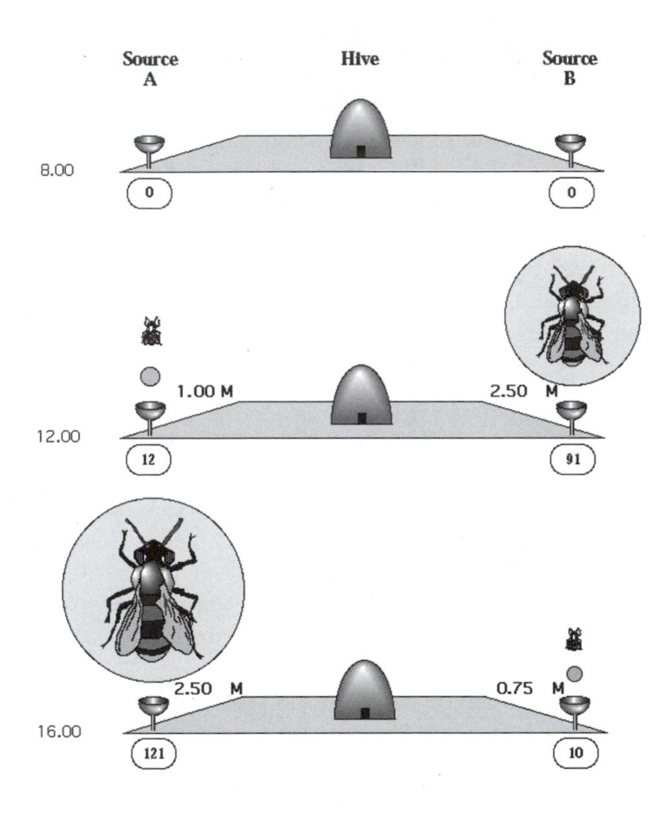

FIGURE 1.8 Schematic representation of the experimental setup.

the sum of the bees dancing for the feeder, the bees at the feeder, and the bees unloading nectar from the feeder) for each feeder obtained from simulations of the simple model depicted in Figure 1.9.

2. Negative feedback counterbalances positive feedback and helps to stabilize the collective pattern: it may take the form of saturation, exhaustion, or competition. In the example of foraging, negative feedback stems from the limited number of available foragers, satiation, food source exhaustion, crowding at the food source, or competition between food sources.
3. SO relies on the amplification of fluctuations (random walks, errors, random task-switching, and so on). Not only do structures emerge despite randomness, but randomness is often crucial, since it enables the discovery of new solutions, and fluctuations can act as seeds from which structures nucleate and grow. For example, foragers may get lost in an ant colony, because they follow trails with some level of error; although such a phenomenon may seem inefficient, lost foragers can find new, unexploited food sources, and recruit nestmates to these food sources.

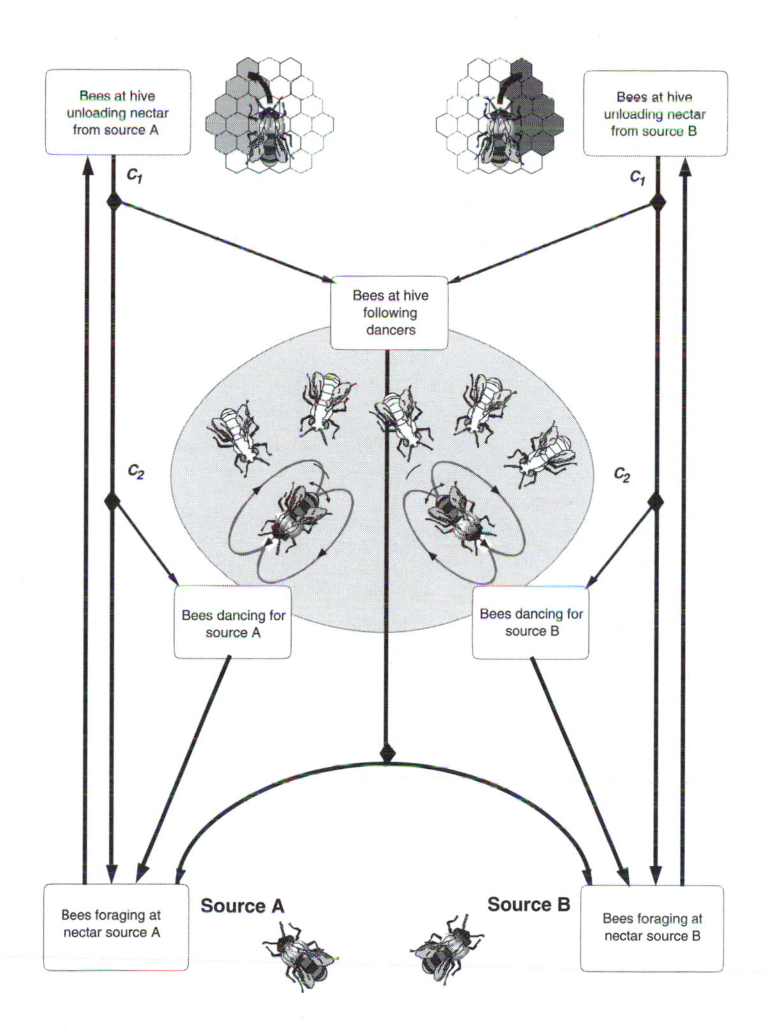

FIGURE 1.9 Schematic representation of foraging activity.

4. All cases of SO rely on multiple interactions. A single individual can generate a self-organized structure such as a stable trail provided pheromonal[3] lifetime is sufficient, because trail-following events can then interact with trail-laying actions. However, SO generally requires a minimal density of mutually tolerant individuals. Moreover, individuals should be able to make use of the results of their own activities as well as of others' activities (although they may perceive

---

[3] A pheromone is a chemical used by animals to communicate. In ants, a pheromone trail is a trail marked with pheromone. Trail-laying trail-following behavior is widespread in ants (see chapter 2)

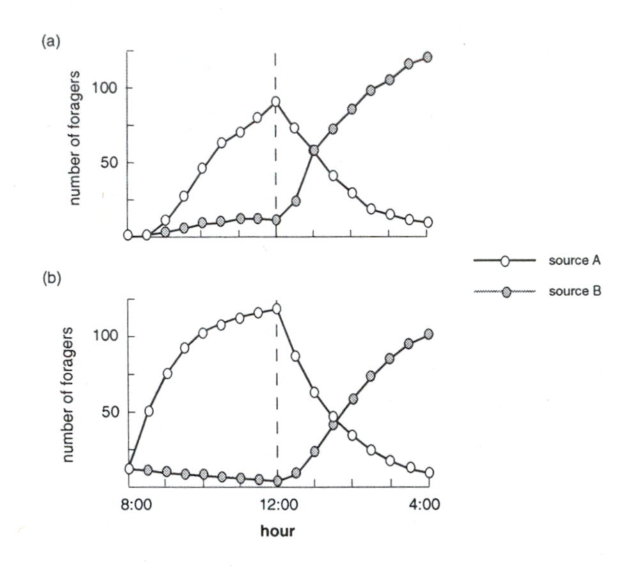

FIGURE 1.10 (a) Experimental number of different individuals that visited each feeder during the previous half hour. (b) Forager group size for each feeder obtained from model.

the difference): for instance, trail networks can self-organize and be used collectively if individuals use others' pheromone. This does not exclude the existence of individual chemical signatures or individual memory which can efficiently complement or sometimes replace responses to collective marks.

When a given phenomenon is self-organized, it can usually be characterized by a few key properties:

1. The creation of spatiotemporal structures in an initially homogeneous medium. Such structures include nest architectures, foraging trails, or social organization. For example, a characteristic well-organized pattern develops on the combs of honeybee colonies. This pattern consists of three concentric regions (a central brood area, a surrounding rim of pollen, and a large peripheral region of honey) (Figure 1.11). It results to a large extent from a self-organized process based on local information [58]. The model described in Camazine [58] relies on the following assumptions, suggested by experimental observations:
   (a) The queen moves more or less randomly over the combs and lays most eggs in the neighborhood of cells already occupied by brood. Eggs remain in place for 21 days.
   (b) Honey and pollen are deposited in randomly selected available cells.
   (c) Four times more honey is brought back to the hive than pollen.

FIGURE 1.11 Comb of a manmade hive: the three concentric regions of cells can be clearly seen.

(d) Typical removal input ratios for honey and pollen are 0.6 and 0.95, respectively.

(e) Removal of honey and pollen is proportional to the number of surrounding cells containing brood.

Simulations of a cellular automaton based on these assumptions were performed [58]. Figure 1.12 shows six successive steps in the formation of the concentric regions of brood (black dot on a white background), pollen (dark grey), and honey (light grey). Assumptions 1a and 1e ensure the growth of a central compact brood area if the first eggs are laid approximately at the center of the comb. Honey and pollen are initially randomly mixed [assumption 1b], but assumptions 1c and 1d imply that pollen cells are more likely to be emptied and refilled with honey, so that pollen located in the periphery is removed and replaced by honey. The only cells available for pollen are those surrounding the brood area, because they have a high turnover rate. The adaptive function of this pattern is discussed by Camazine [58].

2. The possible coexistence of several stable states (multistability). Because structures emerge by amplification of random deviations, any such deviation can be amplified, and the system converges to one among several possible stable states, depending on initial conditions. For example, when two identical food sources, A and B, are presented at the same distance from the nest to an ant colony that resorts to mass recruitment,[4] one of them is eventually massively exploited while the other is neglected: both sources have the same chance of being exploited, but only one of them is, and the colony could choose either one. There are therefore two possible attractors in this example: massive exploitation of A, or

[4]Mass recruitment in ants is based solely on trail-laying trail-following.

massive exploitation of B. Which attractor the colony will converge to depends on random initial events.

3. The existence of bifurcations when some parameters are varied. The behavior of a self-organized system changes dramatically at bifurcations. For example, the termite *Macrotermes* uses soil pellets impregnated with pheromone to build pillars. Two successive phases take place [157]. First, the noncoordinated phase is characterized by a random deposition of pellets. This phase lasts until one of the deposits reaches a critical size. Then, the coordination phase starts if the group of builders is sufficiently large: pillars or strips emerge. The existence of an initial deposit of soil pellets stimulates workers to accumulate more material through a positive feedback mechanism, since the accumulation of material reinforces the attractivity of deposits through the diffusing pheromone emitted by the pellets [48]. This autocatalytic, "snowball effect" leads to the coordinated phase. If the number of builders is too small, the pheromone disappears between two successive passages by the workers, and the amplification mechanism cannot work; only the noncoordinated phase is observed. Therefore, there is no need to invoke a change of behavior by the participants in the transition from the noncoordinated to the coordinated phase: it is merely the result of an increase in group size. A more detailed account of pillar construction in *Macrotermes* can be found in chapter 5 (section 5.2).

1.2.3 STIGMERGY

Self-organization in social insects often requires interactions among insects: such interactions can be direct or indirect. Direct interactions are the "obvious" interactions: antennation, trophallaxis (food or liquid exchange), mandibular contact, visual contact, chemical contact (the odor of nearby nestmates), etc. Indirect interactions are more subtle: two individuals interact indirectly when one of them modifies the environment and the other responds to the new environment at a later time. Such an interaction is an example of stigmergy. In addition to, or in combination with, self-organization, stigmergy is the other most important theoretical concept of this book. Grassé [157, 158] introduced stigmergy (from the Greek *stigma*: sting, and *ergon*: work) to explain task coordination and regulation in the context of nest reconstruction in termites of the genus *Macrotermes*. Grassé showed that the coordination and regulation of building activities do not depend on the workers themselves but are mainly achieved by the nest structure: a stimulating configuration triggers the response of a termite worker, transforming the configuration into another configuration that may trigger in turn another (possibly different) action performed by the same termite or any other worker in the colony. Nest reconstruction consists of first building strips and pillars with soil pellets and stercoral mortar; arches are then thrown between the pillars and finally the interpillar space is filled to make walls. Figure 1.13 sketches how Grassé's [157, 158] notion of stigmergy can be applied to pillar construction.

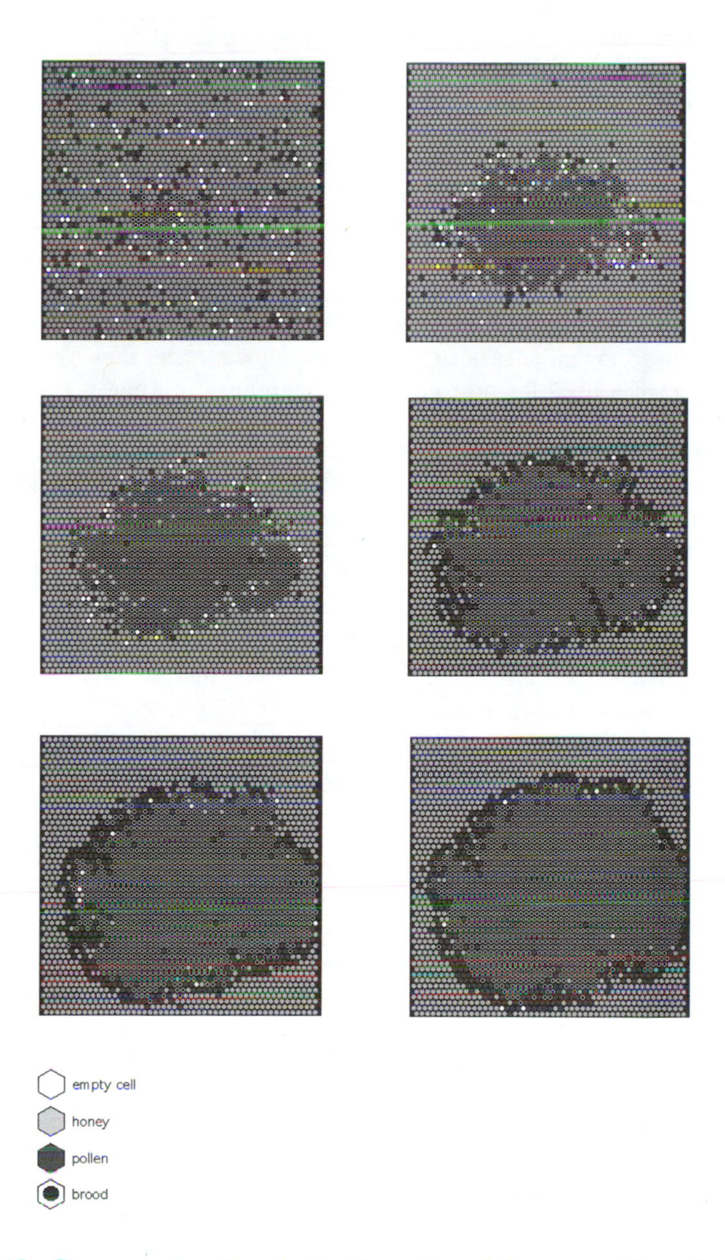

FIGURE 1.12 Six successive steps in the formation of the concentric regions of brood (black dot in a white background), pollen (dark grey), and honey (light grey). Empty cells are represented in white.

Stigmergy is easily overlooked, as it does not explain the detailed mechanisms by which individuals coordinate their activities. However, it does provide a general mechanism that relates individual and colony-level behaviors: individual behavior modifies the environment, which in turn modifies the behavior of other individuals.

The case of pillar construction in termites shows how stigmergy can be used to coordinate the termites' building activities by means of self-organization. Another illustration of how stigmergy and self-organization can be combined is recruitment in ants, described in more detail in chapter 2: self-organized trail laying by individual ants is a way of modifying the environment to communicate with nestmates that follow such trails. In chapter 3, we will see that task performance by some workers decreases the need for more task performance: for instance, nest cleaning by some workers reduces the need for nest cleaning. Therefore, nestmates communicate to other nestmates by modifying the environment (cleaning the nest), and nestmates respond to the modified environment (by not engaging in nest cleaning): that is stigmergy. In chapters 4 and 5, we describe how ants form piles of items such as dead bodies, larvae, or grains of sand. There again, stigmergy is at work: ants deposit items at initially random locations. When other ants perceive deposited items, they are stimulated to deposit items next to them. In chapter 6, we describe a model of nest building in wasps, in which wasp-like agents are stimulated to deposit bricks when they encounter specific configurations of bricks: depositing a brick modifies the environment and hence the stimulatory field of other agents. Finally, in chapter 7, we describe how insects can coordinate their actions to collectively transport prey. When an ant changes position or alignment, it modifies the distribution of forces on the item. Repositioning and realignment cause other transporting ants to change their own position or alignment. The case for stigmergy is a little more ambiguous there, because the item could be considered a medium of "direct" communication between ants. Yet, the same mechanism as in the other examples is at work: ants change the perceived environment of other ants. In every example, the environment serves as a medium of communication.

What all these examples have in common is that they show how stigmergy can easily be made operational. That is a promising first step to design groups of artificial agents which solve problems: replacing coordination through direct communications by indirect interactions is appealing if one wishes to design simple agents and reduce communication among agents. Another feature shared by several of the examples is incremental construction: for instance, termites make use of what other termites have constructed to contribute their own piece. In the context of optimization, incremental improvement is widely used: a new solution is constructed from previous solutions. Finally, stigmergy is often associated with flexibility: when the environment changes because of an external perturbation, the insects respond *appropriately* to that perturbation, as if it were a modification of the environment caused by the colony's activities. In other words, the colony can *collectively* respond to the perturbation with individuals exhibiting the same behavior. When it comes to artificial agents, this type of flexibility is priceless: it means that the agents can respond to a perturbation without being reprogrammed to deal with that particular perturbation.

1.3 MODELING AS AN INTERFACE

As was discussed in section 1.2.1, understanding nature and designing useful systems are two very different endeavors. Understanding nature requires observing, carrying out experiments, and making models that are constrained by observations and experiments. Designing, on the other hand, requires making models that are only limited by one's imagination and available technology. But, in both cases, models are of utmost importance, although they are sometimes implicitly, rather than explicitly, and verbally, rather than mathematically, formulated. Discussing the essence of modeling is definitely beyond the scope of this book. However, it is worth giving a brief outline of what it means to model a phenomenon because models are central to the work presented here. Indeed, models of natural phenomena in social insects set the stage for artificial systems based on them. For the purpose of this book, a model is a simplified picture of reality: a usually small number of observable quantities, thought to be relevant, are identified and used as variables; a model is a way of connecting these variables. Usually, a model has parameters, that, ultimately, should be measurable. Hidden variables, that is, additional assumptions which are not based on observable quantities, are often necessary to build a connection among variables that is consistent with observed behavior. Varying the values of the parameters usually modifies the output or behavior of the model. When the model's output is consistent with the natural system's behavior, parameter values should be compared with the values of their natural counterparts whenever possible. Sources of hidden variables are looked for in the natural system. A good model has several qualities: parsimony, coherence, refutability, etc.

Models channel imagination in the sense that they usually contain the ingredients necessary to explain classes of phenomena: it might be worth exploring models which are known to (plausibly) explain phenomena of interest rather than start from scratch. Exploring a model beyond the constraints of reality amounts to using a tool beyond its initial purpose. Knowing that social insects have been particularly successful at solving problems which can be abstracted away and formulated in, say, algorithmic language, makes models of problem solving in social insects particularly attractive: models that help explain how social insects solve problems serve as a starting point because they can be explored beyond their initial boundaries. For instance, optimization algorithms based on models of foraging often resort to pheromone decay over time scales that are too short to be biologically plausible: in this case, models have been explored beyond what is biologically relevant to generate an interesting new class of algorithms. The underlying principles of these algorithms are strongly inspired by the functioning of social insect colonies, but some of the parameter values do not lie within empirical boundaries. In the context of this book, modeling serves, therefore, as an interface between understanding

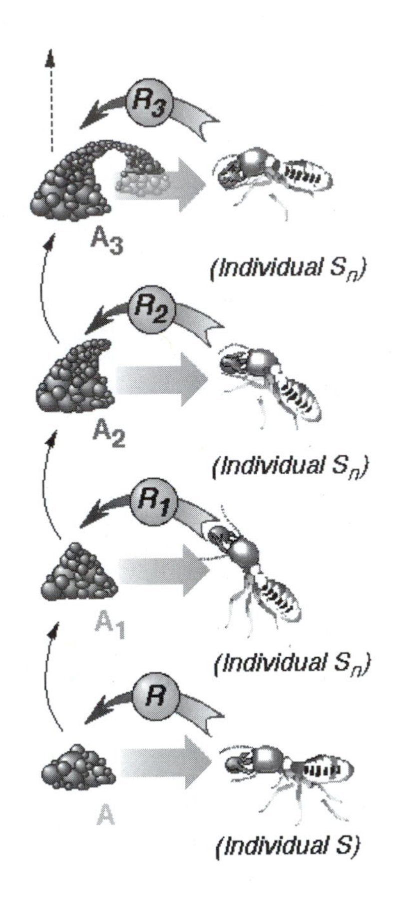

FIGURE 1.13 Assume that the architecture reaches state A, which triggers response R from worker S. A is modified by the action of S (for example, S may drop a soil pellet), and transformed into a new stimulating configuration A_1, that may in turn trigger a new response R_1 from S or any other worker S_n and so forth. The successive responses R_1, R_2, R_n may be produced by any worker carrying a soil pellet. Each worker creates new stimuli in response to existing stimulating configurations. These new stimuli then act on the same termite or any other worker in the colony. Such a process, where the only relevant interactions taking place among the agents are indirect, through the environment which is modified by the other agents, is also called *sematectonic* communication [329]. After Grassé [158]. Reprinted by permission © *Masson*.

nature and designing artificial systems: one starts from the observed phenomenon, tries to make a biologically motivated model of it, and then explores the model without constraints.

1.4 FROM ALGORITHMS TO ROBOTICS

Swarm-based robotics is growing so rapidly that it is becoming very difficult to keep track of what is going on and to keep up with novelty. A recent issue of the journal *Autonomous Robots* was dedicated to "colonies" of robots. Other major journals in the field of robotics have already published, or are in the process of publishing special issues on this topic. Our aim is not to provide the reader with an extensive survey of the field,[5] but rather to present a few selected examples in which researchers have clearly made use of the social insect metaphor, or the swarm-based approach, through one or all of the following: the design of distributed control mechanisms, the type of coordination mechanisms implemented, the tasks that the robots perform. This will be sufficient to have an idea of the advantages and pitfalls of reactive distributed robotics, which can also be called, in the present context, swarm-based robotics.

Why is swarm-based robotics, that we would define loosely as reactive collective robotics, an interesting alternative to classical approaches to robotics? Some of the reasons can be found in the characteristic properties of problem solving by social insects, which is flexible, robust, decentralized, and self-organized. Some tasks may be inherently too complex or impossible for a single robot to perform [62] (see, for example, chapter 7, where pushing a box requires the "coordinated" efforts of at least two individuals). Increased speed can result from using several robots, but there is not necessarily any cooperation: the underlying mechanism that allows the robots to work together is a minimum interference principle. Designing, building, and using several simple robots may be easier because they use, for example, a simpler sensorimotor apparatus, cheaper because they are simple, more flexible without the need to reprogram the robots, and more reliable and fault-tolerant because one or several robots may fail without affecting task completion—although completion time may be affected by such a perturbation—than having a powerful complex robot. Furthermore, theories of self-organization [31] teach us that, sometimes, collective behavior results in patterns which are qualitatively different from those that could be obtained with a single agent or robot. Randomness or fluctuations in individual behavior, far from being harmful, may in fact greatly enhance the system's ability to explore new behaviors and find new "solutions." In addition, self-organization and decentralization, together with the idea that interactions among agents need not be direct but can rather take place through the environment, point to the possibility of significantly reducing communication between robots: explicit robot-to-robot communication rapidly becomes a big issue when the number of robots increases; this

[5]Such reviews—thorough at the time they were released and already somewhat obsolete—can be found, for example, in Cao et al. [61, 62].

issue can be, to a large extent, eliminated by suppressing such communication! Also, central control is usually not well suited to dealing with a large number of agents (this is also true for a telecommunications network: see section 2.7 in chapter 2), not only because of the need for robot-to-controller-and-back communications, but also because failure of the controller implies failure of the whole system.

Of course, using a swarm of robots has some drawbacks. For example, stagnation is one (see chapter 7): because of the lack of a global knowledge, a group of robots may find itself in a deadlock, where it cannot make any progress. Another problem is to determine how these so-called "simple" robots should be programmed to perform user-designed tasks. The pathways to solutions are usually not predefined but emergent, and solving a problem amounts to finding a trajectory for the system and its environment so that the states of both the system and the environment constitute the solution to the problem: although appealing, this formulation does not lend itself to easy programming. Until now, we implicitly assumed that all robots were identical units: the situation becomes more complicated when the robots can perform different tasks, respond to different stimuli, or respond differently to the same stimuli, and so forth; if the body of theory that roboticists can use for homogeneous groups of robots is limited, there is virtually no theoretical guideline for the design of heterogeneous swarms.

The current success of collective robotics is the result of several factors:

1. The relative failure of the Artificial Intelligence (AI) program, which "classical" robotics relied upon, has forced many computer scientists and roboticists to reconsider their fundamental paradigms; this paradigm shift has led to the advent of connectionism, and to the view that sensorimotor "intelligence" is as important as reasoning and other higher-level components of cognition. Swarm-based robotics relies on the anti-classical-AI idea that a group of robots may be able to perform tasks without explicit representations of the environment and of the other robots; finally, planning is replaced by reactivity.
2. The remarkable progress of hardware during the last decade has allowed many researchers to experiment with real robots, which have not only become more efficient and capable of performing many different tasks, but also cheap(er).
3. The field of Artificial Life, where the concept of emergent behavior is emphasized as essential to the understanding of fundamental properties of the living, has done much to propagate ideas about collective behavior in biological systems, particularly social insects; theories that had remained unknown to roboticists have eventually reached them; of course, the rise of the internet-based "global information society" enhanced this tendency.
4. Finally, collective robotics has become fashionable by positive feedback. The field is booming, but not that many works appear to be original, at least conceptually, which is the only level the authors of this book can judge.

Interesting directions for the future include ways of enhancing indirect communications among robots. For example, Deveza et al. [95] introduced odor sensing for robot guidance (which, of course, is reminiscent of trail following in ants), Russell [278] used a short-lived heat trail for the same purpose. Prescott and Ibbotson [264] used a particularly original medium of communication to reproduce the motion of prehistoric worms: bathroom tissue. A dispenser on the back of the robot releases a stream of paper when the robot is moving; two light detectors on each of the side arms measure reflected light from the floor and control the thigmotaxis (toward track) and phobotaxis (away from track) behaviors. Although their work is not about collective robotics, it sets the stage for studying and implementing indirect communication in groups of robots.

Finally, what are the areas of potential application for swarm-based multirobot systems? "In aerospace technology, it is envisioned that teams of flying robots may effect satellite repair, and aircraft engine maintenance could be performed by thousands of robots built into the engine eliminating the need for costly disassembly for routine preventive maintenance. Environmental robots are to be used in pipe inspection and pest eradication. While industrial applications include waste disposal and micro cleaners. Ship maintenance and ocean cleaning could be performed by hundreds of underwater robots designed to remove debris from hulls and ocean floors.... Some researchers envision microsurgical robots that could be injected into the body by the hundreds designed to perform specific manipulation tasks without the need for conventional surgical techniques," writes Kube [208] (p. 17). Most of these applications require miniaturization. Very small robots, micro- and nanorobots, which will by construction have severely limited sensing and computation, may need to "operate in very large groups or swarms to affect the macroworld [237]." Approaches directly inspired or derived from swarm intelligence may be the only way to control and manage such groups of small robots. As the reader will perhaps be disappointed by the "simplicity" of the tasks performed by state-of-the-art swarm-based robotic systems such as those presented in chapters 4, 6, and 7, let us remind him that it is in the perspective of miniaturization that swarm-based robotics becomes meaningful.

In view of these great many potential applications, it seems urgent to work at the fundamental level of what algorithms should be put into these robots: understanding the nature of coordination in groups of simple agents is a first step toward implementing useful multirobot systems. The present book, although it deals marginally with collective robotics, provides a wealth of ideas that, we hope, will be useful in that perspective.

1.5 READING GUIDE

It may be useful at this point to issue a series of warnings to the reader. This book is not intended to provide the reader with recipes for solving problems. It is full of wild speculations and statements that may (or may not) turn out to be just plain wrong. This book is not describing a new kind of algorithmic theory. Nor does it provide an overview of social insect behavior. Finally, this book is not a textbook, although it could be used to propagate our excitement to new generations of researchers.

Knowing what this book is not does not necessarily help define what it is. Well, this book is an educated guess about the future of artificial intelligence. It provides the reader, not with a synthesis of a mature and well-defined field, but rather with new ideas from a booming field with no clear-cut boundaries. It provides the reader, not with proofs of theorems, but with *intuitions of emergent phenomena* [268]. We suggest that the social insect metaphor may go beyond superficial considerations. At a time when the world is becoming so complex that no single human being can really understand it, when information (and not the lack of it) is threatening our lives, when software systems become so intractable that they can no longer be controlled, perhaps the scientific and engineering world will be more willing to consider another way of designing "intelligent" systems, where autonomy, emergence and distributed functioning replace control, preprogramming and centralization. Of course, such an approach raises fundamental issues: how do we design such systems so that they eventually solve the problems we want them to solve? How do we make sure that such systems will not exhibit catastrophic behavior when confronted with pathological situations (note that the same question applies to large pieces of "classical" software)? And many other questions, some of which are waiting to be formulated. To these formulated and unformulated questions, we hope that this book will offer an implicit, rather than explicit, response.

Each chapter deals with a particular behavior observed in social insects: foraging (chapter 2), division of labor (chapter 3), clustering and sorting (chapter 4), building (chapter 5 and chapter 6), and cooperative transport (chapter 7). The first part of each chapter provides a brief description of the phenomenon followed by a more or less thorough description of models developed by ethologists to understand the phenomenon. Engineering-oriented applications, that make use of the emergent behavior of social insects, are then presented. It is worth noting that not all subfields of swarm intelligence are equally advanced. Therefore, although each chapter emphasizes different aspects of swarm intelligence, all of them present an outline of the underlying biology before proceeding to resulting applications, chapters 2, 4, and 7 present more applications than biology or modeling, whereas chapters 3 and 5 are more biology oriented. Chapter 6 is more difficult to classify: it does not contain any actual application, nor does it present a lot of biology. Chapter 6 is speculative at two levels: the model described is highly speculative, whereas potential applications remain to be explored. But we expect all types of chapters to be useful: those emphasizing applications because they show very clearly how our understanding of how social insects collectively solve problems can be applied to designing algorithms and distributed artificial problem-solving devices; those emphasizing the biology because, we believe, they provide new, largely untested, ideas to design new types of algorithms and distributed artificial devices. Chapter 6 belongs to this latter category, although a complete understanding of the underlying biology currently seems out of reach of modeling.

Because this book is about designing problem-solving devices inspired by social insects, the models presented throughout the book need not be "perfect" models of social insect behavior. In fact, all the models presented in the book, including those introduced earlier in this chapter, remain speculative to a varying extent. Carefully designed empirical tests are required to validate them. Some of the models may even turn out to be false—although in most cases they will likely reflect at least part of the truth. But algorithms do not have to be designed after *accurate* or even *true* models of biological systems: efficiency, robustness, and flexibility are the driving criteria, not biological accuracy. This warning is extremely important because the models of social insect behavior presented in the book are still being developed and tested, whereas our writing style may leave the reader with the impression that everything is 100% certain. But the study of self-organization in social insects is almost as new as artificial swarm intelligence!

CHAPTER 2

Ant Foraging Behavior, Combinatorial Optimization, and Routing in Communications Networks

2.1 OVERVIEW

This chapter is dedicated to the description of the collective foraging behavior of ants and to the discussion of several computational models inspired by that behavior—ant-based algorithms or ant colony optimization (ACO) algorithms. In the first part of the chapter, several examples of cooperative foraging in ants are described and modeled. In particular, in some species a colony self-organizes to find and exploit the food source that is closest to the nest.

A set of conveniently defined artificial ants, the behavior of which is designed after that of their real counterparts, can be used to solve combinatorial optimization problems. A detailed introduction to ant-based algorithms is given by using the traveling salesman problem (TSP) as an application problem. Ant-based algorithms have been applied to other combinatorial optimization problems such as the quadratic assignment problem, graph coloring, job-shop scheduling, sequential ordering, and vehicle routing. Results obtained with ant-based algorithms are often as good as those obtained with other general-purpose heuristics. Application to the quadratic assignment problem is described in detail. Coupling ant-based algorithms with local optimizers obtains, in some cases, world-class results. Parallels are drawn between ant-based optimization algorithms and other nature-inspired optimization techniques, such as neural nets and evolutionary computation.

All the combinatorial problems mentioned above are static, that is, their characteristics do not change over time. In the last part of the chapter, the application of ant-based algorithms to a class of stochastic time-varying problems is investigated: routing in telecommunications networks. Given the adaptive capabilities built into the ant-based algorithms, they may be more competitive in stochastic time-varying domains, in which solutions must be adapted online to changing conditions, than in static problems. The performance of AntNet, an ant-based algorithm designed to adaptively build routing tables in packet-switching communications networks, is the best of a number of state-of-the-art algorithms compared on an extensive set of experimental conditions.

2.2 FORAGING STRATEGIES IN ANTS

Many ant species have trail-laying trail-following behavior when foraging: individual ants deposit a chemical substance called *pheromone* as they move from a food source to their nest, and foragers follow such pheromone trails. The process whereby an ant is influenced toward a food source by another ant or by a chemical trail is called *recruitment*, and recruitment based solely on chemical trails is called *mass recruitment*. In this section, two phenomena based on mass recruitment are described. These phenomena inspired the ACO algorithms described later in this chapter.

2.2.1 NATURAL OPTIMIZATION (1): THE BINARY BRIDGE EXPERIMENT

Using a simple and elegant experimental setup, Deneubourg et al. [87] showed that path selection to a food source in the Argentine ant *Linepithema humile* (formerly *Iridomyrmex humilis*) is based on self-organization. In this experiment, a food source is separated from the nest by a bridge with two equally long branches A and B (Figure 2.1). Initially, there is no pheromone on the two branches, which have, therefore, the same probability of being selected by the ants. Nevertheless, random fluctuations will cause a few more ants to randomly select one branch, say A, over the other. Because ants deposit pheromone while walking, the greater number of ants on branch A determines a greater amount of pheromone on A, which in turn stimulates more ants to choose A, and so on [87].

Deneubourg et al. [87] developed a model of this phenomenon, the behavior of which closely matches the experimental observations. Let us assume that the amount of pheromone on a branch is proportional to the number of ants that used the branch to cross the bridge. With this assumption, pheromone evaporation is not taken into account: this is a plausible assumption, because the experiments typically last of the order of an hour, a time scale that may not be sufficient for the amount of pheromone to be reduced significantly. In the model, the probability of choosing a branch at a certain time depends on the total number of ants that used

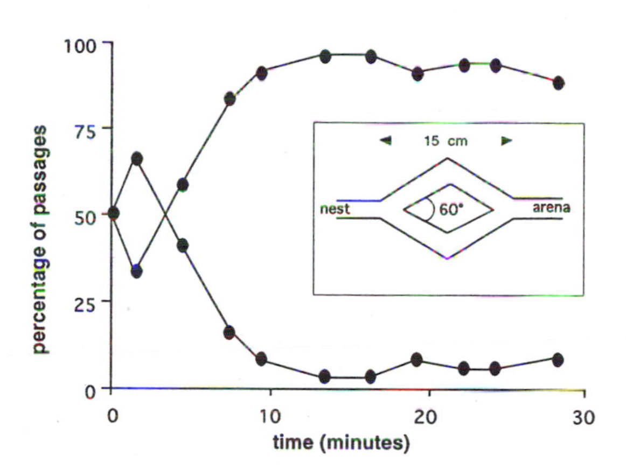

FIGURE 2.1 Percentage of all passages per unit time on each of the two branches as a function of time: one of the branches is eventually used most of the time. Note that the winning branch was not favored by the initial fluctuations, which indicates that these fluctuations were not strong enough to promote exploitation of the other branch. The inset is a schematic representaton of the experimental setup. After Deneubourg et al. [87]. Reprinted by permission © *Plenum Publishing.*

the branch until that time. More precisely, let A_i and B_i be the numbers of ants that have used branches A and B after i ants have used the bridge. The probability $P_A(P_B)$ that the $(i+1)$th ant chooses branch $A(B)$ is

$$P_A = \frac{(k + A_i)^n}{(k + A_i)^n + (k + B_i)^n} = 1 - P_B \,. \tag{2.1}$$

Equation (2.1) quantifies the way in which a higher concentration on branch A gives a higher probability of choosing branch A, depending on the absolute and relative values of A_i and B_i. The parameter n determines the degree of nonlinearity of the choice function: when n is large, if one branch has only slightly more pheromone than the other, the next ant that passes will have a high probability of choosing it. The parameter k quantifies the degree of attraction of an unmarked branch: the greater k, the greater the amount of pheromone to make the choice nonrandom. This particular form of P_A was obtained from experiments on trail following [259]. The values of the parameters k and n that give the best fit to the experimental measures are $n \approx 2$ and $k \approx 20$ [87]. If $A_i \gg B_i$ and $A_i \gg 20$, $P_A \approx 1$; if $A_i \gg B_i$ but $A_i < 20$, then $P_A \approx 0.5$. The same holds true for P_B. The choice dynamics follows from Eq. (2.1):

$$A_{i+1} = \begin{cases} A_i + 1 & \text{if } \delta \leq P_A; \\ A_i & \text{if } \delta > P_A \,, \end{cases} \tag{2.2}$$

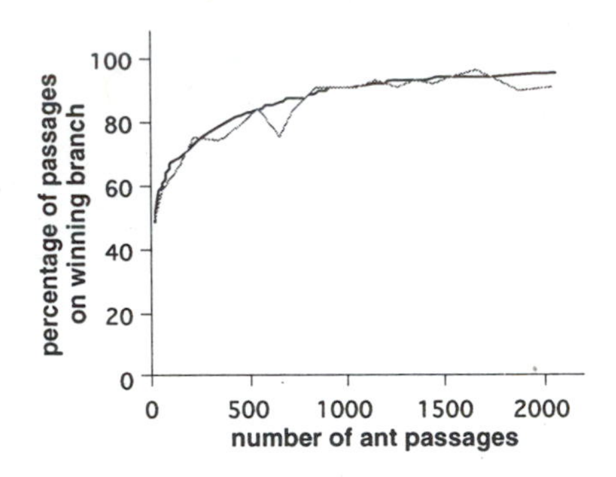

FIGURE 2.2 Percentage of passages on the dominant branch as a function of the total number of ants that took the bridge. Solid black curve: average result of 200 Monte Carlo simulations of the model given by Eqs. (2.1) to (2.4), with 2,000 ant passages in each simulation. Gray broken line: percentage of passages on the dominant branch measured every 100 ant passages, average over 20 experiments of 30 minutes each. Nine colonies of 150–1200 workers each were used for the experiments. After Deneubourg et al. [87]. Reprinted by permission © *Plenum Publishing.*

$$B_{i+1} = \begin{cases} B_i + 1 & \text{if } \delta > P_A; \\ B_i & \text{if } \delta \leq P_A\,, \end{cases} \tag{2.3}$$

$$A_i + B_i = i\,, \tag{2.4}$$

where δ is a random variable uniformly distributed over $[0, 1]$.

The model expressed in Eqs. (2.1) to (2.4) was analyzed by means of Monte Carlo simulations. The results of these simulations are in perfect agreement with the experiments (Figure 2.2).

The experiment where the bridge's branches are the same length can be extended to the case where one branch is longer than the other [154]. Likewise, the model can be modified to apply to this situation. By the same mechanism as in the previous situation, that is, the amplification of initial fluctuations, the shortest branch is most often selected: the first ants coming back to the nest are those that took the shortest path twice (to go from the nest to the source and to return to the nest), so that, immediately after these ants have returned, more pheromone is present on the short branch than on the long branch, stimulating nestmates to choose the short branch (Figure 2.3(a)). Experiments have shown that the chance of the short branch being eventually selected by amplification of initial fluctuations increases with the length ratio r of the two branches. Figure 2.3(b) on the left shows that short branch selection is only statistically verified, when $r = 2$ (the longer branch is twice as long as the short branch): in some cases, more ants initially choose the long branch, so that this branch will be more strongly marked, leading to the preferential exploitation of the long branch. This indicates that recruitment based primarily on pheromone may not be particularly flexible. Figure 2.3(b) on the right shows what happens when, given the same experimental setup, the short branch is presented to the colony 30 minutes after the long branch: the short branch is not selected and the colony remains trapped on the long branch.

With *Lasius niger*, another species of ants resorting to mass recruitment, another mechanism allows the selection of the shorter path, even when it is presented 30 minutes after the longer path. When it finds itself in the middle of the long branch, this ant often realizes that it is heading almost perpendicularly to the required direction: this induces it to make U-turns on the long branch [11]. In this case, the combination of individual memory for the direction to the nest or food source plus collective trail following allows the systematic selection of the short branch (Figure 2.4). In other words, the colony is more flexible.

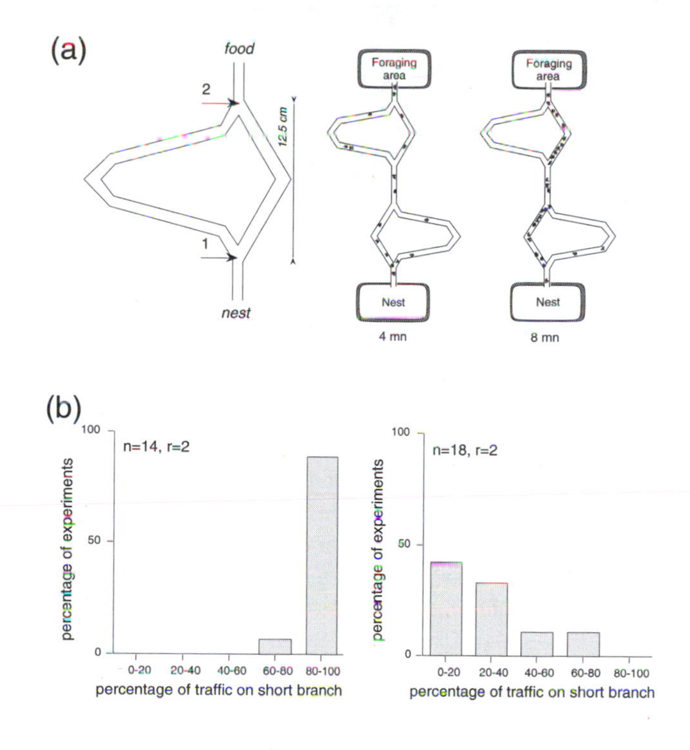

FIGURE 2.3 (a) Experimental setup and drawings of the selection of the short branches by a colony of *Linepithema humile* respectively 4 and 8 minutes after the bridge was placed. After Goss et al. [154]. Reprinted by permission © *Springer-Verlag.* (b) Distribution of the percentage of ants that selected the shorter path over n experiments (r is the length ratio between the two branches). The longer branch is r times longer than the short branch. Left: short and long branches are presented from beginning of the experiment. Right: the short branch is presented to the colony 30 minutes after the long branch. After Goss et al. [154]. Reprinted by permission © *Springer-Verlag. continued.*

(c)

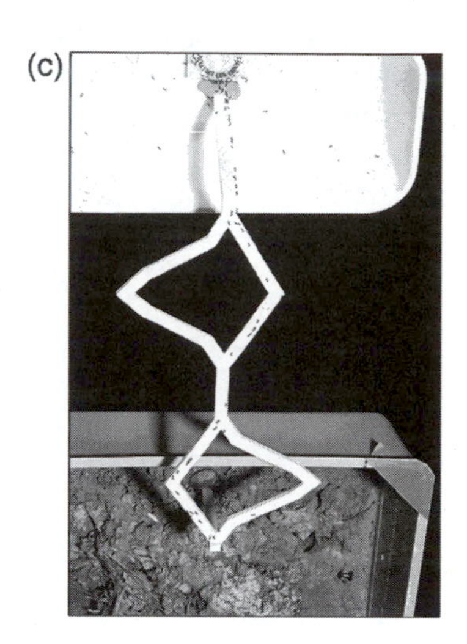

FIGURE 2.3 *continued.* (c) Image of the real experiment. The nest is located in the lower arena and the food source in the upper arena.

In the above-mentioned ant species, which base their recruitment on pheromone(s), it would be interesting to know the physico-chemical properties of the pheromones. It is fair to say that not much is known about these chemicals. "Indirect" experiments, which test for the behavior of ants that encounter different amounts of pheromone on different substrates and/or after varying amounts of time, have provided a rough approximation of quantities such as evaporation rate, absorption rate, diffusion constant, etc. One consistent finding about the above-mentioned species is that their pheromone trails usually persist for a long time, from at least several hours up to several months (depending on the species, the substrate, colony size, weather conditions, etc.), indicating that the lifetimes of pheromones have to be measured on long time scales.

In the field of ant colony optimization and swarm intelligence, common wisdom holds that pheromone evaporation, when occurring on a sufficiently short time scale, allows ant colonies to avoid being trapped on a "suboptimal" solution, as is the case for *Linepithema humile*. This is because maintaining a well-marked pheromone trail is more difficult on a longer trail (Figure 2.5) (see Bonabeau [27, 28]).

In reality, however, this does not seem to be the case, because pheromones are persistent. Also, the notion of optimality is a slippery one in biology: optimality is defined with respect to many constraints, including ecological constraints, such as predation or competition with other colonies. In some situations, it may be to the advantage of the colony to focus its foraging activity on a given site or area, even if

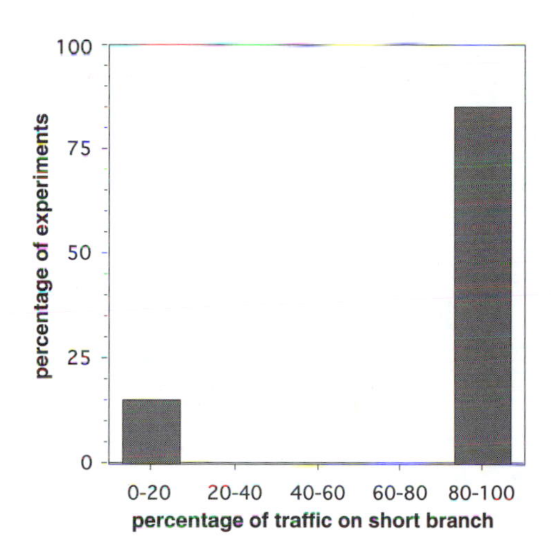

FIGURE 2.4 Same experimental setup as in Figure 2.3(a), but with *Lasius niger*. In this case, the combination of individual memory for the direction to the nest or food source, plus collective trail following allows a more systematic selection of the short branch when the short branch is presented after the long branch. After Beckers et al. [11]. Reprinted by permission © *Academic Press.*

not the best one, because one single trail is easier to protect; switching to other sites in an "opportunistic" way could induce costs, such as a reduction in the colony's level of protection, which should not be neglected. This is an interesting example where there is clear divergence between real and artificial ants: the ant colony algorithms that are described in the following sections heavily rely on techniques similar to pheromone evaporation, which is indeed quite a powerful method for optimization.

2.2.2 NATURAL OPTIMIZATION (2): THE CASE OF INTER-NEST TRAFFIC

Aron et al. [5], in a remarkable experiment involving several nests and bridges between them, have shown that the Argentine ant *Linepithema humile* can solve the minimal spanning tree problem. Although this problem is not combinatorially hard in the formulation used in the experiment, some of its variants are,[1] and it is interesting to have an example of collective problem solving for the simplest variant, since this could inspire similar techniques with which to attack harder versions.

[1] Such as the Steiner problem [144].

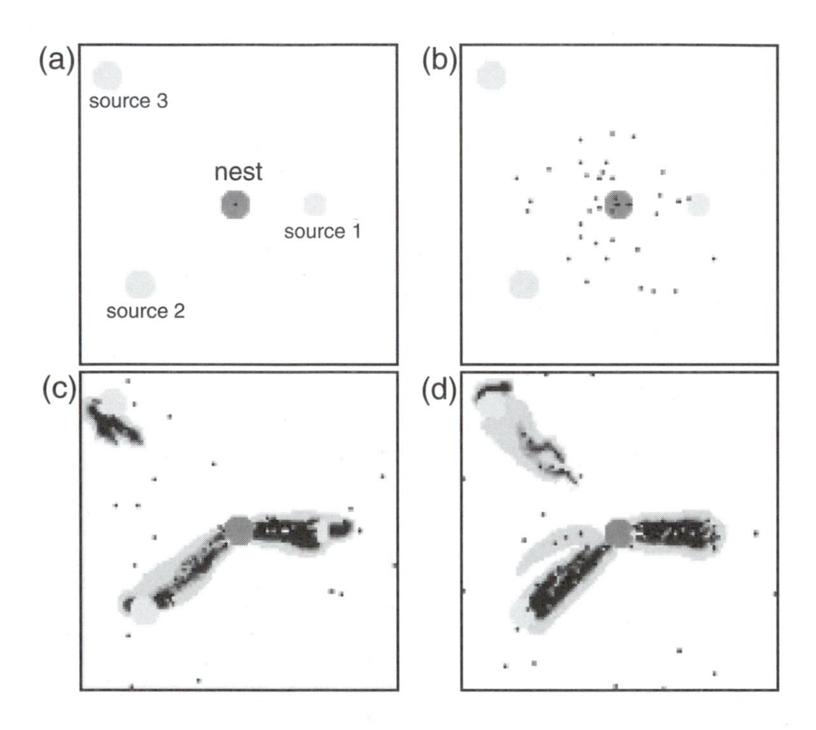

FIGURE 2.5 Simulation of a situation where 3 food sources (S_1, S_2, and S_3) of identical quality are presented to the colony at various distances from the nest (Figure 2.5(a)). Ants returning from a food source to the nest deposit pheromone, and ants that go out of the nest follow pheromone trails, or walk randomly in the absence of pheromone. Trail following is probabilistic: errors can occur and ants may lose the trail. Pheromone decay is assumed to occur over short time scales. At time t_1, ants, represented by black dots, explore their environment randomly (Figure 2.5(b)). At time t_2, trails that connect the nest to the food sources are being established (Figure 2.5(c)). At time t_3, only the trails that connect the nest to the closest food sources are maintained, leading to the exploitation of these sources (Figure 2.5(d)). The next closest source will be exploited later, when the first two exploited sources are exhausted, and so forth. The simulation has been written in the StarLogo programming language, developed by the Epistemology and Learning Group at the MIT Media Lab. The program can be downloaded from the MIT StarLogo website at http://starlogo.www.media.mit.edu. More details about StarLogo and these simulations can be found in Resnick [268].

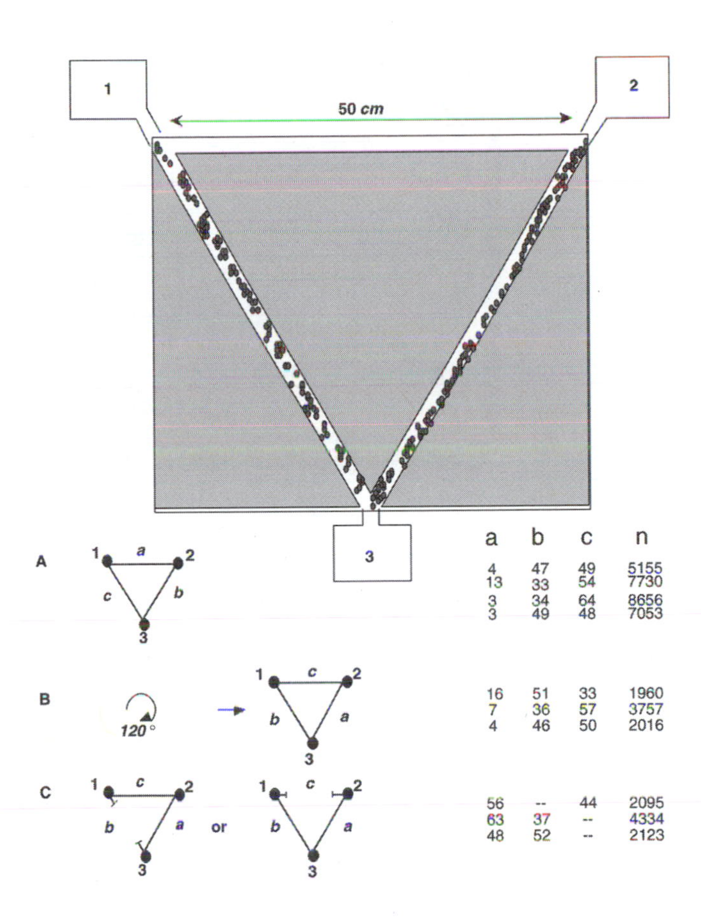

FIGURE 2.6 Top: Experimental design and results for a triangular network connecting three nests. Bottom: Drawings represent the qualitative solutions adopted. A solid line indicates heavy traffic, and an interrupted line a cut branch. The numbers indicate the quantitative results for each experiment, with the percentage of traffic on each branch (a, b, and c) and the total traffic (n) on the branches. (a) The triangular network was left for two weeks before counting the traffic (4 experiments). This experiment indicates that chemical, rather than visual, cues play a key role in the selection of branches. (b) The entire bridge system was rotated 120 degrees, the nests not being moved (3 experiments). (c) The most frequented branch was then cut (3 experiments). After Aron et al. [5]. Reprinted by permission © *Springer-Verlag.*

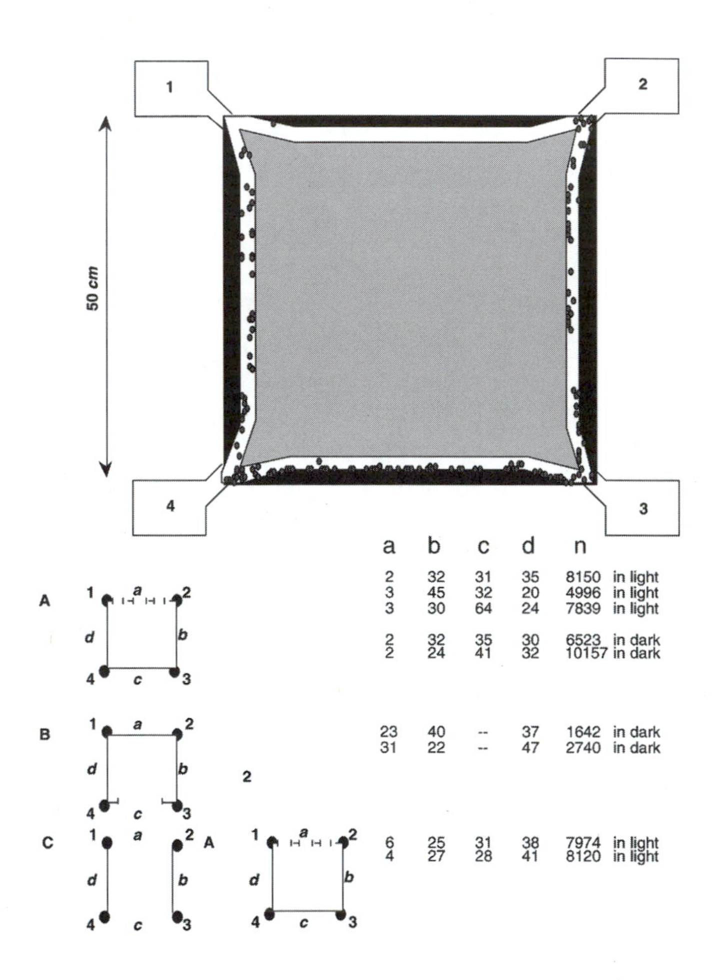

FIGURE 2.7 Top: Experimental design and results for a square network connecting four nests. Bottom: Same as Figure 2.6. "In light" means that the experiment has been performed with light. "In dark" means that the experiment has been performed in the dark. The absence of significant difference between experiments performed with and without light suggests that visual cues are not essential. (a) The square network was left for two weeks before counting the traffic; branch a is not exploited. (b) Branch c (the base of the U-shaped solution in A) was then cut. (c) Branches b and d were presented to the colony for two weeks, and then branches a and c were added; branch a is not exploited. After Aron et al. [5]. Reprinted by permission © *Springer-Verlag.*

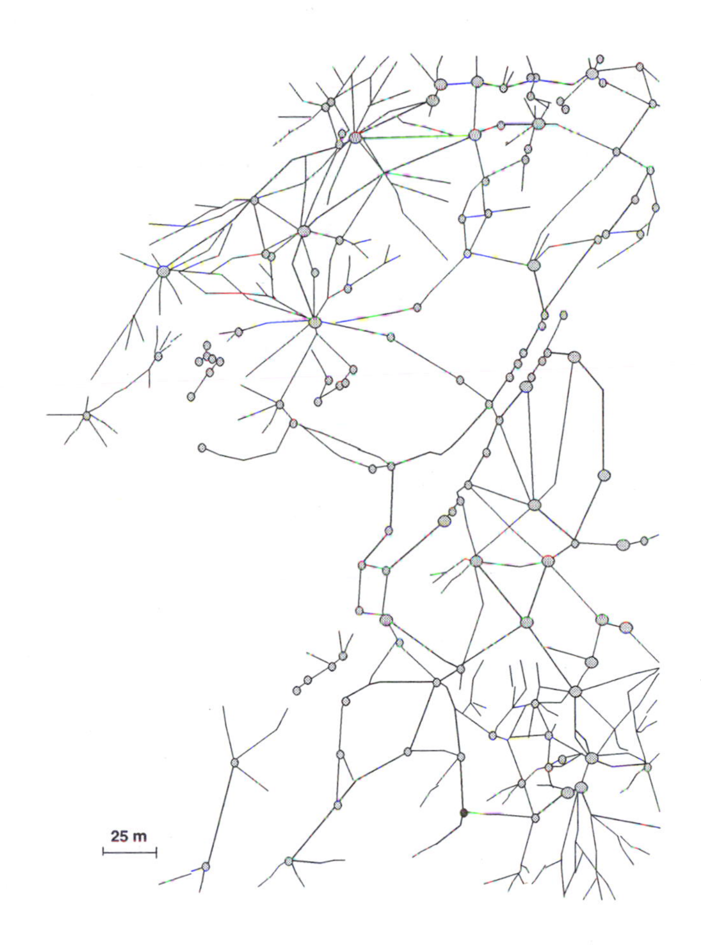

FIGURE 2.8 Network of interconnected nests of a super-colony of *Formica lugubris*. Nests are represented by circles. After Cherix [68]. Reprinted by permission © *Springer-Verlag.*

Societies of *Linepithema humile* are composed of subsocieties connected by a permanent network of chemical trails. Workers, larvae and even queens are continuously exchanged between the nests of these subsocieties. Such exchanges enable a flexible allocation of the work force in the foraging area in response to environmental cues. Moreover, inter-nest trails are extended to include trails to permanent, long-lasting or rich food sources. In the laboratory experiments carried out by Aron et al. [5], three or four nests were connected by cardboard bridges (Figures 2.6 and 2.7). The resulting traffic of ants was such that the ants were travelling on a set of paths connecting all the nests. The set of paths formed a minimal spanning tree, that is, ant did not use redundant bridges. Besides, cutting a frequented bridge caused the traffic to transfer to a previously unused branch (Figures 2.6 and 2.7).

Aron et al. [5] have shown that chemical cues play a crucial role in this process, whereas visual cues were not essential. They were able to reproduce most of the experimental observations with a model similar to the one used to describe the binary bridge experiment of Figure 2.3.

This example may seem trivial, not only because the minimal spanning tree problem is not a hard optimization problem, but also because this particular instance involves only three or four nodes. But some ant colonies build huge networks of connected nests that span hundreds of meters. For example, Figure 2.8 shows a network of interconnected nests of a super-colony of *Formica lugubris* in Switzerland. Although no study has been done in relation with the graph-theoretic properties of this network, it would not be surprising to find that the network is close to a minimum spanning tree that connects all nests, represented by circles in Figure 2.8.

2.2.3 THE RAID PATTERNS OF ARMY ANTS

In this section, we examine the foraging patterns of army ants. Army ants are among the largest and most cohesive societies [86]. Their foraging systems coordinate hundreds of thousands of individuals and cover a thousand square meters in a single day. The largest swarm raids of neotropical army ants are those of *Eciton burchelli*, which may contain up to 200,000 workers raiding in a dense phalanx 15m or more wide [129]. These swarm raids, comprised of individuals that are virtually blind, are fascinating examples of powerful, totally decentralized control. "The raid system is composed of a swarm front, a dense carpet of ants that extends for approximately 1m behind the leading edge of the swarm, and a very large system of anastomosing trails. These trails, along which the ants run out to the swarm front and return with prey items, characteristically form loops that are small near the raid front and get ever bigger and less frequent away from it. The final large loop leads to a single principal trail[2] that provides a permanent link between the army ants' raid and their bivouac" [129]. Raid patterns are dynamic but always exhibit the same basic structure. Figure 2.9 shows the swarm raid structures of three species of army ants, *Eciton hamatum*, *E. rapax*, and *E. burchelli*.

These three species have different "diets": *E. hamatum* feeds mainly on dispersed social insect colonies, *Eciton burchelli* feeds largely on scattered arthropods, and *E. rapax* has an intermediary diet. These different diets correspond to different spatial distributions of food items: for *E. hamatum*, food sources are rare but large, whereas for *E. burchelli*, food can easily be found but in small quantities each time. Could these different spatial distributions explain the different foraging patterns observed in Figure 2.9? More precisely, is it possible to explain the differences in the foraging patterns of the three army ant species without invoking a different behavior, but a different *environment*? This question is interesting from the viewpoint of the biology of army ants: if all three species have common ancestors, it is

[2]The principal trail is often a mostly straight line [122]. A mathematical model which explains how ants build straight trails was given by Bruckstein [47].

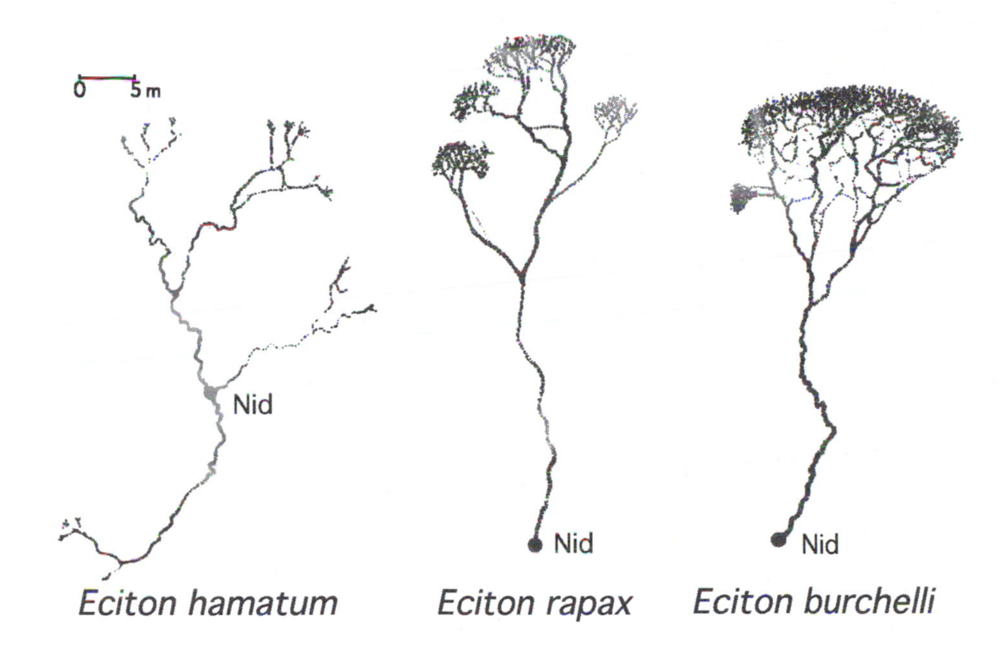

FIGURE 2.9 Foraging patterns of three army ant species: *Eciton hamatum*, *E. rapax*, and *E. burchelli*. After Burton and Franks [52]. Reprinted by permission © *Kansas Science Bulletin*.

not unlikely that their behavior is similar—only their diet may be different, simply because they adapted to different niches. But the question is also interesting from a problem-solving perspective: it means that the same program (the computer science equivalent of behavior) can solve different problems, or can adapt to different environments.

Deneubourg et al.'s [86] self-organization model of army ant raid patterns tells us that the answer to the question is yes. The assumptions of the model are as follows:

(a) For modeling purposes, the environment is represented as a bidimensional grid, represented in Figure 2.10. Each node of the grid is called a site. The system is updated at discrete time steps.

(b) The ants lay pheromone trails both on the way out to the raid front and when returning to the nest or bivouac. Ants deposit one unit of pheromone per unit area (that is, per visited site) on their way out to the raid front; if the amount of pheromone at the site exceeds 1,000 units, no additional pheromone is laid. Returning ants deposit ten units of pheromone per visited site; if the amount of pheromone at the site exceeds 300 units, no additional pheromone is laid. A fixed fraction $e = 1/30$ of the pheromone at each site evaporates at each time step.

(c) Ants return to the nest after finding prey; an ant that returns to the nest is laden with a prey item.

(d) At each time step, an ant decides whether to advance or stay at its current site. Let ρ_l and ρ_r be the amounts of pheromone on the left and right sites, respectively. The probability of moving is given by:

$$p_m = \frac{1}{2}\left[1 + \tanh\left(\frac{\rho_l + \rho_r}{100} - 1\right)\right].$$

Therefore, the more pheromone on the sites the ant is facing, the more likely the ant is to move. This corresponds to the empirical observation that the ants move more and more quickly as the pheromone concentration increases, and move more slowly in unmarked areas.

(e) When an ant chooses to move, it selects the left or right site according to its absolute and relative pheromone concentrations. More precisely, the ant selects the left site with probability

$$p = \frac{(5 + \rho_l)^2}{(5 + \rho_l)^2 + (5 + \rho_r)^2}.$$

This expression is similar to Eq. (2.1) with $n = 2$ and $k = 5$.

(f) Ten ants leave the nest per time step. There cannot be more than 20 ants per site. If the ant has decided to move, and selected a site that is full, it moves to the other site; if both sites are full, the ant stays where it is.

(g) The food distribution is represented by a probability of finding a food source per site. The food source can be small or large. Each ant that finds a food source returns to the nest with one unit of food.

Figure 2.11 shows two patterns obtained from Monte Carlo simulations of Deneubourg et al.'s [86] model, with two different spatial distributions of food. The pattern on the right has been obtained when each site contains 1 unit of food with probability 1/2. Such a food distribution would represent the prey distribution of *E. burchelli*. A well-defined raid front can be observed. Returning ants cause the central trail to split into lateral trails which branch out. The pattern on the right has been obtained when each site has a small probability (1/100) of containing a large number (400) of food units, a food distribution that would represent the prey distribution of *E. hamatum* or *E. rapax*. Indeed, the pattern obtained from the simulation is similar to the swarm raid patterns of these two species: the swarm splits up into a number of small columns.

The similarity between the patterns obtained in Figure 2.11 and those of Figure 2.9 is not perfect. There are, however, some important common characteristics, which suggest that individual foragers with similar behavior operate in all three

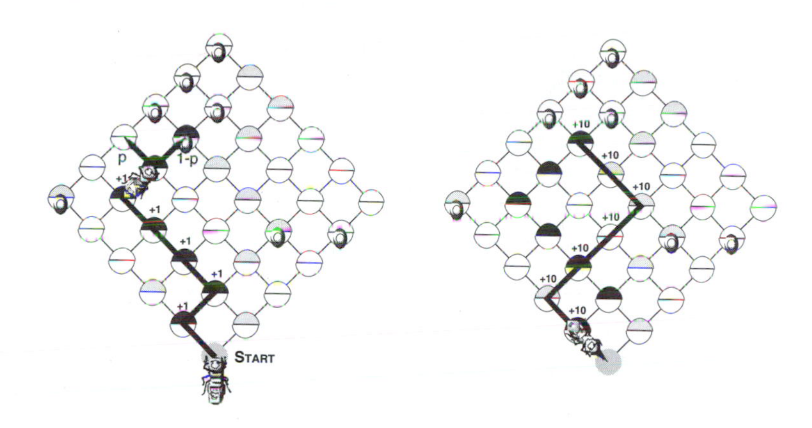

FIGURE 2.10 Grid used to run the Monte Carlo simulations. Left: an ant starts from the nest and moves forward at each time step with probability p_m. When the ant moves, it selects the left site with probability p, or the right site with probability $1-p$. One unit of pheromone is deposited at each visited site. Some sites contain food (small circles). Shaded areas at each site represent pheromone concentrations; a darker grey represents a higher pheromone concentration. Right: an ant is returning from a site where it found food. Ten units of pheromone are deposited at each visited site.

species. The main factor that influences the structure of swarm raid patterns seems to be the environment, here, the food distribution. This prediction of Deneubourg et al.'s [86] model was tested in the field by Franks et al. [129]. By manipulating the prey distributions for *E. burchelli* swarms, they have made them raid in a form more typical of other army ant species, where the swarm breaks up into subswarms.

Although the question has never been formally studied, it would not be surprising, given the structure of their patterns, to find that the swarm raids are optimal "distribution networks," in the sense that they maximize the amount of retrieved prey for a given energy expense, or minimize the energy expense for a given amount of food brought back to the nest. Each pattern seems to be adapted, in that respect, to the species' food distribution. If this is the case, it means that a single algorithm based on self-organization is capable of generating optimal foraging structures, or distribution networks, in different conditions, without any central control. A desirable feature indeed for, say, a routing algorithm in a communications network.

2.3 ANT COLONY OPTIMIZATION: THE TRAVELING SALESMAN PROBLEM

An important aspect of the foraging strategies described in the previous section is that, as an emergent result of the actions of many ants, the shortest path between two or more locations (often between a food source and a nest) is discovered and

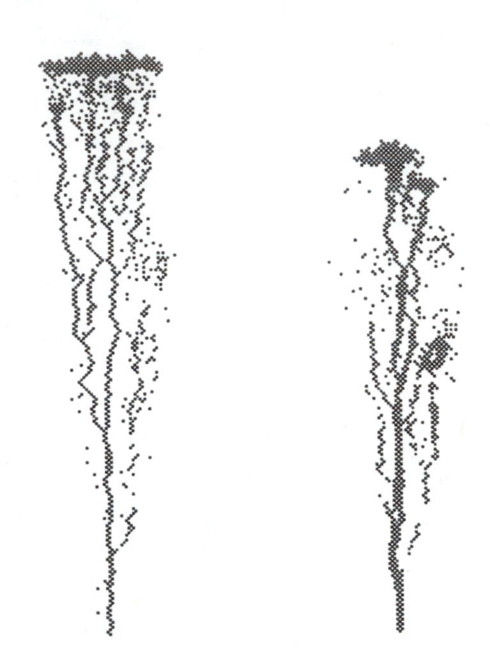

FIGURE 2.11 Patterns obtained from Monte Carlo simulations of the swarm raid model, with two different food distributions. Left: each site contains 1 unit of food with probability 1/2. Right: each site has a probability 1/100 of containing 400 of food units. Insets show prey distributions. After Deneubourg et al. [86] Reprinted by permission © *Plenum Publishing*.

maintained. It is, therefore, natural that the first application of an ant colony algorithm was to a path optimization problem: the traveling salesman problem (TSP).

The TSP was chosen for several reasons:

- It is a shortest path problem to which the ant colony metaphor is easily adapted.
- It is a very difficult (NP-hard) problem [145].
- It has been studied a lot (it is considered to be "the benchmark problem" in combinatorial optimization, see for example Lawler et al. [214] and Reinelt [267]) and therefore many sets of test problems are available, as well as many algorithms with which to run comparisons.
- It is a didactic problem: it is very easy to understand and explanations of the algorithm behavior are not obscured by too many technicalities.

The ant colony optimization approach was initiated by Dorigo [108], in collaboration with Colorni and Maniezzo [76, 77, 107], and has turned out to be more than just a fun metaphor to be presented at exotic conferences: although the first results, obtained with an algorithm called Ant System,[3] were a little bit disappointing, recent developments, which combine the ant colony approach with local searches and/or other optimization methods, are promising.

On the traveling salesman problem (TSP) ant colony optimization algorithms augmented with local search have performance similar to that of the best heuristic approaches (such as Johnson and McGeoch's [191] Iterated Lin-Kernighan and Freisleben and Merz's [133] genetic algorithm), a very encouraging sign.

What is the basic idea underlying all ant-based algorithms? It is to use a *positive feedback* [107, 108] mechanism, based on an analogy with the trail-laying trail-following behavior of some species of ants and some other social insects, to reinforce those portions of good solutions that contribute to the quality of these solutions, or to directly reinforce good solutions. A virtual pheromone, used as reinforcement, allows good solutions to be kept in memory, from where they can be used to make up better solutions. Of course, one needs to avoid some good, but not very good, solutions becoming reinforced to the point where they constrain the search too much, leading to a premature convergence (*stagnation*) of the algorithm. To avoid that, a form of negative feedback is implemented through pheromone evaporation, which includes a time scale into the algorithm. This time scale must not be too large, otherwise suboptimal premature convergence behavior can occur. But it must not be too short either, otherwise no *cooperative behavior* can emerge. Cooperative behavior is the other important concept here: ant colony algorithms make use of the simultaneous exploration of different solutions by a collection of identical ants. Ants that perform well at a given iteration influence the exploration of ants in future iterations. Because ants explore different solutions, the resulting pheromone trail is the consequence of different perspectives on the space of solutions. Even when only the best performing ant is allowed to reinforce its solution, there is a cooperative effect across time because ants in the next iteration use the pheromone trail to guide their exploration.

2.3.1 ANT SYSTEM (AS)

In the TSP the goal is to find a closed tour of minimal length connecting n given cities. Each city must be visited once and only once. Let d_{ij} be the distance between cities i and j. The problem can either be defined in Euclidean space, in which case

$$d_{ij} = \left[(x_i - x_j)^2 + (y_i - y_j)^2\right]^{1/2}, \tag{2.5}$$

[3] Ant System, the first ant colony optimization algorithm, had a performance similar to that of general purpose heuristics, such as simulated annealing or genetic algorithms, on small problems, but did not scale up well.

where x_i and y_i are the coordinates of city i, or can be more generally defined on a graph (N, E) where the cities are the nodes N and the connections between the cities are the edges of the graph E. The graph need not be fully connected. Note that the "distance" matrix need not be symmetric: if it is asymmetric (the corresponding problem is called the asymmetric TSP or ATSP) the length of an edge connecting two cities i and j depends on whether one goes from i to j or from j to i (that is, in general $d_{ij} \neq d_{ji}$). But whether the problem is symmetric or asymmetric does not change how it is being solved by Ant System (AS).

In AS ants build solutions to the TSP by moving on the problem graph from one city to another until they complete a tour. During an iteration of the AS algorithm each ant k, $k = 1, \ldots, m$, builds a tour executing $n = |N|$ steps in which a probabilistic transition rule is applied. Iterations are indexed by t, $1 \leq t \leq t_{\max}$, where $t_{\max}$ is the user defined maximum number of iterations allowed.

For each ant, the transition from city i to city j at iteration t of the algorithm depends on:

1. Whether or not the city has already been visited. For each ant, a memory (also called *tabu list*) is maintained: it grows within a tour, and is then emptied between tours. The memory is used to define, for each ant k, the set J_i^k of cities that the ant still has to visit when it is on city i (at the beginning, J_i^k contains all the cities but i). By exploiting J_i^k an ant k can avoid visiting a city more than once.
2. The inverse of the distance $\eta_{ij} = 1/d_{ij}$, called *visibility*. Visibility is based on strictly local information and represents the *heuristic desirability* of choosing city j when in city i. Visibility can be used to direct ants' search, although a constructive method based on its sole use would produce very low quality solutions. The heuristic information is static, that is, it is not changed during problem solution.
3. The amount of virtual *pheromone trail* $\tau_{ij}(t)$ on the edge connects city i to city j. Pheromone trail is updated on-line and is intended to represent the *learned desirability* of choosing city j when in city i. As opposed to distance, a pheromone trail is a more global type of information. The pheromone trail information is changed during problem solution to reflect the experience acquired by ants during problem solving.

The *transition rule*, that is, the probability for ant k to go from city i to city j while building its tth tour is called *random proportional transition rule* and is given by:

$$p_{ij}^k(t) = \frac{\left[\tau_{ij}(t)\right]^{\alpha} \cdot \left[\eta_{ij}\right]^{\beta}}{\sum_{l \in J_i^k} \left[\tau_{il}(t)\right]^{\alpha} \cdot \left[\eta_{il}\right]^{\beta}}, \tag{2.6}$$

if $j \in J_i^k$, and 0 if $j \notin J_i^k$, where α and β are two adjustable parameters[4] that control the relative weight of trail intensity, $\tau_{ij}(t)$, and visibility, η_{ij}. If $\alpha = 0$, the closest cities are more likely to be selected: this corresponds to a classical stochastic greedy algorithm (with multiple starting points since ants are initially randomly distributed on the nodes). If, on the contrary, $\beta = 0$, only pheromone amplification is at work: this method will lead to the rapid selection of tours that may not be optimal. A tradeoff between tour length and trail intensity therefore appears to be necessary. It is important to note that, although the form of Eq. (2.6) remains constant during an iteration, the value of the probability $p_{ij}^k(t)$ can be different for two ants on the same city i, since $p_{ij}^k(t)$ is a function of J_i^k, that is, of the partial solution built by ant k.

After the completion of a tour, each ant k lays a quantity of pheromone $\Delta\tau_{ij}^k(t)$ on each edge (i,j) that it has used; the value $\Delta\tau_{ij}^k(t)$ depends on how well the ant has performed. At iteration t (the iteration counter is incremented by 1 when all ants have completed a tour), ant k lays $\Delta\tau_{ij}^k(t)$ on edge (i,j):

$$\Delta\tau_{ij}^k(t) = \begin{cases} Q/L^k(t) & \text{if } (i,j) \in T^k(t); \\ 0 & \text{if } (i,j) \notin T^k(t), \end{cases} \tag{2.7}$$

where $T^k(t)$ is the tour done by ant k at iteration t, $L^k(t)$ is its length, and Q is a parameter (although the value of Q only weakly influences the final result, it should be set so that it has a value of the same order of magnitude as that of the optimal tour length, for example, found running a simple constructive heuristic like the nearest neighbor heuristic[5]).

This method could not perform well without pheromone decay: in effect, it would lead to the amplification of the initial random fluctuations, which very probably would not be optimal. In order to ensure efficient solution space exploration, trail intensity must be allowed to decay, otherwise all ants will end up doing the same tour (*stagnation*): because of the additivity of trail intensity, the probabilities of transitions between cities would be dominated by the pheromone term. Trail decay is implemented by introducing a coefficient of decay $\rho, 0 \leq \rho < 1$. The resulting pheromone update rule, which is applied to all edges, is then:

$$\tau_{ij}(t) \leftarrow (1-\rho) \cdot \tau_{ij}(t) + \Delta\tau_{ij}(t)\,, \tag{2.8}$$

where $\Delta\tau_{ij}(t) = \sum_{k=1}^{m} \Delta\tau_{ij}^k(t)$, and m is the number of ants. The initial amount of pheromone on edges is assumed to be a small positive constant τ_0 (that is, there is an homogeneous distribution of pheromone at time $t = 0$).

The total number of ants m, assumed constant over time, is an important parameter: too many ants would quickly reinforce suboptimal trails and lead to early convergence to bad solutions, whereas too few ants would not produce the

[4]The values of all the parameters as used in the experiments are given in the algorithms.

[5]The nearest-neighbor heuristic builds a solution starting from a randomly selected city and repeatedly selecting as the next city the nearest one among those not yet visited.

expected synergistic effects of cooperation because of the (otherwise necessary) process of pheromone decay. Dorigo et al. [109] suggest that to set $m = n$, that is, using as many ants as there are cities in the problem, provides a good tradeoff. At the beginning of each tour, ants are either placed randomly on the nodes (cities), or one ant is placed on each city (no significant difference in performance was observed between the two choices).

Dorigo et al. [109], in an effort to improve AS performance, also introduced "elitist ants" (the term "elitist ant" was chosen due to its similarity with the elitist strategy used in genetic algorithms [171]). An elitist ant is an ant which reinforces the edges belonging to T^+, the best tour found from the beginning of the trial, by a quantity Q/L^+, where L^+ is the length of T^+.

At every iteration e elitist ants are added to the usual ants so that the edges belonging to T^+ get an extra reinforcement $e \cdot Q/L^+$. The idea is that the pheromone trail of T^+, so reinforced, will direct the search of all the other ants in probability toward a solution composed of some edges of the best tour itself. Experiments have shown that a small number of elitist ants can improve the algorithm's performance.

Algorithm 2.1 reports a high-level description of the AS-TSP algorithm, together with parameter values as set by Dorigo et al. [107, 109] in their experiments.

The time complexity of AS is $O(t \cdot n^2 \cdot m)$, where t is the number of iterations done. If $m = n$, that is, if the number of ants is equal to the number of cities, the time complexity becomes $O(t \cdot n^3)$.

Ant System has been compared with other general purpose heuristics on some relatively small traveling salesman problems (these were problems ranging from 30 to 70 cities). The results [109] were very interesting and disappointing at the same time. AS-TSP[6] was able to find and improve the best solution found by a genetic algorithm for Oliver30 [327], a 30-city problem, and it had a performance similar to or better than that of the general purpose heuristics with which it was compared (see Table 2.1). Unfortunately, for problems of growing dimensions (e.g., Eilon50 and Eilon75 [119], see also TSPLIB, a library of traveling salesman problems at: http:// www.iwr.uni-heidelberg.de/ iwr/ comopt/ soft/ TSPLIB95/ TSPLIB.html), AS-TSP never reached the best known solutions within a number of iterations bounded by $t_{\max} = 3,000$, although it exhibited quick convergence to good solutions. Also, its performance level was much lower than that of specialized algorithms for the TSP. The problem here was that all general purpose heuristics are outperformed by specialized algorithms. Current wisdom says that high-performance heuristic algorithms for the TSP, as well as for many other combinatorial problems, must couple good generators of initial solutions with rapid local optimizers [191]. We will see in the next section that an improved version of Ant System augmented with a simple local search reaches outstanding performance.

---

[6]Because Ant System (AS) has been used to attack different problems, we will often add a suffix, such as AS-TSP, to make clear to which problem AS was applied.

Algorithm 2.1 High-level description of AS-TSP

/* **Initialization** */
For every edge (i, j) **do**
  $\tau_{ij}(0) = \tau_0$
End For
For $k = 1$ to m **do**
  Place ant k on a randomly chosen city
End For
Let T^+ be the shortest tour found from beginning and L^+ its length
/* **Main loop** */
For $t = 1$ to $t_{\max}$ **do**
  For $k = 1$ to m **do**
    Build tour $T^k(t)$ by applying $n-1$ times the following step:
    Choose the next city j with probability

$$p_{ij}^k(t) = \frac{[\tau_{ij}(t)]^\alpha \cdot [\eta_{ij}]^\beta}{\sum_{l \in J_i^k} [\tau_{il}(t)]^\alpha \cdot [\eta_{il}]^\beta},$$

    where i is the current city
  End For
  For $k = 1$ to m **do**
    Compute the length $L^k(t)$ of the tour $T^k(t)$ produced by ant k
  End For
  If an improved tour is found **then**
    update T^+ and L^+
  End If
  For every edge (i, j) **do**
    Update pheromone trails by applying the rule:
    $\tau_{ij}(t) \leftarrow (1-\rho) \cdot \tau_{ij}(t) + \Delta\tau_{ij}(t) + e \cdot \Delta\tau_{ij}^e(t)$ where

$$\Delta\tau_{ij}(t) = \sum_{k=1}^{m} \Delta\tau_{ij}^k(t),$$

$$\Delta\tau_{ij}^k(t) = \begin{cases} Q/L^k(t) & \text{if } (i,j) \in T^k(t); \\ 0 & \text{otherwise}, \end{cases}$$

    and

$$\Delta\tau_{ij}^e(t) = \begin{cases} Q/L^+ & \text{if } (i,j) \in T^+; \\ 0 & \text{otherwise}, \end{cases}$$

  End For
  For every edge (i, j) **do**
    $\tau_{ij}(t+1) = \tau_{ij}(t)$
  End For
End For
Print the shortest tour T^+ and its length L^+
Stop
/* **Values of parameters used in experiments** */
$\alpha = 1, \beta = 5, \rho = 0.5, m = n, Q = 100, \tau_0 = 10^{-6}, e = 5$

TABLE 2.1 Comparison of the average best solution generated by Ant System (AS-TSP), tabu search (TS) and simulated annealing (SA). TS is a class of procedures in which a set of solutions recently obtained by the algorithm is kept in memory so as to restrict local choices by preventing some moves to neighboring solutions [150, 151, 168]. SA is a physics-inspired local search method in which a slowly decreasing temperature-like parameter determines a contraction of the search step until the system reaches a state of minimum energy [1, 199]. Each algorithm was given one hour of computing time on an IBM-compatible PC with an 80386 processor. Averages are over 10 runs. Distances between cities are computed using integer numbers. Test problem: Oliver30. From Dorigo et al. [109].

	Best tour	Average	Std. Dev.
AS-TSP	420	420.4	1.3
TS	420	420.6	1.5
SA	422	459.8	25.1

Dorigo et al. [109] also ran a number of experiments aimed at understanding the way AS functions. Among others, a set of experiments showed that the population of solutions generated by AS does not converge to a single common solution. On the contrary, the algorithm continues to produce new, possibly improving, solutions. Figure 2.12 shows the evolution of the best tour length in these experiments, Figure 2.13 the evolution of the population standard deviation, and Figure 2.14 the evolution of the average node branching number. (The branching number of a node is the number of edges that exit from the node, the pheromone concentration of which exceeds a small threshold.) It is clear that, although the best solution produced by AS converges toward the optimal solution within 500 iterations, the population maintains a high diversity, as shown by the relatively high standard deviation value and by the fact that the average branching number has a value greater than 5 after 3,000 iterations. If all generated solutions were identical, the average branching number would quickly drop to 2. This "nonconvergence" property of the population of solutions is interesting because:

1. It tends to avoid the algorithm getting trapped in local optima.
2. It makes AS promising for applications to dynamical problems, that is, problems the characteristics of which change at run time.

2.3.2 ANT COLONY SYSTEM (ACS)

The Ant Colony System (ACS) algorithm has been introduced by Dorigo and Gambardella [111, 112] to improve the performance of Ant System, that was able to find good solutions within a reasonable time only for small size problems. ACS is based on four modifications of Ant System: a different transition rule, a different pheromone trail update rule, the use of local updates of pheromone trail to favor exploration, and the use of a candidate list to restrict the choice of the next city to visit. These are described in the following.

Transition rule. The transition rule is modified to allow explicitly for exploration. An ant k on city i chooses the city j to move to following the rule:

$$j = \begin{cases} \arg\max_{u \in J_i^k} \left\{ [\tau_{iu}(t)] \cdot [\eta_{iu}]^\beta \right\} & \text{if } q \le q_0; \\ J & \text{if } q > q_0, \end{cases} \tag{2.9}$$

where q is a random variable uniformly distributed over $[0, 1]$, q_0 is a tunable parameter $(0 \le q_0 \le 1)$, and $J \in J_i^k$ is a city that is randomly selected according to probability

$$p_{iJ}^k(t) = \frac{\left[\tau_{iJ}(t)\right] \cdot \left[\eta_{iJ}\right]^\beta}{\sum_{l \in J_i^k} \left[\tau_{il}(t)\right] \cdot \left[\eta_{il}\right]^\beta}, \tag{2.10}$$

which is very similar to the transition probability used by Ant System (see Eq. (2.6)). We see therefore that the ACS transition rule is identical to Ant System's when $q > q_0$, and is different when $q \le q_0$. More precisely, $q \le q_0$ corresponds to an exploitation of the knowledge available about the problem, that is, the heuristic knowledge about distances between cities and the learned knowledge memorized in the form of pheromone trails, whereas $q > q_0$ favors more exploration. Cutting exploration by tuning q_0 allows the activity of the system to concentrate on the best solutions instead of letting it explore constantly. It is clear that tuning q_0 is similar to tuning temperature in simulated annealing: when q_0 is close to 1, only the locally optimal solution is selected (but a combination of locally optimal solutions may not result in a globally optimal solution), whereas when q_0 is close to 0,

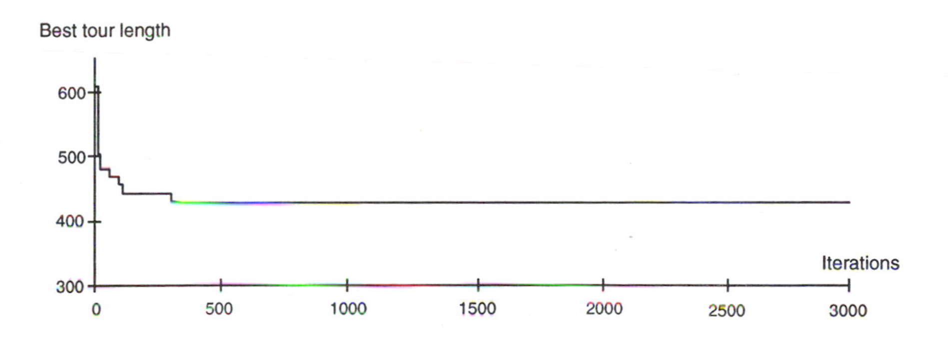

FIGURE 2.12 Evolution of best tour length (Test problem: Oliver30). Typical run. After Dorigo et al. [109]. Reprinted by permission © *IEEE Press*.

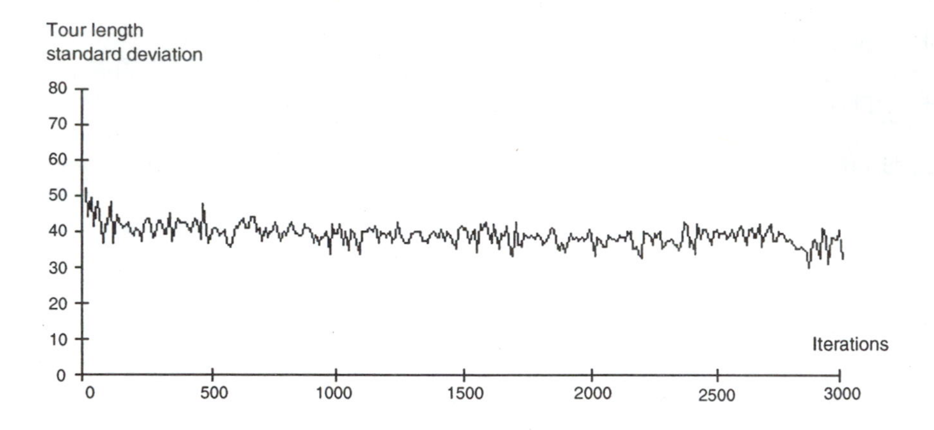

FIGURE 2.13 Evolution of the standard deviation of the population's tour lengths (Test problem: Oliver30). Typical run. After Dorigo et al. [109]. Reprinted by permission © *IEEE Press.*

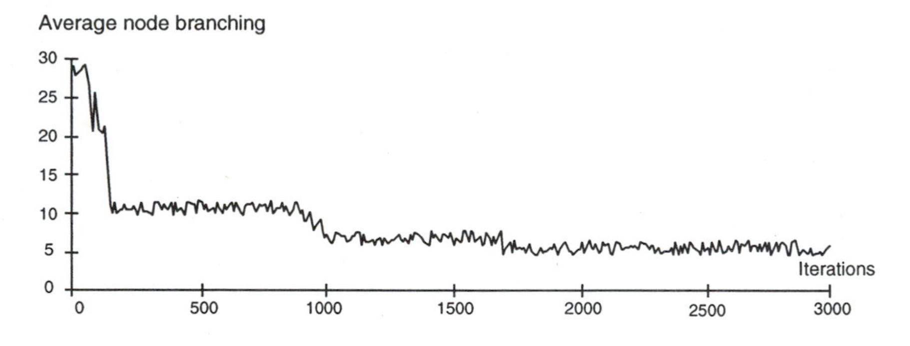

FIGURE 2.14 Evolution of the average node branching of the problem's graph (Test problem: Oliver30). Typical run. After Dorigo et al. [109]. Reprinted by permission © *IEEE Press.*

all local solutions are examined, although a larger weight is given to locally optimal solutions (unlike simulated annealing, where all states have similar weights at high temperature). It is therefore possible in principle to progressively freeze the system by tuning q_0 from 0 to 1, in order to favor exploration in the initial part of the algorithm and then favor exploitation. This possibility has not been explored yet.

Pheromone trail update rule. In Ant System all ants are allowed to deposit pheromone after completing their tours. By contrast, in ACS only the ant that generated the best tour since the beginning of the trial is allowed to globally update the concentrations of pheromone on the branches. The ants therefore are encouraged to search for paths in the vicinity of the best tour found so far. In other words, exploration is more directed. Another difference is that in Ant System the pheromone trail updating rule was applied to all edges, while in ACS the global pheromone trail updating rule is applied only to the edges belonging to the best tour since the beginning of the trial. The updating rule is:

$$\tau_{ij}(t) \leftarrow (1-\rho) \cdot \tau_{ij}(t) + \rho \cdot \Delta\tau_{ij}(t)\,, \tag{2.11}$$

where (i, j)'s are the edges belonging to T^+, the best tour since the beginning of the trial, ρ is a parameter governing pheromone decay, and

$$\Delta\tau_{ij}(t) = 1/L^+\,, \tag{2.12}$$

where L^+ is the length of T^+. We see that this procedure allows only the best tour to be reinforced by a global update. Local updates are however also performed, so that other solutions can emerge.

Local updates of pheromone trail. The local update is performed as follows: when, while performing a tour, ant k is in city i and selects city $j \in J_i^k$, the pheromone concentration of (i, j) is updated by the following formula:

$$\tau_{ij}(t) \leftarrow (1-\rho) \cdot \tau_{ij}(t) + \rho \cdot \tau_0\,. \tag{2.13}$$

The value τ_0 is the same as the initial value of pheromone trails and it was experimentally found that setting $\tau_0 = (n \cdot L_{nn})^{-1}$, where n is the number of cities and L_{nn} is the length of a tour produced by the nearest neighbor heuristic, produces good results.[7]

When an ant visits an edge, the application of the local update rule makes the edge pheromone level diminish. This has the effect of making the visited edges less and less attractive as they are visited by ants, indirectly favoring the exploration of not yet visited edges. As a consequence, ants tend not to converge to a common path. This fact, which was observed experimentally, is a desirable property given that if ants explore different paths then there is a higher probability that one of them will find an improved solution than there is in the case that they all converge to the same tour (which would make the use of m ants pointless). In other words, the role of the ACS local updating rule is to shuffle the tours, so that the early cities in one ant's tour may be explored later in other ants' tours. The effect of local-updating is, therefore, to make the learned desirability of edges change dynamically: every time an ant uses an edge this becomes slightly less desirable (since it loses some of its pheromone). In this way ants will make better use of pheromone information: without local updating all ants would search in a narrow neighborhood of the best previous tour.

---

[7] Gambardella and Dorigo [138] and Dorigo and Gambardella [110] also tested a variant of ACS in which the local update rule was inspired by Q-learning [323], but, despite the increased amount of computation, this did not produce significantly better results than using the fixed value τ_0.

Use of a candidate list. ACS exploits a candidate list, that is, a data structure commonly used when trying to solve large TSP instances. A candidate list is a list of preferred cities to be visited from a given city: instead of examining all the possibilities from any given city, unvisited cities in the candidate list are examined first, and only when all cities in the candidate list have been visited are other cities examined. The candidate list of a city contains *cl* cities (*cl* being a parameter of the algorithm), which are the *cl* closest cities. Cities in the candidate list are ordered by increasing distance, and the list is scanned sequentially. ACS-TSP with candidate list works as follows: an ant first restricts the choice of the next city to those in the list, and considers the other cities only if all the cities in the list have already been visited. If there are cities in the candidate list, then the next city j is chosen according to Eqs. (2.9) and (2.10). Otherwise j is set to the closest of the yet unvisited cities.

Algorithm 2.2 reports a high-level description of ACS-TSP, the ACS algorithm applied to the TSP.

ACS-TSP was tested on problems of various sizes and compared with other algorithms, such as the elastic net algorithm (EN) [117], self-organizing maps (SOM), and simulated annealing (SA) [262], genetic algorithms (GA) [20, 327], and evolutionary programming (EP) [125]. Table 2.2 gives a comparison of results (average tour length over 25 runs) obtained by ACS-TSP on randomly generated sets of symmetric 50-city problems with those obtained by SA, EN, and SOM. ACS-TSP was run for 2500 iterations with 10 ants, which leads to approximately the same number of tours being generated by every algorithm. It can be seen that ACS-TSP produces the best results in most cases.

Table 2.3 shows a comparison of results obtained with ACS-TSP with those obtained with GA, EP, and SA on a set of geometric problems (geometric problems are problems taken from the real world, like finding the shortest tour that visits all the European Union states capitals). ACS-TSP was run for 1250 iterations with 20 ants, which, again, leads to approximately the same number of tours being generated by every algorithm.

In order to apply ACS-TSP to larger problems, Dorigo and Gambardella [111] introduced a local search procedure to be performed in combination with ACS. This means combining ACS with a procedure used for iteratively improving a solution. In the case of the TSP, the most widespread local search procedures are 2-opt and 3-opt [220], and Lin-Kernighan (LK) [221], whereby two, three, and a variable number of edges are exchanged iteratively until a local minumum is reached. Dorigo and Gambardella [111] used 3-opt, because 3-opt can be used for both symmetric and asymmetric instances of TSP. They resorted to a variant of 3-opt, a restricted 3 -opt procedure, in which only 3-opt moves that do not revert the order in which the cities are visited are applied: this amounts to changing three edges (k, l), (p, q), and (r, s) into three other edges (k, q), (p, s), and (r, l) while maintaining the orientations of all other subtours. This 3-opt procedure avoids unpredictable tour length changes due to the inversion of a subtour.

Algorithm 2.2 High-level description of ACS-TSP

```
/* Initialization */
For every edge (i, j) do
  τ_ij(0) = τ_0
End For
For k = 1 to m do
  Place ant k on a randomly chosen city
End For
Let T+ be the shortest tour found from beginning and L+ its length
/* Main loop */
For t = 1 to t_max do
  For k = 1 to m do
    Build tour T^k(t) by applying n − 1 times the following steps:
    If exists at least one city j ∈ candidate list then
      Choose the next city j, j ∈ J_i^k, among the cl cities in the candidate list as follows
```

$$j = \begin{cases} \arg\max_{u \in J_i^k} \{[\tau_{iu}(t)] \cdot [\eta_{iu}]^\beta\} & \text{if } q \leq q_0; \\ J & \text{if } q > q_0, \end{cases}$$

```
      where J ∈ J_i^k is chosen according to the probability:
```

$$p_{ij}^k(t) = \frac{[\tau_{ij}(t)] \cdot [\eta_{ij}]^\beta}{\sum_{l \in J_i^k} [\tau_{il}(t)] \cdot [\eta_{il}]^\beta},$$

```
      and where i is the current city
    Else
      choose the closest j ∈ J_i^k
    End If
    After each transition ant k applies the local update rule:
    τ_ij(t) ← (1 − ρ) · τ_ij(t) + ρ · τ_0
  End For
  For k = 1 to m do
    Compute the length L^k(t) of the tour T^k(t) produced by ant k
  End For
  If an improved tour is found then
    update T+ and L+
  End If
  For every edge (i, j) ∈ T+ do
    Update pheromone trails by applying the rule:
    τ_ij(t) ← (1 − ρ) · τ_ij(t) + ρ · Δτ_ij(t) where Δτ_ij(t) = 1/L+
  End For
  For every edge (i, j) do
    τ_ij(t + 1) = τ_ij(t)
  End For
End For
Print the shortest tour T+ and its length L+
Stop
/* Values of parameters used in experiments */
β = 2, ρ = 0.1, q_0 = 0.9, m = 10, τ_0 = (n · L_nn)^−1, cl = 15
```

TABLE 2.2 Comparison of ACS-TSP with other algorithms on randomly generated problems. Results for EN are from Durbin and Willshaw [117] and results for SA and SOM are from Potvin [262]. For ACS-TSP, averages are over 25 runs. Best results are in boldface. From Dorigo and Gambardella [111].

	ACS-TSP	SA	EN	SOM
City set 1 (50-city problem)	**5.88**	**5.88**	5.98	6.06
City set 2 (50-city problem)	6.05	**6.01**	6.03	6.25
City set 3 (50-city problem)	**5.58**	5.65	5.70	5.83
City set 4 (50-city problem)	**5.74**	5.81	5.86	5.87
City set 5 (50-city problem)	**6.18**	6.33	6.49	6.70

TABLE 2.3 Comparison of ACS-TSP with GA, EP, and SA on four test problems (available at TSPLIB: http://www.iwr.uni-heidelberg.de/iwr/comopt/soft/TSPLIB95/TSPLIB.html). Results for ACS-TSP are from Dorigo and Gambardella [111]. ACS-TSP is run for 1250 iterations using 20 ants, which amounts to approximately the same number of tours searched by the other heuristics; averages are over 15 runs. Results for GA are from Bersini et al. [20] for the KroA100 problem, and from Whitley et al. [327] for the Oliver30, Eil50 and Eil75 problem; results for EP are from Fogel [125]; and results for SA from Lin et al. [222]. For each algorithm results are given in two columns: the first one gives the best integer tour length (that is, the tour length obtained when distances among cities are given as integer numbers), and, in parenthesis, the best real tour length (that is, the tour length obtained when distances among cities are given as real numbers); the second one gives the number of tours that were generated before the best integer tour length was discovered. Best results are in boldface. (N/A: Not available.)

	ACS-TSP best	ACS-TSP # iter.	GA best	GA # iter.	EP best	EP # iter.	SA best	SA # iter.
Eil50 (50-city problem)	**425** (427.96)	**1830**	428 (N/A)	25000	426 **(427.86)**	100000	443 (N/A)	68512
Eil75 (75-city problem)	**535** **(542.37)**	**3480**	545 (N/A)	80000	542 (549.18)	325000	580 (N/A)	173250
KroA100 (100-city problem)	**21282** **(21285.44)**	**4820**	21761 (N/A)	103000	N/A (N/A)	N/A	N/A (N/A)	N/A

Algorithm 2.3 The local optimization step in ACS-3-opt

For $k = 1$ to m **do**
$T^k(t) \leftarrow$ 3-opt $(T^k(t))$ {apply the local optimizer to each $T^k(t)$}
End For

/* **Values of parameters used in experiments** */
$\beta = 2, \rho = 0.1, q_0 = 0.98, m = 10, \tau_0 = (n \cdot L_{nn})^{-1}, cl = 20$

TABLE 2.4 Comparison of results, over 10 runs, obtained by ACS-3-opt (the results are from Dorigo and Gambardella [111]) on symmetric TSP problems of various sizes with those obtained by the genetic algorithm (STSP—the results are the best between those published by Freisleben and Merz [133, 134]) which won the First International Contest on Evolutionary Optimization [21]. Best results are in boldface.

	ACS-3-opt best	ACS-3-opt average	STSP best	STSP average
d198 (198-city problem)	**15780**	15781.7	**15780**	**15780**
lin318 (318-city problem)	**42029**	**42029**	**42029**	**42029**
att532 (532-city problem)	27693	27718.2	**27686**	**27693.7**
rat783 (783-city problem)	8818	8837.9	**8806**	**8807.3**

The ACS algorithm with restricted 3-opt procedure, ACS-3-opt, is the same as ACS except for the fact that the tour $T^k(t)$ produced by each ant k is taken to its local optimum. The "for cycle" reported in Algorithm 2.3 is applied at every iteration of the algorithm, after each ant k has completed the construction of its tour $T^k(t)$ and before the best tour T^+ and its length L^+ are updated. Note that, as indicated in Algorithm 2.3, some algorithm parameters were set to different values from those used in ACS.

Table 2.4 shows results of simulations of ACS-3-opt (with candidate list) performed on symmetric problems, compared with results obtained by the genetic algorithm (STSP) [133, 134] that won the First International Contest on Evolutionary Optimization [21]. Table 2.5 shows results obtained on asymmetric TSP instances. The performance of the two algorithms is similar, with the genetic algorithm having a slightly better performance on symmetric problems and ACS-3-opt on the asymmetric ones. Note that in both cases the genetic algorithm operators were finely tuned to the TSP application and that a local optimizer took the solutions generated by the GA to the local optimum.

TABLE 2.5 Comparison of results, over 10 runs, obtained by ACS-3-opt (the results are from Dorigo and Gambardella [111]) on asymmetric TSP problems of various sizes with those obtained by the genetic algorithm (ATSP—the results are from Freisleben and Merz [133, 134]) which won the First International Contest on Evolutionary Optimization [21]. Best results are in boldface.

	ACS-3-opt best	ACS-3-opt average	ATSP best	ATSP average
p43 (43-city problem)	**2810**	**2810**	**2810**	**2810**
ry48p (48-city problem)	**14422**	**14422**	**14422**	14440
ft70 (70-city problem)	**38673**	**38679.8**	**38673**	38683.8
kro124p (124-city problem)	**36230**	**36230**	**36230**	36235.3
ftv170 (170-city problem)	**2755**	**2755**	**2755**	2766.1

As further, not implemented improvements of ACS, Dorigo and Gambardella [111] suggest the following possible additions:

1. Allow the two best ants (or more generally, the best r ants) to update the pheromone trail, instead of using a single ant. This should make the probability of getting trapped in local minima smaller.
2. Remove pheromone from edges that belong to the worst tours. Giving negative reinforcement to bad tours should increase the convergence speed toward good solutions.
3. Use a more powerful local search procedure.

Finally, the "nonconvergence" property of AS (see section 2.3.1), which is common to many swarm-based systems, can also be observed in ACS. As an example, Figure 2.15 shows that the pheromone level on different edges of a problem graph can greatly differ. The level of pheromone for a run of ACS is represented by line thickness in Figure 2.15(a). Some edges are strongly marked, while others are weakly marked. Strongly marked edges are most likely to be part of the best solution found by the system. However, more weakly marked edges point to potential "alternative" solutions. The continued availability of these weakly marked edges helps ACS maintain a diverse population of solutions, thereby avoiding convergence to a single solution. That is a nice and desirable property because it can help avoid getting trapped in local optima. But there is more: imagine for example that the problem, instead of being static, is dynamic with nodes and/or edges being

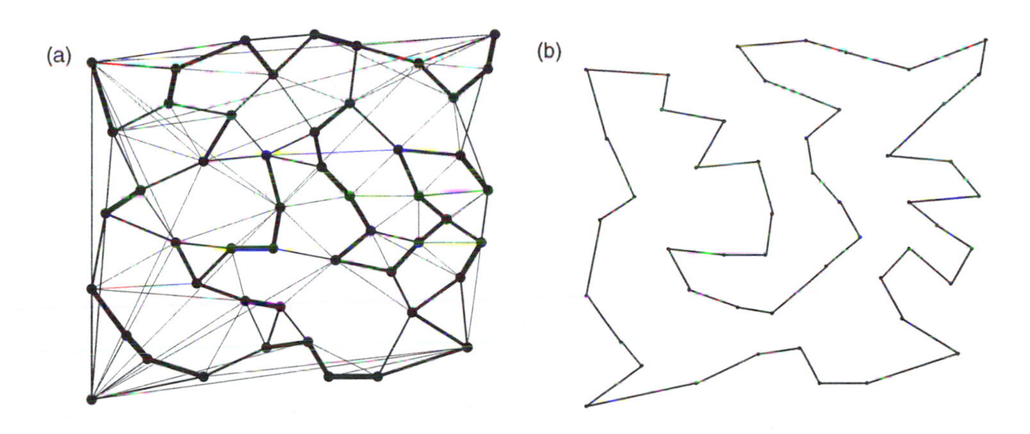

FIGURE 2.15 (a) An example of a trail configuration found by ACS in a 50-city problem (Eil50). Line thickness reflects pheromone concentration. (b) Best solution found by ants.

dynamically added to, or removed from, the graph. In this case, the trail configuration provides a set of alternative pathways which can be used to quickly produce a new solution. Although this type of problem would need to be more precisely formulated by allowing for the design of dynamic benchmark problems, we believe it is on dynamic problems that swarm-based algorithms might exhibit particularly good performance. This latter conjecture remains to be confirmed.

2.3.3 OTHER METHODS BASED ON AS OR ACS

Several suggestions have been made to modify the original AS algorithm in order to improve the quality of its solutions, but not necessarily its convergence time.

Stützle and Hoos [298, 299] have introduced Max-Min AS (hereafter MMAS), which is basically similar to AS, but (i) only one ant (the best in the current iteration) is allowed to update pheromone trails, (ii) pheromone trail values are restricted to an interval $[\tau_{\min}, \tau_{\max}]$, and (iii) trails are initialized to their maximum value $\tau_{\max}$. Putting a bound, $\tau_{\max}$, on the maximum value a pheromone trail can have avoids some edges being reinforced to the point where it is impossible to make a path without them. This helps to avoid stagnation, which was one of the reasons why AS performed poorly (in terms of solution quality) on large problems. Stützle and Hoos [298, 299] have also added what they called a "trail smoothing mechanism" to further avoid stagnation. They implement this mechanism through a proportional update mechanism: $\Delta\tau_{ij}(t) \propto (\tau_{\max} - \tau_{ij}(t))$. In this way stronger pheromone trails are proportionally less reinforced than weaker ones, a procedure that obviously favors the exploration of new paths. They found that, when applied to the TSP, MMAS finds significantly better tours than AS, but comparable to those obtained with ACS. Stützle and Hoos [298, 299] have also proposed adding a local search procedure (2-opt or 3-opt for ATSP) to their system, in a way similar to Dorigo and Gambardella [111].

Bullnheimer et al. [49] proposed yet another modification of AS, called AS-rank. Their modification is based on (i) using elitist ants, as was done in Dorigo et al. [109] (see also section 2.3.1), and (ii) ranking the m ants by tour length $(L^1(t), L^2(t), \ldots, L^m(t))$ and making ants update the edges with a quantity of pheromone proportional to their rank. They used σ elitist ants, and only the best $\sigma - 1$ ants were allowed to deposit a pheromone trail. The weight of a given ant's update was set to $\max\{0, \sigma - \mu\}$, where μ is the ant's rank; in this way no ant could give a contribution higher than that given by the elitist ants.

The dynamics of the amount of pheromone $\tau_{ij}(t)$ on edge (i, j) is given by:

$$\tau_{ij}(t) \leftarrow (1 - \rho) \cdot \tau_{ij}(t) + \sigma \cdot \Delta\tau_{ij}^{+}(t) + \Delta\tau_{ij}^{r}(t)\,, \tag{2.14}$$

where $\Delta\tau_{ij}^{+}(t) = Q/L^{+}(t)$, L^{+} being, as usual, the length of the best solution from trial beginning, and $\Delta\tau_{ij}^{r}(t) = \sum_{\mu=1}^{\sigma-1} \Delta\tau_{ij}^{\mu}(t)$, $\Delta\tau_{ij}^{\mu}(t) = (\sigma - \mu) \cdot Q/L^{\mu}(t)$ if ant μ uses (i, j) and $\Delta\tau_{ij}^{\mu}(t) = 0$ otherwise. $L^{\mu}(t)$ is the length of the tour performed by ant μ at iteration t, and Q was set to the value $Q = 100$. Bullnheimer et al. [49] found that this new procedure significantly improves the quality of the results obtained with AS, but did not try to compare them with those obtained with ACS.

2.4 ANT COLONY OPTIMIZATION: THE QUADRATIC ASSIGNMENT PROBLEM

The quadratic assignment problem (QAP) has been introduced by Koopmans and Beckman [201] in an economic context. The problem can be stated as follows. Consider a set of n activities that have to be assigned to n locations (or vice versa). A matrix $D = [d_{ij}]_{n,n}$ gives distances between locations, where d_{ij} is the Euclidean distance between location i and location j, and a matrix $F = [f_{hk}]_{n,n}$ characterizes flows among activities (transfers of data, material, humans, etc.), where f_{hk} is the flow between activity h and activity k. An assignment is a permutation π of $\{1, \ldots, n\}$, where $\pi(i)$ is the activity that is assigned to location i. The problem is to find a permutation π_m such that the product of the flows among activities be minimized by the distances between their locations. Formally, the QAP can be formulated as the problem of finding the permutation π which minimizes the following objective function:

$$C(\pi) = \sum_{i,j=1}^{n} d_{ij} f_{\pi(i)\pi(j)}\,. \tag{2.15}$$

The optimal permutation π_{opt} is defined by

$$\pi_{opt} = \arg \min_{\pi \in \Pi(n)} C(\pi)\,, \tag{2.16}$$

where $\Pi(n)$ is the set of all permutations of $\{1, \ldots, n\}$.

The QAP has been shown to be NP-hard [279], and the problem rapidly becomes intractable with growing instance size. Heuristics have, therefore, been developed to find good solutions in a reasonably short computation time. In the following sections we will illustrate how this problem can be attacked by ant colony optimization algorithms.

2.4.1 ANT SYSTEM FOR THE QAP

AS-QAP, the first application of an ant colony algorithm to the QAP, was proposed by Maniezzo et al. [233]. In their application of ant colony optimization to the QAP, the graph representation is similar to the one used in the TSP case: the nodes of the graph are locations, and the ants' goal is to visit all the locations and to couple one activity with each location. The pheromone trail τ_{ij}, which in the QAP case represents the learned desirability of setting $\pi(i) = j$ in the solution, is deposited on edges and used to build solutions as in AS-TSP. Also the constraint satisfaction method is very similar to the one used for the TSP: while building a permutation, ants memorize the activities still to be assigned and the locations which have already been assigned an activity.

The only tricky point was then to devise a heuristic that allows a constructive definition of the solutions. This was done as follows. Consider the vectors $\bar{d}$ and $\bar{f}$ called *distance potentials* and *flow potentials*. Each of these vectors is obtained by summing the elements of each line of the two matrices of distances D and of flows F:

$$d_i = \sum_{j=1}^{n} d_{ij} \quad \{i = 1, \ldots, n\}, \tag{2.17}$$

$$f_h = \sum_{k=1}^{n} f_{hk} \quad \{h = 1, \ldots, n\}. \tag{2.18}$$

The potential values indicate, respectively, the sum of the distances between a particular node and all the others, and the total flow exchanged between an activity and all the others. The lower the value d_i (distance potential) of the ith node, the more this node is barycentric in the network; the higher the value f_h (flow potential) of activity h, the more important this activity is in the system of flows exchanged.

As an example, consider a randomly generated problem of order 4 characterized by the following matrices D and$F$:

$$D = \begin{bmatrix} 0 & 10 & 4 & 2 \\ 10 & 0 & 6 & 4 \\ 4 & 6 & 0 & 1 \\ 2 & 4 & 1 & 0 \end{bmatrix}$$

$$F = \begin{bmatrix} 0 & 3 & 8 & 3 \\ 3 & 0 & 2 & 4 \\ 8 & 2 & 0 & 5 \\ 3 & 4 & 5 & 0 \end{bmatrix}$$

The vectors of the potentials $\bar{d}$ and $\bar{f}$ are:

$$\bar{d} = \begin{bmatrix} 16 \\ 20 \\ 11 \\ 7 \end{bmatrix}$$

$$\bar{f} = \begin{bmatrix} 14 \\ 9 \\ 15 \\ 12 \end{bmatrix}$$

Location 4 (with $d_4 = 7$) is in a central position with respect to all the others, while location 2 (with $d_2 = 20$) is in a peripheric position. Concerning the activities, the third of these (with $f_3 = 15$) is the one which has most exchanges with the others, while the second ($f_2 = 9$), is the least active. It is possible to obtain good, although probably not optimal, permutations via a *min-max coupling rule*: activities with high flow potentials are coupled with high probability to locations with low distance potentials (and vice versa). This is explained in the following.

Let $B = \bar{d} \cdot \bar{f}^T$ be the coupling matrix in which the element b_{ij} is given by the product $d_i \cdot f_j$. In the above example, this corresponds to the following values:

$$B = \begin{bmatrix} 224 & 144 & 240 & \underline{192} \\ 280 & \underline{180} & 300 & 240 \\ \underline{154} & 99 & 165 & 132 \\ 98 & 63 & \underline{105} & 84 \end{bmatrix}$$

A solution is constructed as follows (we assume, for simplicity, that ants act in a deterministic way and not probabilistically, as in reality happens in the algorithm). Consider the row of matrix B corresponding to the location with lowest distance potential (row 4 in the example). The activity with highest flow potential is assigned to this location according to the min-max rule. In the example, activity 3 is assigned to location 4 because the element b_{43} is the greatest of row 4. The procedure is repeated considering the location with lowest distance potential among those not yet coupled (location 3 in the example). This location is coupled with activity 1 (the coupling location 3—activity 3 is inhibited because activity 3 was already assigned to location 4); the procedure is iterated to obtain the couplings 1–4 and 2–2. In the above matrix the costs of the assignments made are underlined.

Note that the values given to location-activity couplings by the above min-max rule, and more in general the values b_{ij} of matrix B, only give an indication of the "potential goodness" of the couplings, and do not represent the effective cost. These values are used to compute the heuristic desirability η_{ij} of assigning activity j to location i. η_{ij} is set to the value of the corresponding element of the coupling matrix:

$$\eta_{ij} = b_{ij} = d_i \cdot f_j \,. \tag{2.19}$$

Once the "heuristic matrix" is built, the algorithm works exactly as in the TSP case. The transition probability $p_{ij}^k(t)$ becomes an "assignment probability," and represents the probability that the kth ant chooses activity j as the activity to assign to location i. Its value is

$$p_{ij}^k(t) = \frac{[\tau_{ij}(t)]^\alpha \cdot [\eta_{ij}]^\beta}{\sum_{l \in J_i^k} [\tau_{il}(t)]^\alpha \cdot [\eta_{il}]^\beta} \,, \tag{2.20}$$

if $j \in J_i^k$, and 0 if $j \notin J_i^k$, where J_i^k, similarly to the TSP case, is the set of feasible activities, that is, the set of activities ant k still has to assign.

A difference from the TSP application is that, in the QAP case, the construction of the permutation starts from the row (index i) corresponding to the location with the lowest distance potential and the coupling is made by choosing probabilistically from all the activities; at the second step the second location is assigned in order of potential, choosing probabilistically an activity that excludes the ones already coupled, and so on. The procedure is repeated for all the n rows. The solution construction is repeated m times, as many times as there are ants in the population.

The parameters α and β allow the user to define the relative importance of the pheromone trail $\tau_{ij}(t)$ with respect to the heuristic desirability η_{ij} (as in the previous section, all the parameters' values as used in the experiments are reported in the algorithms). Thus the probability $p_{ij}^k(t)$ is a compromise between the heuristic desirability of a coupling (activities with high flow potential should be assigned to locations with low distance potential) and the pheromone trail intensity (if coupling (i, j) received a lot of pheromone in the recent past, then this coupling is probably very desirable).

The pheromone trail update rule is the same as in the TSP case:

$$\tau_{ij}(t) \leftarrow (1 - \rho) \cdot \tau_{ij}(t) + \sum_{k=1}^{m} \Delta\tau_{ij}^k(t) \,, \tag{2.21}$$

where $\Delta\tau_{ij}^k$ is set to $Q/C^k(t)$ if the kth ant has chosen coupling (i, j), to 0 otherwise; m is the number of ants, Q and $0 \leq \rho < 1$ are two parameters, while $C^k(t)$ is the value of the objective function obtained by the kth ant at iteration t. The pheromone trail update rule causes the couplings which determine a low value of the objective function (i.e., a low $C^k(t)$ value), to have a stronger pheromone trail.

Algorithm 2.4 reports a high-level description of AS-QAP, the AS algorithm applied to the QAP, as well as the values of parameters as set by Maniezzo et al. [233].

Algorithm 2.4 High-level description of AS-QAP

/* **Initialization** */
For every coupling (i, j) **do**
$\tau_{ij}(0) = \tau_0$
End For
Compute the distance and flow potentials and the coupling matrix B
Set the heuristic desirability to $\eta_{ij} = b_{ij}$
For $k = 1$ to m **do**
Place ant k on a randomly chosen city
End For

/* **Main loop** */
For $t = 1$ to $t_{\max}$ **do**
For $k = 1$ to m **do**
Build solution $A^k(t)$ by applying $n - 1$ times the following two steps:
(1) Choose the location i with the lowest distance potential among those not yet assigned
(2) Choose the activity j to assign to location i by the probabilistic rule

$$p_{ij}^k(t) = \frac{[\tau_{ij}(t)]^\alpha \cdot [\eta_{ij}]^\beta}{\sum_{l \in J_i^k} [\tau_{il}(t)]^\alpha \cdot [\eta_{il}]^\beta}$$

End For
For $k = 1$ to m **do**
Compute the cost $C^k(t)$ of the assignment $A^k(t)$ produced by ant k
End For
If an improved assignment is found **then**
update best assignment found
End If
For every coupling (i, j) **do**
Update pheromone trails by applying the rule:
$\tau_{ij}(t) \leftarrow (1 - \rho) \cdot \tau_{ij}(t) + \Delta\tau_{ij}(t)$ where

$$\Delta\tau_{ij}(t) = \sum_{k=1}^{m} \Delta\tau_{ij}^k(t),$$

and

$$\Delta\tau_{ij}^k(t) = \begin{cases} Q/C^k(t) & \text{if } (i, j) \in A^k(t); \\ 0 & \text{otherwise}. \end{cases}$$

End For
For every coupling (i, j) **do**
$\tau_{ij}(t + 1) = \tau_{ij}(t)$
End For
End For
Print the best assignment
Stop
/* **Values of parameters used in experiments** */
$\alpha = 1, \beta = 1, \rho = 0.9, m = n, Q = 10, \tau_0 = 10^{-6}$

TABLE 2.6 A comparison between simulated annealing (SA), tabu search (TS), genetic algorithms (GA), evolution strategies (ES), sampling and clustering (SC), Ant System (AS-QAP), and two versions of Ant System enriched with local search (AS-LS and AS-SA). Best results are in boldface.

	Nugent (7)	Nugent (12)	Nugent (15)	Nugent (20)	Nugent (30)	Elshafei (19)	Krarup (30)
SA	**148**	**578**	**1150**	**2570**	6128	17937024	89800
TS	**148**	**578**	**1150**	**2570**	**6124**	**17212548**	90090
GA	**148**	588	1160	2688	6784	17640584	108830
ES	**148**	598	1168	2654	6308	19600212	97880
SC	**148**	**578**	**1150**	**2570**	6154	**17212548**	**88900**
AS-QAP	**148**	**578**	**1150**	2598	6232	18122850	92490
AS-LS	**148**	**578**	**1150**	**2570**	6146	**17212548**	89300
AS-SA	**148**	**578**	**1150**	**2570**	6128	**17212548**	**88900**

The results obtained with the above-described algorithm were compared with some other heuristics with good, although not extraordinary, results (see Table 2.6): on a set of standard problems of dimensions in the range 5 to 30, AS-QAP outperformed two versions of evolutionary algorithms (the genetic algorithm GA and evolution strategies ES) but had a performance worse than that of simulated annealing (SA), tabu search (TS), and sampling and clustering (SC) [25]. AS-QAP was also enriched with two different types of local search procedure (AS-LS and AS-SA in Table 2.6), which increased its performance in a nondramatic way (AS-QAP plus local search performed at approximately the same level as SC and TS). Results reported in Table 2.6 were obtained with a PC-386, 16 MHz, with a constant computation time of one hour.

2.4.2 HYBRID ANT SYSTEM

Gambardella et al. [140] proposed a new ant colony optimization algorithm, called Hybrid Ant System (HAS-QAP), which departs radically from the previously described ACO algorithms. The main novelties are: (i) ants modify solutions, as opposed to building them, and (ii) pheromone trails are used to guide solutions modifications, and not as an aid to direct their construction. In the initialization phase, each ant k, $k = 1, \ldots, m$, is given a permutation π^k which consists of a randomly generated permutation taken to its local optimum by the local search procedure explained in the following. The trail values $\tau_{ij}(t)$ are initialized by the value $\tau_0 = 1/(100 \cdot C(\pi^+))$, where π^+ is the best among the m initial permutations, and $C(\pi^+)$ is its cost.

The algorithm is then composed of three procedures, called (i) pheromone-trail-based modification, (ii) local search, and (iii) pheromone trail updating.

Pheromone-trail-based modification. Each ant modifies its permutation π^k by applying a set of R swaps, where R is an algorithm parameter: an index i is chosen randomly in $\{1,\ldots,n\}$, a second index $j \neq i$ is selected in $\{1,\ldots,n\}$, and $\pi^k(i)$ and $\pi^k(j)$ are swapped. The selection of j is done in one of two ways: (i) with probability q, j is selected such that $\tau_{i\pi^k(j)} + \tau_{j\pi^k(i)}$ is the maximum (which corresponds to an exploitation of the knowledge memorized in the pheromone trails), and (ii) with probability $1-q$, j is selected with probability

$$p_{ij}^k = \frac{\tau_{i\pi^k(j)} + \tau_{j\pi^k(i)}}{\sum_{l=1,l\neq i}^{n} \left(\tau_{i\pi^k(l)} + \tau_{l\pi^k(i)}\right)}, \tag{2.22}$$

which corresponds to a biased exploratory strategy. These R swaps generate a solution $\tilde{\pi}^k$.

Local search. The solution $\tilde{\pi}^k$ is transformed into a new solution $\hat{\pi}^k$ by a local search procedure. The local search procedure examines all possible swaps in a random order and accepts any swap that improves the current solution. The difference $\Delta(\pi, i, j)$ in the objective function due to the swap of $\pi(i)$ and $\pi(j)$ in a given permutation π can be evaluated in $O(n)$ steps with the formula:

$$\begin{aligned}\Delta(\pi,i,j) = {} & (d_{ii} - d_{jj})\left(f_{\pi(j)\pi(j)} - f_{\pi(i)\pi(i)}\right) \\ & + (d_{ij} - d_{ji})\left(f_{\pi(j)\pi(i)} - f_{\pi(i)\pi(j)}\right) \\ & + \sum_{\substack{k=1\\k\neq i\\k\neq j}}^{n} \Big\{ (d_{ki} - d_{kj})\left(f_{\pi(k)\pi(j)} - f_{\pi(k)\pi(i)}\right) \\ & + (d_{ik} - d_{jk})\left(f_{\pi(j)\pi(k)} - f_{\pi(i)\pi(k)}\right) \Big\}.\end{aligned} \tag{2.23}$$

Because there are $n(n-1)$ possible swaps, the local search procedure is performed in $O(n^3)$ steps. Notice that the resulting $\hat{\pi}^k$ is usually better than $\tilde{\pi}^k$, but need not be a local optimum.

Pheromone trail updating. When all ants have performed pheromone-trail-based modifications and local search, the pheromone trail matrix is updated. The pheromone trail update rule is similar to that of AS-TSP and AS-QAP, while the updating term is similar to that of ACS-TSP. In fact, as in AS, all the edges are updated, but only the ones belonging to π^+, the best assignment from the beginning of the trial, receive a positive reinforcement. The pheromone trail updating rule is:

$$\tau_{i\pi(i)}(t) \leftarrow (1-\rho)\cdot\tau_{i\pi(i)}(t) + \rho\cdot\Delta\tau_{i\pi(i)}(t), \tag{2.24}$$

where $\Delta\tau_{i\pi(i)}(t) = 1/C(\pi^+)$ if $(i,\pi(i)) \in \pi^+$, and 0 otherwise. As usual, ρ, $0 \leq \rho < 1$, represents the pheromone evaporation rate.

A problem with this schema is that, since trail updating is governed by the value $C(\pi^+)$ of solution π^+ at a given iteration, the trail distribution is determined by past solutions π^+'s. When a new solution π^{k^+} is found such that $\pi^{k^+} < \pi^+$, π^{k^+}

becomes the new π^+. In general, it will take some iterations before the influence of the new π^+ on pheromone trail distribution will have all its effect. Therefore, a procedure called *intensification* is activated when the best solution so far has been improved: the best permutation found between π^k and $\hat{\pi}^k$ is kept in memory, and $\hat{\pi}^k$ is replaced by that best permutation. Intensification allows a more intense exploration of the neighborhood of newly found good solutions.

Finally, to avoid stagnation, a *diversification* procedure is activated if no improvement has been made to the current best solution in the last S iterations. When diversification is activated, all pheromone trail values are reinitialized (i.e., they are all set to the value $\tau_0 = 1/(100 \cdot C(\pi^+))$, where π^+ is the best solution so far), $m-1$ ants are given a new randomly generated solution, and one ant is given the best solution π^+ so far.

Algorithm 2.5 presents a high-level description of the HAS-QAP algorithm.

The algorithm has been tested on two types of problems: in the QAP case, the performance of an algorithm usually depends on the structure of the distance and flow matrices. Gambardella et al. [140] divided their set of problems into problems where flow and/or distance matrices have "regular" entries, that is, entries that follow the same statistical distribution, and those that are characterized by at least one of the matrices having irregular and highly variable entries.[8] The latter class of problems includes most real-world problems. Many heuristics perform well on the former class of problems, where a lot of good solutions are "easily" accessible because they are spread in the space of feasible solutions, and perform poorly on the latter. The quality of the solutions found by the algorithm has been compared with the quality of solutions found by the best performing algorithms for the QAP, as well as with a general heuristic, simulated annealing (SA). The QAP-specific algorithms considered are two tabu searches, the reactive tabu search (RTS) of Battiti and Tecchiolli [9] and the robust tabu search of Taillard (TT) [306], and the genetic hybrid tabu search approach introduced by Fleurent and Ferland [123]. All problem instances can be found in the QAPLIB library compiled by Burkard et al. [51] (http://fmtbhpl.tu-graz.ac.at/~karisch/qaplib).

Two types of experiments were run for each problem typology: "short runs" and "long runs." In short runs the Hybrid Ant System (HAS-QAP) was run for 10 iterations, which, from a computing time point of view, is equivalent to about $100 \cdot n$ iterations of the RTS and TT algorithms, where n is problem size, $1250 \cdot n^2$ iterations of SA and 25 iterations of GH. In long runs HAS-QAP was run for 100 iterations, TT or RTS for $1000 \cdot n$ iterations, SA for $1250 \cdot n^2$ iterations, and GH for 250 iterations.

On the irregular problems, for both short and long runs, HAS-QAP always provided the best solutions or solutions which do not differ significantly from the best solution. GH also performs well but not as well as HAS-QAP. Table 2.7 reports some of the results obtained on long runs.

---

[8] Real world, irregular, and structured problems are distinguished from randomly generated, regular, and unstructured problems by means of the value assumed by a statistics called flow-dominance [140, 219, 284].

Algorithm 2.5 High-level description of HAS-QAP

```
/* Initialization */
For k = 1 to m do
   generate a random initial permutation π^k
   π^k = loc-opt (π^k)
End For
Let π^+ be the best solution
For For every coupling (i, j) do
   τ_ij(0) = 1/(Q · C(π^+))
End For
Activate intensification
/* Main loop */
For t = 1 to t_max do
   /* solution manipulation */
   For k = 1 to m do
      apply R pheromone-trail-based swaps to π^k(t) to obtain π̃^k(t)
      π̂^k(t) = loc-opt (π̃^k(t))
   End For
   /* Intensification */
   For k = 1 to m do
      If intensification is active then
         π^k(t + 1) = best-of {π^k(t), π̂^k(t)}
      Else
         π^k(t + 1) = π̂^k(t)
      End If
   End For
   If π^k(t + 1) = π^k(t) for each ant k then
      deactivate intensification
   End If
   If C(π̂^k(t)) < C(π^+) for at least one ant k then
      π^+ = π̂^k
      activate intensification
   End If
   /* Pheromone trail updating */
   For every coupling (i, j) do
      Update pheromone trails by applying the rule:
      τ_ij(t) ← (1 − ρ) · τ_ij(t) + Δτ_ij(t) where
```

$$\Delta\tau_{ij}(t) = \begin{cases} 1/C(\pi^+) & \text{if } (i,j) \in \pi^+; \\ 0 & \text{otherwise}. \end{cases}$$

```
   End For
   For every coupling (i, j) do
      τ_ij(t + 1) = τ_ij(t)
   End For
   /* Diversification */
   If S iterations have been performed without improving π^+ then
      For every coupling (i, j) do
         τ_ij(t) = 1/(Q · C(π^+))
      End For
   End If
End For
/* Values of parameters used in experiments */
ρ = 0.1, m = 10, q = 0.9, Q = 100, R = n/3, S = n/2
```

TABLE 2.7 Comparison of HAS-QAP with other four heuristics on irregular QAP problems, long runs. The figures in the problem names give the problem dimensions (i.e., the number of cities). The second column gives the best known solution. Following columns give the average quality of solutions produced by different methods (over 10 runs), expressed in percentage above best known solution. Best solutions are represented in italic boldface. Solutions which do not differ significantly from best solution ($P > 0.1$ in a Mann-Withney U-test) are represented in boldface. The last column gives the average CPU time (in seconds) to run HAS-QAP for $100 \cdot n$ iterations; when in all 10 runs HAS-QAP found the best known solution, then the CPU time is in parentheses and indicates the average computing time to reach the best known solution. From Gambardella et al. [140].

Problem name	Best known solution	TT $1000n$	RTS $1000n$	SA $12500n^2$	GH 250	HAS-QAP 100	HAS-QAP 100 CPU (sec)
bur26a	5426670	0.0004	-	0.1411	0.0120	***0***	(10)
bur26b	3817852	0.0032	-	0.1828	0.0219	***0***	(17)
bur26c	5426795	0.0004	-	0.0742	***0***	***0***	(3.7)
bur26d	3821225	0.0015	-	0.0056	0.0002	***0***	(7.9)
bur26e	5386879	***0***	-	0.1238	***0***	***0***	(9.1)
bur26f	3782044	0.0007	-	0.1579	***0***	***0***	(3.4)
bur26g	10117172	0.0003	-	0.1688	***0***	***0***	(7.7)
bur26h	7098658	0.0027	-	0.1268	0.0003	***0***	(4.1)
chr25a	3796	6.9652	9.8894	12.4973	***2.6923***	**3.0822**	40
els19	17212548	***0***	0.0899	18.5385	***0***	***0***	(1.6)
kra30a	88900	0.4702	2.0079	1.4657	***0.1338***	0.6299	76
kra30b	91420	**0.0591**	0.7121	0.1947	***0.0536***	**0.0711**	86
tai20b	122455319	***0***	-	6.7298	***0***	0.0905	27
tai25b	344355646	0.0072	-	1.1215	***0***	***0***	(12)
tai30b	637117113	0.0547	-	4.4075	0.0003	***0***	(25)
tai35b	283315445	0.1777	-	3.1746	0.1067	***0.0256***	147
tai40b	637250948	0.2082	-	4.5646	0.2109	***0***	(51)
tai50b	458821517	**0.2943**	-	0.8107	**0.2142**	***0.1916***	480
tai60b	608215054	0.3904	-	2.1373	0.2905	***0.0483***	855
tai80b	818415043	1.4354	-	1.4386	**0.8286**	***0.6670***	2073

Very different results were obtained when these algorithms were applied to regular problems. For both short and long runs TT and RTS were found to be the best algorithms, while HAS-QAP did not perform well on these problems. A feature of these problems is that they lack "structure." As discussed before, ant colony algorithms have the ability to detect the structure of good solutions by reinforcement of good feasible solutions, which is an advantage when the vast majority of the good solutions are located in the neighborhood of the best solution. This advantage is lost when it comes to regular problems, where good solutions are spread all over the set of feasible solutions: in this case reinforcing a good solution does not ensure that the algorithm is exploring the vicinity of a very good solution.

In conclusion, the HAS-QAP algorithm is a very good heuristic for real-world problems in which the space of solutions is structured. For such problems, it compares favorably with the best known heuristics. For regular problems, which lack structure, the HAS-QAP does not perform as well because it cannot take advantage of the underlying structure of the space of solutions. This difference in performance when HAS-QAP is applied to these different types of problems might be insightful, from a wider perspective, for understanding why and how ant colony algorithms work. As we already said, solution spaces which have some kind of organization, where the best solution is surrounded by local optima, which themselves are surrounded by other more shallow local optima, and so forth, favor ant colony algorithms: this is certainly true for problems other than the QAP.

2.4.3 EXTENSIONS OF AS-QAP

Stützle and Hoos [300] have implemented an improved version of AS-QAP based on their Max-Min Ant System (MMAS) (see section 2.3.3). They used the same heuristic as in Maniezzo et al. [233] (see Eq. 2.19 in section 2.4.1) and they added two types of local search, one based on 2-opt, and the other on short tabu search runs. On structured real-world problems, both the resulting algorithms have had a level of performance similar to that of HAS-QAP. As it was the case with HAS-QAP, MMAS-QAP performance was not as good on regular, randomly generated problems.

More recently, Maniezzo and Colorni [232], and Maniezzo [231] have proposed the following variations of AS-QAP. First, they defined slightly different heuristic measures of goodness (based on the Gilmore [149] and Lawler [213] bound, in Maniezzo and Colorni, and on a specially designed bound in Maniezzo) used to direct the construction of solutions. Second, they slightly changed the assignment probability to the following form:

$$p_{ij}^k(t) = \frac{\gamma \cdot \tau_{ij}(t) + (1-\gamma) \cdot \eta_{ij}}{\sum_{r \in J_i^k} \gamma \cdot \tau_{ir}(t) + (1-\gamma) \cdot \eta_{ir}}, \tag{2.25}$$

if $j \in J_i^k$, and 0 otherwise, with $0 \leq \gamma \leq 1$.

Third, they also introduced a local search procedure by which solutions generated by ants were taken to their local optima.

Maniezzo and Colorni [232] compared their version of the Ant System with GRASP [218], a well-known heuristic for the QAP, with very good results. All experiments were run on a Pentium PC 166 MHz, and comparisons were made on a set of 46 test problems (most of them available at the QAPLIB [51], http://fmatbhp1.tu-graz.ac.at/~karisch/qaplib/). Each algorithm was given the same CPU time (10 minutes). Under these experimental conditions, the ant colony algorithm outperformed GRASP: it found a greater number of best known solutions, it had a smaller percentage of error and a smaller maximum error (both with respect to the best known solution). Moreover, the time required to find the best solution was on average slightly higher for GRASP. Maniezzo [231] compared his extended version of the Ant System to both GRASP [218] and Tabu Search [306]. Extensive experiments were run (using a Pentium PC 166 MHz) on all the QAPLIB test problems. He found that the ant colony algorithm was the best performing both in terms of the quality of the best and average solutions produced, although he did not differentiate between structured and unstructured problems.

These results confirm the results obtained by the previous versions of ant colony algorithms: ant colony optimization is currently the best approach for structured QAP problems.

2.5 OTHER APPLICATIONS OF ANT COLONY ALGORITHMS TO OPTIMIZATION

Although the methods developed in the two previous sections can look rather specific, they can in fact be readily modified and applied to many other combinatorial optimization problems. In Algorithm 2.6 we give a high-level description of a basic version of ACO algorithms. An ACO algorithm alternates, for $t_{\max}$ iterations, the application of two basic procedures: (i) a parallel solution construction/modification procedure in which a set of m ants builds/modifies in parallel m solutions to the considered problem, and (ii) a pheromone trail updating procedure by which the amount of pheromone trail on the problem graph edges is changed. The solution construction/modification is done probabilistically, and the probability with which each new item is added to the solution under construction is a function of the item heuristic desirability η and of the pheromone trail τ deposited by previous ants.

Pheromone trail modifications are a function of both the evaporation rate ρ and the quality of the solutions produced.

Given the above-described algorithm, it is clear that an ACO algorithm can be applied to any combinatorial problem as far as it is possible to define:

(i) An appropriate **problem representation** which allows ants to incrementally build/modify solutions by using a construction/modification **probabilistic transition rule** that makes use of pheromone trail (which provides compact information on the past history performance of construction/modification moves) and of a local heuristic (which informs on the desirability of construction/modification moves).
(ii) The **heuristic desirability** η of edges.
(iii) A **constraint satisfaction** method which forces the construction of feasible solutions.
(iv) A **pheromone updating rule** which specifies how to modify pheromone trail τ on the edges of the graph.

Algorithm 2.6 High-level description of a basic ACO algorithm

```
/* Initialization */
For every edge (i, j) do
  τ_ij(0) = τ_0
End For
For k = 1 to m do
  Place ant k on a randomly chosen city
End For
/* Main loop */
For t = 1 to t_max do
  For k = 1 to m do
    Build a solution S^k(t) by applying n − 1 times a probabilistic construc-
    tion/modification rule where choices are a function of a pheromone trail τ and
    of an heuristic desirability η
  End For
  For k = 1 to m do
    Compute the cost C^k(t) of the solution S^k(t) built by ant k
  End For
  If an improved solution is found then
    update best solution found
  End If
  For every edge (i, j) do
    Update pheromone trails by applying a pheromone trail update rule
  End For
End For
Print the best solution
Stop
```

(v) A **probabilistic transition rule** function of the heuristic desirability and of pheromone trail.

Let us summarize how this was done in two of the ant systems that we discussed in the previous two sections: AS-TSP and AS-QAP. Similar considerations hold for the other ACO algorithms which were presented.

In the AS-TSP case:

AS-TSP: Problem representation. The problem is represented as a graph $G = (N, E)$ where the set of nodes N represents the cities and the set of edges E the connections between the cities; ants build their solutions by sequentially visiting all the cities.

AS-TSP: Heuristic desirability. A very simple heuristic is used: the inverse of the distance between two cities.

AS-TSP: Constraint satisfaction. The constraint satisfaction method is implemented as a simple, short-term memory of the visited cities, in order to avoid visiting a city more than once.

AS-TSP: Pheromone updating rule. It is the one typical of Ant System (see Eq. (2.8)): pheromone evaporates on all edges and new pheromone is deposited by all ants on visited edges; its value is proportional to the quality of the solution built by the ants.

AS-TSP: Probabilistic transition rule. The probabilistic transition rule, called random proportional, is the one typical of Ant System (Eq. (2.6)).

The AS-QAP case presents very similar characteristics:

AS-QAP: Problem representation. The problem is represented as in the TSP, and ants build their solutions by sequentially assigning activities to locations.

AS-QAP: Heuristic desirability. The heuristic is based on a flow-distance matrix used to implement the well-known min-max heuristic.

AS-QAP: Constraint satisfaction. Similarly to the AS-TSP case, the constraint satisfaction method is implemented as a simple, short-term memory, this time of the assigned couplings.

AS-QAP: Pheromone updating rule. It is the one typical of Ant System (Eq. (2.8)): pheromone evaporates on all edges and new pheromone is deposited by all ants on visited couplings; its value is proportional to the quality of the assignment built by the ants.

AS-QAP: Probabilistic transition rule. It is the random proportional probabilistic transition rule, the one typical of Ant System (Eq. (2.6)).

It is interesting that there are currently applications of various versions of the basic ACO algorithm to a number of combinatorial optimization problems. In Table 2.8 we summarize the most successful applications, which are briefly discussed in the following.

2.5.1 AS-JSP: AN ACO ALGORITHM FOR THE JOB-SHOP SCHEDULING PROBLEM

AS-JSP: Problem definition and representation. In the job-shop scheduling problem (JSP) [156], a set of M machines and a set of J jobs are given. The jth job, $j = 1, \ldots, J$, consists of an ordered sequence of operations from a set $O = \{\ldots o_{jm} \ldots\}$. Each operation $o_{jm} \in O$ belongs to job j and has to be processed on machine m for d_{jm} consecutive time instants. $N = |O|$ is the total number of operations. The problem is to assign the operations to time intervals in such a way that the maximum of the completion times of all operations is minimized and no two jobs are processed at the same time on the same machine. JSP is an NP-hard prob-

TABLE 2.8 A chronologically ordered summary of most successful applications of ACO algorithms to combinatorial optimization problems.

Problem		Algorithm name	Authors	Year
TSP	- Traveling salesman problem	AS-TSP	Dorigo et al.	1991
QAP	- Quadratic assignment problem	AS-QAP	Maniezzo et al.	1994
JSP	- Job-shop scheduling problem	AS-JSP	Colorni et al.	1994
TSP	- Traveling salesman problem	Ant-Q	Gambardella et al.	1995
TSP	- Traveling salesman problem	ACS-TSP	Dorigo et al.	1997
TSP	- Traveling salesman problem	MaxMin-AS	Stützle et al.	1997
TSP	- Traveling salesman problem	Rank AS	Bullnheimer et al.	1997
GCP	- Graph coloring problem	ANTCOL	Costa et al.	1997
QAP	- Quadratic assignment problem	HAS-QAP	Gambardella et al.	1997
VRP	- Vehicle routing problem	AS-VRP	Bullnheimer et al.	1997
SOP	- Sequential ordering problem	HAS-SOP	Gambardella et al.	1997
QAP	- Quadratic assignment problem	MMAS-QAP	Stützle et al.	1998
SCP	- Shortest common supersequence problem	AS-SCS	Michel et al.	1998

lem [143]. Colorni et al. [78] suggest the following representation to be used with the ACO approach: a JSP with M machines, J jobs, and an operation set O containing N operations can be represented by a directed weighted graph $G = (O', A)$, where $O' = O \cup \{o_0\}$, o_0 is an extra vertex added in order to specify which job will be scheduled first, in case several jobs have their first operation on the same machine, and A is the set of arcs that connects o_0 with the first operation of each job and that completely connects the nodes of O except for the nodes belonging to the same job. Operations that do not belong to the same job are connected with a bidirectional link. Two operations that belong to the same job are connected by a unidirectional link only if one is the successor of the other (the unidirectional link reflects the order in which operations have to be completed within the job); otherwise, two such operations are not connected. There are, therefore, $N+1$ nodes and $J + [N(N-1)/2]$ edges in this graph.

Figure 2.16 presents an example graph with 3 jobs and 2 machines, in which one machine processes operations o_{11}, o_{21}, and o_{31}, while the other processes operations o_{12}, o_{22}, and o_{32}. The first operation of jobs 1 and 3 as well as the second operation of job 2 have to be performed on machine 1, while the first operation of job 2 and the second operation of jobs 1 and 3 have to be performed on machine 2. Ants build a solution by visiting all the nodes o_{jm} of the graph: the resulting permutation specifies the orientation of the edges of the cliques representing the machines.

AS-JSP: Heuristic desirability. The heuristic desirability η_{ij} to have edge (i, j) in the solution is computed by the simple greedy heuristic known as Longest Remaining Processing Time (that is: what is the operation o_j, connected to o_i, that has the longest remaining processing time?).

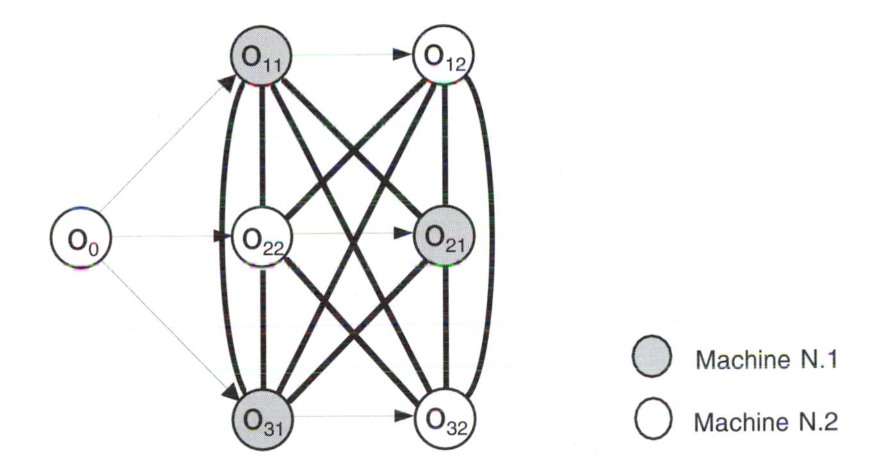

FIGURE 2.16 Representation graph for an example problem with 3 jobs and 2 machines. Bold edges represent bidirectional connections.

AS-JSP: Constraint satisfaction. Because the graph is not fully connected, the set of allowed nodes is defined with a procedure adapted to the problem. First, a node that has been selected cannot be visited anymore. Second, if the selected node is not the last in its job, its immediate successor in the job is added to the set of allowed nodes (in which it could not be before). As an example, consider the problem represented in Figure 2.16: the set of nodes to be visited at $t = 0$ is $\{o_{11}, o_{22}, o_{31}, o_{12}, o_{21}, o_{32}\}$, but the set of allowed nodes is $\{o_{11}, o_{22}, o_{31}\}$.

AS-JSP: Pheromone updating rule. It is the one typical of Ant System (Eq. (2.8)): pheromone evaporates on all edges and new pheromone is deposited by all ants on visited edges; its value is proportional to the inverse of the total completion time of the solutions built by the ants.

AS-JSP: Probabilistic transition rule. The probabilistic transition rule is the random proportional, typical of Ant System (Eq. (2.6)).

AS-JSP: Results. This method has been applied by Colorni et al. [78] and by Dorigo et al. [109] to problems with up to 10 jobs and 15 machines with reasonable success. Further study is necessary to understand how AS-JSP compares to other heuristic approaches on this type of problem.

2.5.2 ANTCOL: AN ACO ALGORITHM FOR THE GRAPH COLORING PROBLEM

ANTCOL: Problem definition and representation. The graph coloring problem (GCP) can be formulated as follows. Let $G = (V, E)$ be a nondirected graph, where $V = \{v_i\}$, $i = 1, \ldots, n$, is the set of n vertices, and E the set of edges. A q-coloring of G, where q is a strictly positive integer, is a mapping $c : V \rightarrow \{1, 2, \ldots, q\}$ such that $c(v_i) \neq c(v_j)$ if $(v_i, v_j) \in E$. $c(v_i)$ is called the color of vertex v_i. There is a minimal q below which no q-coloring exists. This value of q is called the chromatic number of the graph and is denoted $\chi(G)$. The graph coloring problem consists of finding an $\chi(G)$-coloring of G. GCP is an NP-hard problem [145]. Ants build their solution by means of the ANT-RLF procedure (see probabilistic transition rule below).

ANTCOL: Heuristic desirability. To define the heuristic desirability, Costa and Hertz [81] tried many different possibilities by taking various specialized heuristics for the coloring problem. The one that gave the best results is the RLF (recursive largest first) heuristic [216].

ANTCOL: Constraint satisfaction. The choice of the color to assign to a vertex is such that neighbor vertices always have different colors.

ANTCOL: Pheromone updating rule. Costa and Hertz [81] define an $n \times n$ "pheromone" matrix P, which is updated by ants after each iteration of the algorithm. Given two nonadjacent vertices v_r and v_s, the corresponding entry P_{rs} reflects the quality of the colorings obtained in previous trials by giving the same color to v_r and v_s. All of P's entries are initially set to 1. Entries, representing pheromone concentration, "evaporate" at a rate ρ. Let m denote the number of ants, $(s^1, \ldots, s^m)$ the colorings obtained by the m ants at the end of an iteration, $S_{rs} \subseteq \{s^1, \ldots, s^m\}$ the subset of those colorings in which vertices v_r and v_s have the same color, and q^k the number of colors used in $s^k, 1 \leq k \leq m$. P_{rs} is updated as follows:

$$P_{rs}(t) \leftarrow (1 - \rho) \cdot P_{rs}(t) + \sum_{s^k \in S^{rs}} \frac{1}{q^k}. \tag{2.26}$$

ANTCOL: Probabilistic transition rule. Here it must be mentioned that Costa and Hertz [81] define their ANTCOL algorithm as a special case of AS-ATP, a generalization of the AS algorithm to assignment type problems. In AS-ATP two probabilities are defined: the probability with which an item to be assigned is chosen and the probability with which a resource is chosen to assign to the chosen item. In their best performing version of ANTCOL (which uses the RLF heuristic) the colors are assigned sequentially (i.e., the first color assigned is $q = 1$, then $q = 2$, and so on until all vertices are colored) and, therefore, the only probability that must be defined is the probability with which resources are chosen. In fact, the probabilistic transition rule is implemented as a procedure called ANT-RLF. ANT-RLF is a stochastic procedure that makes use of the RLF heuristic and of the pheromone matrix P to build solutions.

ANTCOL: Results. ANTCOL was tested on a set of random graphs and results have shown that its performance is comparable to that obtained by the best heuristics: it found an average q of 15.05 colors for 20 randomly generated graphs of 100 vertices with any two vertices connected with probability 0.5, whereas the best known result, obtained by Costa et al. [80] and Fleurent and Ferland [124], is 14.95. Costa and Hertz [81] found that the performance of the algorithms increases with "colony size" up to a certain size (that depends on the problem size and on the heuristic used to guide the constructive part of the algorithm). They also found that, when compared with the best graph coloring heuristics, ANTCOL tends to become less competitive for increasing problem dimensions.

2.5.3 AS-VRP: AN ACO ALGORITHM FOR THE VEHICLE ROUTING PROBLEM

Bullnheimer et al. [50] have proposed an application of an extended version of the Ant System (AS) algorithm presented in section 2.3.1 to a vehicle routing problem (VRP) defined as follows.

AS-VRP: Problem definition and representation. Let $G = (V, A, d)$ be a complete weighted directed graph, where $V = \{v_0, \ldots, v_n\}$ is the set of vertices, $A = \{(i, j) : i \neq j\}$ is the set of arcs, and a weight $d_{ij} \geq 0$ representing the distance between v_i and v_j is associated to arc (i, j). Vertex v_0 represents a depot, while the other vertices represent customers' locations. A demand $d_i \geq 0$ and a service time $\delta_i \geq 0$ are associated to each customer v_i ($d_0 = 0$ and $\delta_0 = 0$). The objective is to find minimum cost vehicle routes such that (i) every customer is visited exactly once by exactly one vehicle, (ii) for every vehicle the total demand does not exceed the vehicle capacity D, (iii) every vehicle starts and ends its tour in the depot, and (iv) the total tour length of each vehicle does not exceed a bound L.

It is easy to see that VRPs and TSPs are closely related: a VRP consists of the solution of many TSPs with common start and end cities. As such, the VRP is an NP-hard problem. As in the TSP, ants build their solutions by sequentially visiting all the cities.

AS-VRP: Heuristic desirability. The heuristic desirability of a city is measured as a parametrical saving function (proposed by Paessens [254]):

$$\eta_{ij} = d_{i0} + d_{0j} - g \cdot d_{ij} + f \cdot |d_{i0} - d_{0j}| \ .$$

AS-VRP: Constraint satisfaction. As it was the case in AS-TSP, an ant in city i chooses the next city j to move to among those not yet visited. Artificial ants construct vehicle routes by successively choosing cities to visit until each city has been visited. Whenever the choice of another city would lead to an unfeasible solution due to vehicle capacity or total route length, the depot is chosen as the next city and a new tour is started.

AS-VRP: Pheromone updating rule. Pheromone trail update is done using the ranked/ elitist scheme that Bullnheimer et al. [50] had already used with the TSP (see section 2.3.3, Eq. (2.14)).

AS-VRP: Probabilistic transition rule. The probabilistic choice rule is the same as in Ant System (Eq. (2.6)).

AS-VRP: Results. In AS-VRP a candidate list is used together with a local optimizer based on the 2-opt swap strategy. A set of experiments was run to compare AS-VRP with other heuristics based on tabu search, simulated annealing, and neural networks. It was found that AS-VRP outperforms simulated annealing and neural networks, while having a slightly lower performance than tabu search.

2.5.4 HAS-SOP: AN ACO ALGORITHM FOR THE SEQUENTIAL ORDERING PROBLEM

Gambardella and Dorigo [139] have developed a Hybrid Ant System (called HAS-SOP) which is a modified version of ACS plus a local optimization procedure specifically designed to solve the sequential ordering problem (SOP).

HAS-SOP: Problem definition and representation. The sequential ordering problem [120] consists of finding a minimum weight Hamiltonian path on a directed graph with weights on the arcs and the nodes, subject to precedence constraints among nodes.

Consider a complete graph $G = (V, A)$ with node set V and arc set A, where nodes correspond to jobs $0, \ldots, i, \ldots, n$ and $|V| = n + 1$. To each arc (i, j) is associated a cost $t_{ij} \in \Re$ with $t_{ij} \geq 0$. This cost represents the required waiting time between the end of job i and the beginning of job j. Each node i has an associated cost $p_i \in \Re$, $p_i \geq 0$, which represents the processing time of job i. The set of nodes V includes a starting node (node 0) and a final node (node n) connected with all the other nodes. The costs between node 0 and the other nodes is equal to the setup time of node i, $\forall_i \; t_{0i} = p_i$ and $t_{in} = 0$. Precedence constraints are given by an additional acyclic digraph $P = (V, R)$ that is defined on the same node set V. An arc $(i, j) \in R$ if job i has to precede job j in any feasible solution. Given the above definitions, the SOP problem can be stated as the problem of finding a job sequence subject to the precedence constraints which minimizes the total makespan. This is equivalent to the problem of finding a feasible Hamiltonian path with minimal cost in G under precedence constraints given by P, and therefore SOP is NP-hard. As in the TSP, ants build their solutions by sequentially visiting all the cities.

HAS-SOP: Heuristic desirability. The heuristic used is the same as in the TSP applications: η_{ij} is the inverse of the distance between nodes i and j.

HAS-SOP: Constraint satisfaction. The set of feasible nodes comprises the nodes j still to be visited such that all nodes that have to precede j, according to precedence constraints, have already been inserted in the sequence.

HAS-SOP: Pheromone updating rule. The same rule as in ACS is used (Eqs. (2.11) and (2.12)): only the ant which generated the best solution from the beginning of the trial is allowed to deposit pheromone. Also, as in ACS, a local updating rule is applied (Eq. (2.13)).

HAS-SOP: Probabilistic transition rule. The probabilistic transition rule used is the same as in ACS (Eq. (2.6)).

HAS-SOP: Results. Results obtained with HAS-SOP are excellent. HAS-SOP, which was tested on an extensive number of problems, outperforms existing heuristics in terms of solution quality and computation time. Also, HAS-SOP has improved many of the best known results on problems maintained in TSPLIB.

2.5.5 AS-SCS: AN ACO ALGORITHM FOR THE SHORTEST COMMON SUPERSEQUENCE PROBLEM

AS-SCS: Problem definition and representation. Given a set L of strings over an alphabet Σ, find a string of minimal length that is a supersequence of each string in L. A string S is a supersequence of a string A if S can be obtained from A by inserting zero or more characters in A. Consider for example the set $L = \{bcab, bccb, baab, acca\}$. The string *baccab* is a shortest supersequence. The shortest common supersequence (SCS) problem is NP-hard even for an alphabet of cardinality two [266]. Ants build solutions by repeatedly adding symbols to the supersequence under construction. Symbols are chosen according to the probabilistic transition rule defined below.

AS-SCS: Heuristic desirability. Michel and Middendorf [239] use two types of heuristic information. The first one is based on a *lookahead function* which takes into account the influence of the choice of the next symbol to append on the next iteration. The second one is based either on the MM heuristic or on the LM heuristic [40], which is a weighted version of the majority merge (MM) heuristic [127]. The MM heuristic looks at the first characters of each string in L and appends the most frequent one to the supersequence. The appended character is then removed by the front of the strings and the procedure is repeated. The LM heuristic is defined by adding a weight to each character in the strings of L and then selecting the one for which the sum of the weights of all its occurences in the front of strings in L is maximal. The weight of a symbol is given by the number of characters that follow it in the string to which it belongs.

AS-SCS: Constraint satisfaction. Symbols to be selected are chosen only from among those in the first position of at least one string in L.

AS-SCS: Pheromone updating rule. The same rule as in AS was used (Eq. (2.8)), with two elitist ants.

AS-SCS: Probabilistic transition rule. The probabilistic transition rule used is similar to the one used in AS (Eq. (2.6)). The usual η value is set to the value returned by the lookahead function. A difference from Eq. (2.9) is that, in the version of AS-SCS using the LM heuristic, the pheromone trail term is replaced by the product of the pheromone trail with the value returned by the LM heuristic.

AS-SCS: Results. The AS-SCS-LM algorithm (i.e., AS-SCS with LM heuristic), further improved by the use of an island model of computation (that is, different populations of ants work on the same problem concurrently using private pheromone trail distributions; every some-fixed-number of iterations they exchange the best solution found), found very good results. AS-SCS-LM was compared with both the MM and LM heuristics, as well as with a recently proposed genetic algorithm specialized for the SCS problem. On almost all the test problems AS-SCS-LM was the best performing algorithm. This is a rather good result since the genetic algorithm is one of the best performing algorithms on this problem.

2.6 SOME CONSIDERATIONS ON ACO ALGORITHMS

Before we discuss a different class of applications for algorithms inspired by the foraging behavior of ants, let us draw some parallels between the ACO algorithms presented above and other approaches to combinatorial optimization which took inspiration from other natural processes. We focus on ACO algorithms applied to the TSP and draw some analogies with neural networks and with evolutionary algorithms.

2.6.1 ACO AND NEURAL NETWORKS

Ant colonies, being composed of numerous interacting units, are "connectionist" systems, the most famous examples of which are neural networks [167]. In artificial neural networks, units are directly connected to each other and exchange information. The strength of a connection between two units, called "synaptic" strength in analogy with real neurons in the brain, measures how tightly two neurons are coupled or, in other words, how strongly two neurons influence each other. Synaptic connections are reinforced or decreased during a learning process: for instance, examples are presented to the neural network, and synaptic strengths adapt to "learn" these examples, to either recognize, classify, discriminate among them, or predict their properties. It is tempting to draw a parallel here between synaptic connections and pheromone trails, which are reinforced or decreased by ants to find good solutions. Indeed, Chen [67] proposed a neural network approach to the TSP which bears important similarities with the ant colony approach:

(i) A tour is selected by choosing cities probabilistically according to the synaptic strength between cities.
(ii) The synaptic strengths of the links that form the tour are then reinforced (reduced) if the tour length is shorter (longer) than the average result of the previous trials.

More precisely, in Chen's [67] algorithm, the synaptic strength w_{ij} between cities i and j starts with a greedy-like value: $w_{ij} = e^{-d_{ij}/T}$, where d_{ij} is the distance between the two cities and T is a (temperature-like) parameter that sets the scale for initial synaptic strengths. Building a tour consists of selecting cities sequentially according to synaptic strengths: first, a city i_1 is selected randomly among all the cities, then a city i_2 is selected among the remaining cities with probability $P_{i_1 i_2} \propto w_{i_1 i_2}$, and then a city i_3 is selected from among the cities which remain unvisited, with probability $P_{i_2 i_3} \propto w_{i_2 i_3}$, and so forth, until a tour, containing all n cities, is created. To enhance computational efficiency, Chen [67] also uses candidate lists (see section 2.3.2). The obtained tour is then improved by applying a local search procedure, here 2-opt [220]. This tour is then compared with other tours obtained in previous r trials. Let $\{i_1 i_2 \ldots i_n\}$ denote the current tour, of length d, and $\{i'_1 i'_2 \ldots i'_n\}$ one of the previous tours, of length d'. Synaptic strengths are modified as follows:

$$w_{i_k i_{k+1}} \leftarrow w_{i_k i_{k+1}} e^{-\frac{\alpha}{r}(d-d')}, k = 1, \ldots, n, \tag{2.27}$$

$$w_{i'_k i'_{k+1}} \leftarrow w_{i'_k i'_{k+1}} e^{-\frac{\alpha}{r}(d'-d)}, k = 1, \ldots, n, \tag{2.28}$$

where, by convention, $i_{n+1} = i_1$ and $i'_{n+1} = i'_1$, and α is a "learning" rate, that is, the modification rate of synaptic strengths. According to this rule, not only are links that belong to the current tour updated, links that belong to the r previous tours are also updated. The similarity between this neural-network formulation and the ant colony approach is striking. Here are some of the correspondences:

pheromone trail $\leftrightarrow$ synaptic strength
add pheromone (amount $\propto 1/d$) $\leftrightarrow$ reinforce synaptic strength (amount $\propto (d-d')$)
evaporate pheromone $\leftrightarrow$ reduce synaptic strength (amount $\propto (d - d')$)
$Q \leftrightarrow \alpha$
ρ (evaporation) $\leftrightarrow$ α (learning rate)
local search $\leftrightarrow$ local search
candidate list $\leftrightarrow$ candidate list

Although there are some differences, the common features are overwhelming. The basic principles are the same: reinforcement of portions of solutions that tend to belong to many good solutions, and dissipation to avoid ending up in local optima. This example illustrates in the clearest way the close relationship between the ant colony approach and connectionist models.

2.6.2 ACO AND EVOLUTIONARY COMPUTATION

To see the similarities between ACO algorithms and evolutionary computation, we focus on population-based incremental learning (PBIL), an algorithm proposed by Baluja and Caruana [6], which takes inspiration from genetic algorithms. PBIL maintains a vector of real numbers, the generating vector, which plays a role similar to that of the population in genetic algorithms. Starting from this vector, a population of binary strings is randomly generated: each string in the population will have the ith bit set to 1 with a probability which is a function of the ith value in the generating vector (in practice, values in the generating vector are normalized to the interval $[0, 1]$ so that they can directly represent the probabilities). Once a population of solutions is created, the generated solutions are evaluated and this evaluation is used to increase (or decrease) the probabilities of each separate component in the generating vector so that good (bad) solutions in the future generations will be produced with higher (lower) probability. When applied to the TSP, PBIL uses the following encoding: each city is assigned a string of length $\lceil \log_2 n \rceil$, where n is the number of cities; a solution is therefore a string of size $n \cdot \lceil \log_2 n \rceil$ bits. Cities are then ordered by increasing integer values; in case of ties the leftmost city in the string comes first in the tour. In ACO algorithms the pheromone trail matrix plays a role similar to Baluja and Caruana's [6] generating vector, and pheromone updating has the same goal as updating the probabilities in the generating vector. Still, the two approaches are different since in ACO the pheromone trail matrix changes while ants build their solutions, while in PBIL the probability vector is modified only after a population of solutions has been generated. Moreover, ACO algorithms use heuristics to direct search, while PBIL does not.

2.6.3 CONTINUOUS OPTIMIZATION

The ACO approach works for discrete optimization problems. How can it be extended to problems in which a continuous space of solutions has to be searched? Such problems cannot be represented by graphs in which nodes are connected by edges that can be marked with virtual pheromone. In order to search a continuous space, some form of discretization must be used. Bilchev and Parmee [24] suggest considering a finite set of regions at each iteration of the algorithm: agents are sent to these regions, from which they explore randomly selected directions within a radius of exploration. Agents reinforce their paths according to their performance. Trails diffuse, evaporate, and recombine.

The first step in Bilchev and Parmee's [24] approach is to place the nest, to be located in a promising region of search space identified by a previous coarse-grained algorithm: Bilchev and Parmee's [24] algorithm concentrates on local search. The radius of exploration decreases over time as agents converge toward a good solution.

Regions are first randomly placed in search space, or may correspond to regularly sampled directions from the nest. An agent selects a region with a probability proportional to the virtual pheromone concentration of the path that goes from the nest to the region. Let $\tau_i(t)$ be the pheromone concentration of the path to region i. Initially, $\tau_i(t=0)=\tau_0$ for all regions. The probability that an agent selects region i is given by

$$p_i(t)=\frac{\tau_i^{\alpha}(t)\cdot\eta_i^{\beta}(t)}{\sum_{j=1}^{N}\tau_j^{\alpha}(t)\cdot\eta_j^{\beta}(t)},$$

where N is the (constant) number of regions, $\eta_i^{\beta}(t)$ is either equal to 1 or incorporates some problem-specific heuristics (it reflects the local "desirability" of a portion of solution, such as the inverse distance between two cities in the traveling salesman problem). The agent then moves to the region's center, chooses a random direction, and moves a short distance from the region's center in that direction. The distance δr hopped by an agent at time t is a time-dependent random variable defined by $\delta r(t,R)=R(1-u^{(1-t/T)^c})$, where R is the maximum search radius, determined by the extent of the region one wishes to explore, u is a uniformly random real number in $[0,1]$, T is the total number of iterations of the algorithm (so that δr eventually converges to 0 as t tends to T), and c is a "cooling" parameter. If the agent finds a better solution than that of the region's center, the region is moved to the agent's position and the agent increases $\tau_i(t)$ by an amount proportional to the improvement made to the region's solution. The trail evaporates after each iteration at a rate $\rho:\tau_i(t+1)=(1-\rho)\cdot\tau_i(t)$. A form of diffusion is also applied to trails: several "parent" trails are selected, and produce an offspring which is a trail resulting from the weighted average of its parent trails. For example, the x-coordinate of a point of curvilinear coordinate s on a new trail produced by two parent trails will be given by $x(s)=w_1x_1(s)+w_2x_2(s)$, where w_i is the weight of trail i (for instance, an increasing function of τ_i) and $x_i(s)$ is the coordinate of a point of curvilinear coordinate s on trail i.

This algorithm operates at three different levels: (1) the individual search strategy of individual agents, (2) the cooperation between local agents using pheromone trails to focus local searches, and (3) the exchange of information between different regions, performed by some kind of "diffusion," similar to a crossover in a genetic algorithm. The algorithm can also be easily modified to handle constrained optimization: Bilchev and Parmee [24] define the level of constraint violation as the euclidian distance from the feasible region; a point with a high level of constraint violation will not be accepted as a "food source" by the algorithm, so that the path leading to it will soon decay.

In order to design a more global approach and to avoid convergence toward local optima, Wodrich [332] introduces two types of agents: local search agents (LSAs), which make about 20% of the population, and global search agents (GSAs) which make the remaining 80%. Local agents are similar to those of Bilchev and Parmee [24]. Global agents are the equivalent of explorer ants in foraging: they perform random walks and find new regions to replace those regions which do not seem to contain good solutions. Wodrich [332] also introduces aging for regions: a region's age is decremented whenever an improvement has been made and incremented when no improvement has been made within an iteration. The search radius

δr is still given by $\delta r(t, R) = R(1 - u^{(1-t/T)^c})$, but t is now the age of the region where the agent is located. This allows the adjustment of δr to the agents' success.

Bilchev and Parmee [24] and Wodrich [332] tested their algorithms on continuous function optimization. They used common benchmark functions to compare the performance of this approach with that of other algorithms, and found that it performs well compared to evolutionary strategies, genetic algorithms, evolutionary programming, hill-climbing, and population-based incremental learning. Bilchev and Parmee [24] also find that this approach produces good results in constrained optimization. Given the success of the ant colony metaphor in discrete optimization problems, this approach should be studied in more depth. However, too few results exist to make a definite statement about whether or not this is a promising approach. Moreover, too many important implementation details are missing from the descriptions given by both Bilchev and Parmee [24] and Wodrich [332] and the details of the functioning of the algorithm cannot be understood, especially its "meta-cooperation" or diffusion aspects.

2.7 APPLICATIONS TO TELECOMMUNICATIONS NETWORKS

All the optimization problems dealt with in the previous sections are static: the problem to be solved does not change over time. One desirable feature of the swarm-based approach is that it may allow for enhanced efficiency when the representation of the problem under investigation is spatially distributed and changing over time. Distributed time-varying problems are, therefore, the next big field to move to for researchers working on swarm-based optimization and problem-solving. This section is mainly devoted to the presentation of the first paradigmatic application to a problem with the above-mentioned characteristics: routing in telecommunications networks. This problem, in addition to being dynamic, is also distributed.

Routing is a mechanism that allows information transmitted over a network to be routed from a source to a destination through a sequence of intermediate switching/buffering stations or nodes. Routing is a necessity because in real networks not all nodes are directly connected: the cost of completely connecting a network becomes prohibitive for more than a few nodes. Routing selects paths that meet the objectives and constraints set by the user traffic and the network technology, and therefore determines which network resources are traversed by which user traffic [23, 195, 291].

The problem to be solved by any routing algorithm is to direct traffic from sources to destinations maximizing network performance while minimizing costs (e.g., rate of call rejection, throughput, packet delay, etc.). There are many possible routing problems differing mainly by the characteristics of the network and of the traffic, which call for different routing algorithms. In real networks traffic conditions are constantly changing, and the structure of the network itself may fluctuate (nodes or links can fail). Because there are usually many possible pathways for one message to go from a given node to another node, it is possible, in principle, to make routing algorithms adaptive enough to overcome local congestion: calls can be rerouted to less congested areas of the network. If there is a sudden burst of activity, or if one node becomes the destination or the emitter of a large number of calls, rerouting becomes crucial. Static routing, whereby routing remains fixed independent of the current states of the network and user traffic, is therefore almost never implemented: most routing schemes respond in some way to changes in network or user traffic states. But there exists a wide spectrum of dynamic routing systems, which vary dramatically in their speed of response and in the types of changes they respond to [291]: some routing systems can be seen as "quasi static," because routing is modified only in response to exceptional events (link or switch failure) and/or on a long time scale, while other routing systems are highly dynamic and autonomously update traffic routing in real time in response to perceived changes in user and network states. Dynamic routing requires more computational resources than static or quasi static routing. It relies on active participation of entities within the network to measure user traffic, network state and performance, and to compute routes.

In the following we will present two routing algorithms for two types of networks. The first one, called ant-based control (ABC) and introduced by Schoonderwoerd et al. [283], was designed with a telephone network application in mind. The second, called AntNet and proposed by Di Caro and Dorigo [97, 103, 104], was primarily conceived for packet-switching, connectionless networks like the Internet, although its extension to connection-oriented data networks is straightforward.

2.7.1 ABC: ROUTING IN TELEPHONE NETWORKS

Schoonderwoerd et al. [283] have proposed a very interesting adaptive routing algorithm based on the use of many simple agents, called ants, that modify the routing policy at every node in a telephone network by depositing a virtual pheromone trail on routing tables entries. The ants' goal is to build, and adapt to load changes at run time, routing tables so that network performance is maximized. The network performance is measured by the rate of incoming calls which are accepted (incoming calls can be either accepted or rejected depending on the network resources available at call set-up time).

In Schoonderwoerd et al.'s [283] work, a telephone network is represented by a graph with N nodes and E bidirectional links. Each node i has k_i neighbors (two nodes directly connected by a link are said to be neighbors), and is characterized by a capacity C_i, a spare capacity S_i, and by a routing table $R_i = [r^i_{n,d}(t)]_{k_i,N-1}$ with k_i rows and $N-1$ columns. Links have infinite capacity, that is, they can carry an infinite number of connections. From a technological point of view, the nodes are crossbar switches with limited connectivity. C_i represents the maximum number of connections node i can establish, while S_i represents the percentage of capacity that is still available for new connections. Each row in the routing table corresponds to a neighbor node and each column to a destination node. The value $r^i_{n,d}(t)$ is used as a probability by ants: in this case it gives the probability that a given ant, the destination of which is node d, be routed from node i to neighbor node n. Calls, differently from ants, use routing tables in a deterministic way: at set-up time a route from node s to node d is built by choosing sequentially and deterministically, starting from node s, the neighbor node with the highest probability value, until node d is reached. Once the call is set up in this way, the capacity of each node on the selected route is decreased by a fixed amount. If at call set up any of the nodes along the route under construction has no spare capacity left, then the call is rejected. When a call terminates, the corresponding reserved capacity of nodes on its route is made available again for other calls.

Ants can be launched from any node in the network at any time. An ant's destination is selected randomly from among all the other nodes in the network. The probability of launching an ant per time unit must be tuned to maximize the performance of the system. It appears that too few ants make it difficult to reach good solutions, whereas too many ants degrade the performance of the system. Ants go from their source node to their destination node by moving from node to node. The next node an ant will move to is selected probabilistically according to the routing table of its current node. When the ant reaches its destination node it is deleted from the network. Each visited node's routing table is updated: more precisely, an ant deposits some pheromone on the routing table entry corresponding to the neighbor node it just came from and to the ant source node, which is viewed as a destination node. Formally, routing tables are updated as follows. Let s be the source node of an ant, $i-1$ be the node it just came from, and i its current node at time t. The entry $r^i_{i-1,s}$ is reinforced while other entries $r^i_{n,s}(t)$, $n \neq i-1$, in the same column decay by probability normalization:

$$r^i_{i-1,s}(t+1) = \frac{r^i_{i-1,s}(t) + \delta r}{1+\delta r}\,, \tag{2.29}$$

$$r^i_{n,s}(t+1) = \frac{r^i_{n,s}(t)}{1+\delta r}\,, n \neq i-1\,, \tag{2.30}$$

where δr is a reinforcement parameter that depends on some ant's characteristics such as its age. Note that this updating procedure conserves the normalization of the values $r^i_{n,s}(t)$

$$\sum_n r^i_{n,s}(t) = 1\,, \tag{2.31}$$

if they are initially normalized, so that the values $r^i_{n,s}(t)$ can always be considered as probabilities. $r^i_{i-1,s}(t)$ is comparatively more reinforced when it is small (that is, when node $i-1$ was not on the preferred route to node s) than when it is large (that is, when node $i-1$ was already on the preferred route to node s). This is an interesting feature (similar in spirit to Stützle and Hoos's [298, 299] trail-smoothing mechanism by proportional update; see section 2.3.3), as it should allow new routes to be discovered quickly when the preferred route gets congested because established routing solutions become unstable more easily. There is, however,

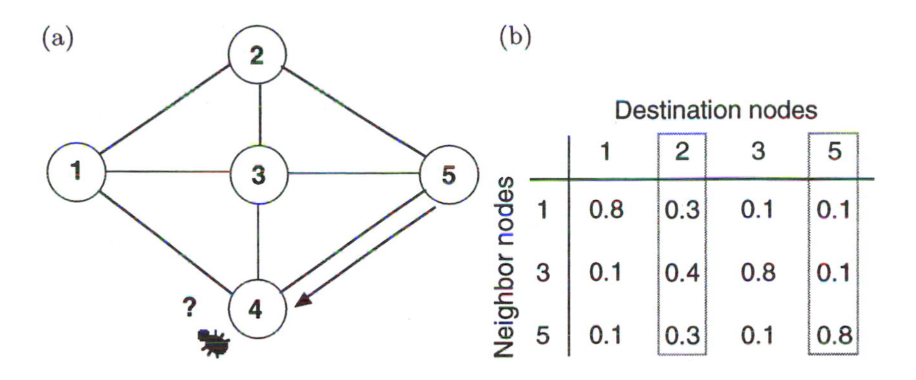

Neighbor nodes	1	2	3	5
1	0.8	0.3	0.1	0.1
3	0.1	0.4	0.8	0.1
5	0.1	0.3	0.1	0.8

FIGURE 2.17 Example of use and update of a routing table. (a) An ant, whose destination is node 2 and whose source is node 5, arrives in node 4 from node 5. (b) The routing table of node 4: the left framed column gives the probabilities with which the ant, which is headed towards node 2, chooses the next node to move to among the neighbors; the right framed column is the column whose values are updated by the ant.

an exploration/exploitation tradeoff, for too much instability might not always be desirable.

An example of the way routing tables are updated is given in Figure 2.17. The source node of the ant in the figure is node 5, while its destination is node 2. In the situation represented in Figure 2.17(a) the ant just arrived in node 4 from node 5. The routing table of node 4 is shown in Figure 2.17(b). The values in the column relative to the ant's destination, that is the left framed column, are used to choose probabilistically the next node to move to. The values in the column relative to the ant's source, that is the right framed column, are updated using the rule given by Eqs. (2.29) and (2.30). The new, updated, values of probabilities for destination 5 are therefore: $0.1/(1+\delta r)$, $0.1/(1+\delta r)$, $(0.8+\delta r)/(1+\delta r)$.

We have seen that ants update routing tables of nodes viewing their node of origin as a destination node. In fact, while moving, an ant collects information about the path it traversed and which connects its source node with its current node: the ant can therefore modify only the entries of the routing tables that influence the routing of ants and calls that have the ant's source node as destination.

It is important to note that the feasibility of this choice is based on an important assumption: that any specific path connecting two nodes s and d must exhibit approximately the same level of congestion in both directions. Although this is a realistic assumption for telephone networks, it becomes questionable in the case of many packet-switching networks and, therefore, as we will see in section 2.7.2, a different routing table strategy must be devised.

The amount of reinforcement δr deposited on the routing table by a given ant depends on how well this ant is performing. Aging can be used to modulate δr: if an ant has been waiting a long time along its route to its destination node, it means that the nodes it has visited and the links it has used are congested, so that δr should decrease with the ant's age (measured in time units spent in the network). Aging should, in principle, be relative to the length (expressed in time units) of the optimal path from an ant's source node to its destination. Schoonderwoerd et al. [283] choose to use the absolute age T, measured in time units spent in the network, and propose

$$\delta r = \frac{a}{T} + b \,, \tag{2.32}$$

where a and b are parameters. Aging is also used in a complementary way: age is manipulated by delaying ants at congested nodes. These delays determine two effects: (1) the flow of ants from a congested node to its neighbors is temporarily reduced, so that entries of the routing tables of neighbors that lead to the congested node cannot be reinforced, and (2) the age of delayed ants increases by definition, so that delayed ants have less influence on the routing tables of the nodes they reach, if δr decreases with age. Schoonderwoerd et al. suggest that the delay D imposed on an ant reaching a node with spare capacity S should be given by

$$D = c \cdot e^{-d \cdot S} \,, \tag{2.33}$$

where c and d are two parameters and S is the spare capacity (expressed in percentage of the node's capacity).

It is important to note that in situations that remain static for a long time and then suddenly change, good routes could become "frozen," that is, probabilities on the routing tables of a good route can become so high that no more exploration is going on. This can make the discovery of new, better routes impossible when the network load status changes. To solve this problem, Schoonderwoerd et al. [283] suggest the addition of a tunable "noise" or "exploration" factor $g, 0 < g < 1$. At every node, an ant chooses a purely random path with probability g and chooses its path according to the node's routing table with probability $(1 - g)$. Noise allows a constant exploration of the network so that it becomes less probable that new, better routes (due, for example, to the release from congestion of a node) go unnoticed.

Overall, the effect of ants on routing tables is such that routes which are visited frequently and by "young" ants will be favored routes when building paths to route new calls. This has the effect that links close to the ant source node get higher reinforcement than links that are far away. It is clear that there is a strong and complex interplay among routing of calls, routing of ants, updating of routing tables by ants, and user-generated traffic. A schematic representation of these relationships is given in Figure 2.18.

Algorithm 2.7 reports a high-level description of the ABC algorithm, together with algorithm parameters values as set by Schoonderwoerd et al. [283] in their experiments. Before the algorithm starts to route calls, an initialization phase during which ants are allowed to explore the network in absence of users' calls takes place.

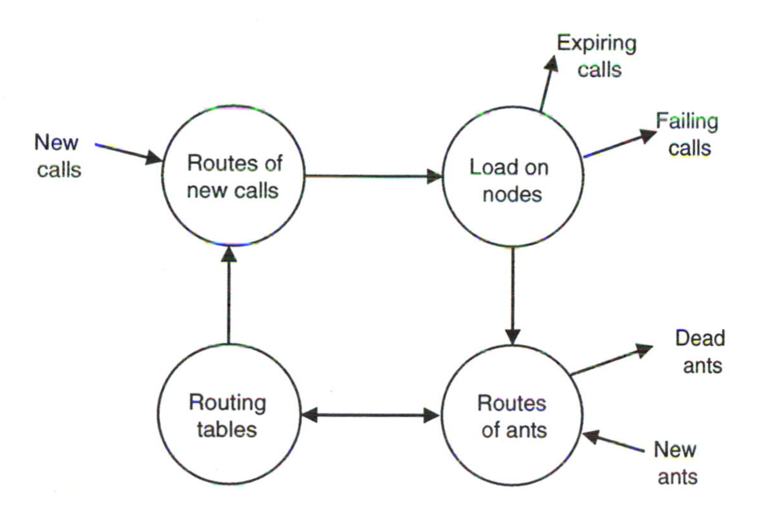

FIGURE 2.18 Schema of the interplay between routing of calls, routing of ants, updating of routing tables by ants, and user-generated traffic.

This initialization phase is intended to let the ants initialize the routing tables. It is interesting that after a short time the highest probabilities in the routing tables define routes very close to those obtainable by using a shortest path algorithm, and circular routes have been eliminated. At this point, calls can be safely put on the network. In their experiments Schoonderwoerd et al. [283] set the initialization phase to either 250 time steps, when ants without noise are used, or to 500 time steps, in the case of ants with noise.

Schoonderwoerd et al. [283] have tested their ABC algorithm on the 30-node interconnection network of British Telecom as an underlying realistic network topology (Figure 2.19). In their simulation, each node has a maximum capacity of 40 calls. Within every time step, an average of one call is generated, with an average duration of 170 time steps. Probabilities that nodes be emitters or receivers are chosen uniformly over $[0.01, 0.07]$ and then normalized. Generation and normalization of emission and reception probabilities for all nodes define a set of call probabilities, and a change in call probabilities means that new emission and reception probabilities have been generated (new nodes become more likely to be emitters or receivers, and others less likely). At the start, before the initialization phase, routing tables are characterized by equiprobable routing (all neighboring nodes are equally likely to be selected as next node), and there are no calls in the network. The performance measure used is the percentage of call failures: the lower the better.

Schoonderwoerd et al. have run experiments to compare five different algorithms: a fixed routing scheme based on a shortest path algorithm [22], an algorithm based on mobile agents [3], an improved version of the mobile agents algo-

Algorithm 2.7 High-level description of ABC

```
/* Initialization */
For t = 1 to t_initialize do
  For node = 1 to n do
    Launch ant
  End For
  For each ant do
    Choose next node j among neighbors using probabilistic routing tables
    Update routing tables according to Eqs. (2.29) and (2.30)
  End For
End For
/* Main loop */
For t = 1 to t_max do
  For node = 1 to n do
    Launch ant
  End For
  For each ant do
    Choose next node j among neighbors using probabilistic routing tables
    Update routing tables according to Eqs. (2.29) and (2.30)
  End For
  Remove expired calls from the network
  Add new calls to the network
  Compute network statistics
End For
/* Values of algorithm parameters used in experiments */
a = 0.08, b = 0.05, c = 80, d = 0.075
```

rithm [283], the ABC algorithm without noise, and the ABC algorithm with noise ($g = 0.05$).

Table 2.9 shows the average percentage and its standard deviation (computed over the last 7,500 time steps) of call failures for a set of ten simulations, each one lasting 15,000 steps, where call probabilities are set at simulation start and left unchanged thereafter. It is clear that the ABC algorithms perform significantly better than the others. In this situation, where call probabilities are unchanged, ABC without noise is slightly more efficient. Table 2.10 shows the average percentage of call failures and its standard deviation (again computed over the last 7,500 time steps) for a set of ten 15,000 steps simulations where call probabilities are unchanged, but the adaptive component is switched off at time 7,500 for three of the routing algorithms: improved mobile agents, ants without noise, and ants with noise ($g = 0.05$). That the ABC algorithms still perform better despite the fact that routing tables have been frozen suggests that the ants have been able to find good routes for the given network conditions. Table 2.11 shows the same as Table 2.9, but now a different set of call probabilities is introduced at time 7,500. Here again,

TABLE 2.9 Average percentage and its standard deviation (time steps 7,500–15,000) of call failures for a series of ten 15,000 steps simulations where call probabilities are unchanged, for five different algorithms. After Schoonderwoerd et al. [283]. Reprinted by permission © *MIT Press.*

	Average call failures	Std. Dev.
Shortest path	12.57	2.16
Mobile agents	9.19	0.78
Improved mobile agents	4.22	0.77
ABC without noise	1.79	0.54
ABC with noise	1.99	0.54

the ABC algorithms perform better than the others, with the noisy version being slightly more efficient in the dynamic environment. In this dynamic environment, if routing tables are frozen at the same time call probabilities are changed, the system's performance degrades significantly but the ABC algorithms remain the best performing (Table 2.12).

At each time step, an ant as described by Schoonderwoerd et al. [283] only updates the column that corresponds to its source node (viewed as a destination node). A recent addition by Guérin [159] introduces updating of all columns corresponding to all the intermediate nodes visited by the ant, in a way reminiscent of dynamic programming's *optimality principle*: the assumption is that if a route between two nodes is optimal (irrespective of the exact definition of optimality), all subroutes that connect intermediate nodes should also be optimal. Or, differently put, optimal routes are made up of optimal subroutes. In order to take advantage of this simple fact in the context of Schoonderwoerd et al.'s ant colony routing, Guérin [159] defines a new species of ants—*smart ants*—which reinforce intermediate routes in addition to their main route. Smart ants reinforce an entry associated with a given destination node by a quantity that depends on the smart ant's age relative to the time it visited that node. Guérin notes that this scheme, although it requires more complex ants, reduces the number of ants to be released in the network because each ant performs more updates: assuming that the average length of an ant's trip, expressed as number of visited nodes, is L, an original ant reinforces positively a total of L entries in all the routing tables of the various nodes it has visited, whereas a smart ant reinforces positively $L(L-1)/2$ entries, that is, a factor $(L-1)/2$ more entries. Therefore, it may, in principle, be possible to reduce the number of ants in the network by a factor $(L-1)/2$ and obtain the same type of performance.

Although Guérin's [159] method relies on ants that perform round trips from their source nodes to their destination nodes and back, it can be readily applied

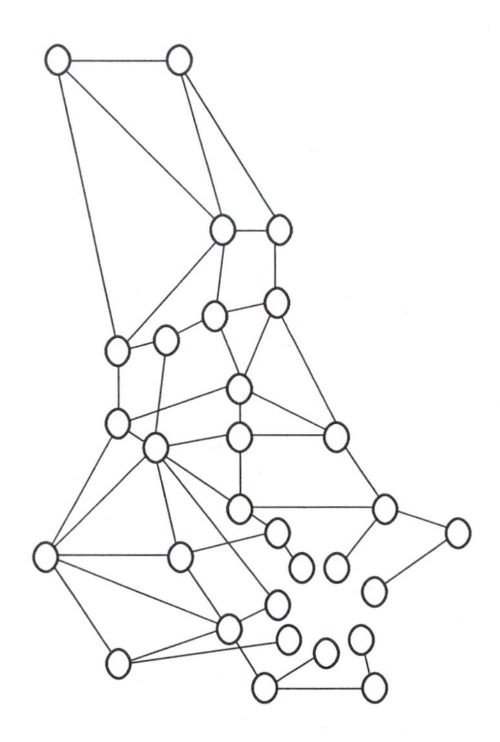

FIGURE 2.19 Schematic representation of the interconnection structure of the British Telecom SDH network. After Schoonderwoerd et al. [283]. Reprinted by permission © *MIT Press.*

TABLE 2.10 Average percentage and its standard deviation (time steps 7,500–15,000) of call failures for a series of ten 15,000 steps simulations where call probabilities are unchanged, but the adaptive mechanism is switched off at time 7,500, for three algorithms. After Schoonderwoerd et al. [283]. Reprinted by permission © *MIT Press.*

	Average call failures	Std. Dev.
Improved mobile agents	6.43	2.17
ABC without noise	2.11	0.60
ABC with noise	2.48	0.69

TABLE 2.11 Average percentage and its standard deviation (time steps 7,500–15,000) of call failures for a series of ten 15,000 steps simulations where call probabilities change at time 7,500, for five different algorithms. After Schoonderwoerd et al. [283]. Reprinted by permission © *MIT Press*.

	Average call failures	Std. Dev.
Shortest path	12.53	2.04
Mobile agents	9.24	0.80
Improved mobile agents	4.41	0.85
ABC without noise	2.72	1.24
ABC with noise	2.56	1.05

TABLE 2.12 Average percentage and its standard deviation (time steps 7,500–15,000) of call failures for a series of ten 15,000 steps simulations where call probabilities change at time 7,500 and the load balancing mechanism is switched off at time 7,500, for three algorithms. After Schoonderwoerd et al. [283]. Reprinted by permission © *MIT Press*.

	Average call failures	Std. Dev.
Improved mobile agents	8.03	2.88
ABC without noise	4.29	2.06
ABC with noise	4.37	2.27

to the case of one-way ants [33]. Let i be a node visited by a smart ant on its way to its destination. The ant updates the columns of i's routing table corresponding to all the intermediate nodes which belong to the path the smart ant is building. The updating of the column corresponding to an intermediate node f is performed using a relative age instead of an absolute age. The entry $r^i_{i-1,f}(t)$, where $i-1$ is the node preceding node i in the smart ant route, is reinforced while other entries $r^i_{l,f}(t)$, $l \neq i-1$, in the same column decay. The updating rules are exactly the same as in ABC (see Eqs. (2.29) and (2.30)), and the value δr is given by Eq. (2.32) in which the absolute age T is substituted by the value $T_i - T_f$:

$$\delta r = \frac{a}{T_i - T_f} + b\,, \tag{2.34}$$

where T_f is the ant's absolute age when reaching node f and T_i its age when reaching node i.

The ant colony algorithm ABC with smart ants yields significantly better performance than without. To evaluate smart ants, simulations have been run on the same experimental testbed as Schoonderwoerd et al. [283], using exactly the same simulation parameters: the 30-node interconnection network of British Telecom was used as an example; each node i has a capacity of $C_i = 40$ calls; during every time step, an average of one call is generated, with an average duration of 170 time steps; the probabilities that nodes be emitters or receivers are uniform in $[0.01, 0.07]$ and normalized; generation and normalization of emission and reception probabilities for all nodes defines a set of call probabilities, and a change in call probabilities means that new emission and reception probabilities have been generated (new nodes become more likely to be emitters or receivers, and others less likely). At initialization, routing tables are characterized by equiprobable routing (all neighboring nodes are equally likely to be selected as next node), and there is no message in the network. Messages are routed independently of the agents' dynamics: when a message reaches a node, it selects the largest entry in the appropriate row in its current table and is routed towards the neighboring node corresponding to this largest entry.

Figure 2.20(a) shows the average number of call failures per 500 steps together with error bars representing the standard deviation observed over 10 simulation trials, the first 500 steps being discarded, after the network has been initialized [33]. Two phases are examined: the first phase is defined by $500 < t \leq 3,000$, and the second phase by $3,000 < t \leq 7,500$. The first phase corresponds to a transient dynamics, while the second phase corresponds to a phase where a stationary regime has been reached. During the first phase following initialization, smart ants perform significantly better than Schoonderwoerd et al.'s [283] ants (t-test: $P < 0.003$). During the second phase, smart ants also perform significantly better (t-test: $P < 0.001$), with a level of significance comparable to the one obtained during the first phase. Figure 2.20(b) shows the average number of call failures per 500 steps together with error bars representing the standard deviation observed over 10 trials of the simulation, the first 500 steps being discarded, after a change in the call probabilities [33]. Results similar to those obtained after network initialization are observed: during both phases, smart ants perform significantly better than simple ants (t-test phase 1: $P < 0.001$; t-test phase 2: $P < 0.025$).

The above results must be taken with some care. The performance of ABC with smart ants depends on the applicability of the optimality principle to the considered problem, which in turn depends on the network topology and the traffic patterns. More experiments will be necessary to better characterize the situations in which smart ants are useful from those in which they are not.

Two other recent approaches which extend the ABC algorithm have been proposed. Subramanian et al. [301] have proposed an application of ABC to packet-switching networks. In the Subramanian et al. model an ant is represented by a triple (s, d, c) where s is the ant source node, d its destination node, and c is a path cost which starts at zero and is incremented while the ant is building its path. Routing tables are probabilistic and they are updated exactly as in ABC. Subramanian et al. also define so called *uniform ants* that choose the next node to visit uniformly randomly (that is, all neighbor nodes have the same probability of being chosen). Unfortunately, both these schemes call for symmetric path costs (i.e., for

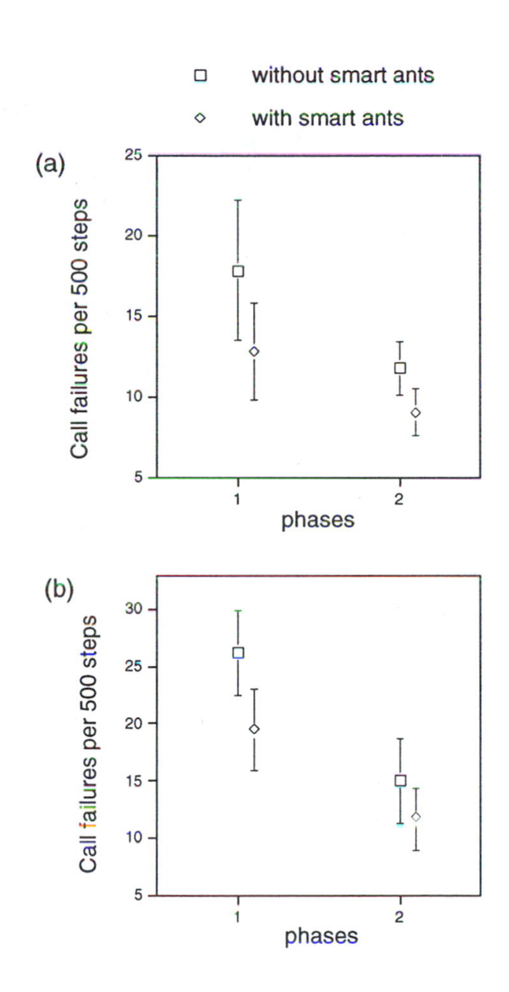

FIGURE 2.20 (a) Adaptation (500 first steps discarded). Phase 1: from $t = 501$ to $t = 3,000$, Phase 2: from $t = 3,001$ to $t = 7,500$. $a = 0.08$, $b = 0.005$, $c = 80$, $d = 0.075$, $g = 0.05$, node capacity $C_i = 40$. (b) Change in call probabilities (500 first steps discarded). Phase 1: from $t = 501$ to $t = 3,000$, Phase 2: from $t = 3,001$ to $t = 7,500$. $a = 0.08$, $b = 0.005$, $c = 80$, $d = 0.075$, $g = 0.05$, node capacity $C_i = 40$.

any traffic situation the cost to traverse the link (i, n) needs to be the same as that of traversing the link (n, i)). Requiring path costs to be symmetric is a strong constraint, hardly met by most of the interesting instances of packet-switching networks, and their ant algorithm can therefore find useful applications only to the small subset of networks whose links can be assumed to have symmetric costs.

Heusse et al. [169] have proposed a model, called cooperative asymmetric forward (CAF) routing, for packet-switching networks with asymmetric path costs that combines the ABC approach with some of the Q-routing ideas [39]. In CAF, differently from what happens in both ABC and AntNet, routing tables are based on cost estimates to destination nodes. In their implementation, packet delay is used as the cost metric, and the ants measure the delay between nodes by sharing the same transmission lines and queues with data packets. They use these cost estimates to update the routing tables on-line. The presence of asymmetric path costs introduces two problems for ant routing algorithms like ABC. The first problem is that an ant originated in node s and moving from node $i-1$ to neighbor node i cannot use the cost information it collects about the cost of link $(i-1,i)$ to change the probability $r^i_{i-1,s}$ of going from $i-1$ to i, (on the contrary, this can be done in case of symmetric path costs like those used by Schoonderwoerd et al.'s and by Subramanian et al.'s algorithms). To solve this problem Heusse et al. [169] maintain on each node i a model of the time $T_{n\to i}$ required to go from any neighbor node n of i to node i. When the ant moves from node i to node n it does two things: (i) it reads the value $T_{n\to i}$ it finds on node i and uses it to update the value $r^n_{i,s}$ of the routing table of node n, and (ii) it updates the model of the time $T_{i\to n}$ maintained at node n. (In fact, Heusse et al. use a very simple model: the value of $T_{i\to n}$ is set to the value written by the last ant which used link (i,n). That is, they do not maintain any memory of what happened in the past.) The second problem is given by the fact that, when path costs are asymmetric, the data traffic moving say from node s to node d can follow different paths than the data traffic moving from node d to node s. When the path costs are symmetric, i.e., traffic from a node s to a node d has the same characteristics as traffic moving from node d to node s, the very same routing tables can be used to route both data packets and ants. In this way a variance reduction effect by which ants explore the neighborhood of the paths that are currently used by data packets can be obtained. This cannot be done in the presence of asymmetric path costs, as can be better understood by looking at Figure 2.21. In this example, data packets from node s to node d follow a different preferred route from that followed by data packets from node d to node s. Ants with source node s and destination node d, since they update routing tables entries used by data packets directed toward s, should follow the same preferred route as that used by data packets directed toward s (lower branch in the figure). Therefore, they cannot use the same routing tables as data packets directed toward d: if they did, they would follow the upper branch.

To solve this problem, Heusse et al. [169] maintain, on each node i, a table $RevR_i = [rev^i_{n,d}(t)]_{k_1,N-1}$ which keeps track of the data traffic toward any possible destination node d via any possible neighbor node n. In practice, each time a data packet originated in node d traverses link (n,i) the value $rev^i_{n,d}(t)$ is incremented by one. Ants are routed probabilistically using the normalized values of the table $RevR_i$. Heusse et al. also implemented the smart ant capability described above, plus some capabilities to manage dynamic networks, that is, networks which change configuration at run time (e.g., link failures, mobile devices, etc.).

In both cases, during the writing of this book, results about these works were not accurate enough to say whether or not they resulted in an improvement over the

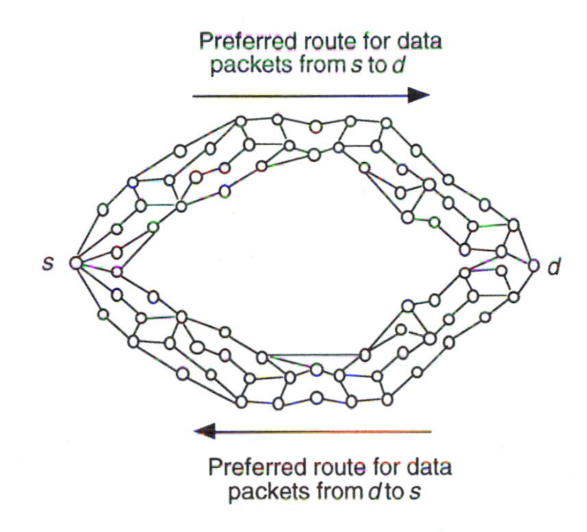

FIGURE 2.21 Data packets from node s to node d and data packets from node d to node s follow different preferred routes. Ants with source node s and destination node d update routing tables entries used by data packets directed towards s. They should therefore follow the same preferred route as that used by data directed towards s. Therefore ants cannot use the same routing tables as data packets.

original ABC and AntNet algorithms. The preliminary results presented by Heusse et al. [169], however, are promising.

2.7.2 ANTNET: ROUTING IN DATA COMMUNICATIONS NETWORKS

Di Caro and Dorigo [97, 103, 104] have introduced an adaptive routing algorithm based on ant colonies that explore the network with the goal of building routing tables and keeping them adapted to traffic conditions. Their algorithm, called AntNet, although based on principles similar to those of Schoonderwoerd et al.'s approach, has some important differences:

1. The way ants collect and distribute information was designed so that AntNet can be applied to both connection-oriented and connectionless types of communication networks.
2. Ants collect information which is used to build local parametric models of the network status. These models are used to compute reinforcements to change the probabilistic routing tables.
3. AntNet has been tested on a packet-switching network model and its performance, in terms of throughput and delivery delays, has been compared to the performance of well-known existing routing algorithms.

This work is a serious attempt at evaluating the efficiency of an ant-based algorithm, although the network model used is still a simplification, albeit a sophisticated one, of a real network. Moreover, Di Caro and Dorigo's tests are based on continuous-time discrete event simulation, which is much more powerful and closer to reality than the sequential discrete iterations simulation system used by Schoonderwoerd et al. [283].

The same notation as in the previous subsection is used. Let s and d be the source and destination nodes, respectively, of an ant, $i-1$ the node it just came from, and i the current node at time t. In the network there are N nodes, and each node i is characterized by its number of neighbors k_i and by a routing table $R_i = [r^i_{n,d}(t)]_{k_i,N-1}$. The entry $r^i_{n,d}(t)$ in the routing table of node i represents the probability at time t of choosing node n (which is a neighbor of node i) as the next node to move to when the destination is node d. Alternatively, $r^i_{n,d}(t)$ can be seen as the *desirability* of going through n when trying to reach d from i. The routing table is not the only relevant data structure present at a node: there is also a set of estimates for trip times from the current node i to all other nodes d. This set of estimates, denoted by $\Gamma_i = \{\mu_{i\to d}, \sigma^2_{i\to d}\}$, comprises average estimated trip times $\mu_{i\to d}$, that is, the estimated average time to go from node i to node d, and their associated variances $\sigma^2_{i\to d}$. Γ_i allows maintenance of a local picture of the global network's status at node i. Both the routing table R_i and the local model Γ_i are updated by ants in a way that is described below.

Two types of ants exist: *forward ants*, denoted by $F_{s\to d}$, that go from a source node s to a destination node d, and *backward ants*, denoted by $B_{d\to s}$, that go from destination nodes to source nodes. $F_{s\to d}$ is transformed into $B_{d\to s}$ when it reaches d. Forward ants move in the network at the same level of priority as data packets, and therefore experience queues and traffic load, which allows them to collect information about the state of the network; backward ants are used to back-propagate this information to nodes on their paths as quickly as possible, which is why they are given a higher priority than forward ants and data.[9] The AntNet algorithm can be informally described as follows.

- Forward ants $F_{s\to d}$ are launched at regular intervals from every node s in the network, with a randomly selected destination d. Destinations are chosen to match the current traffic patterns.
- Each forward ant selects the next hop among the not yet visited neighbor nodes following a random scheme. The choice probability is made proportional to the *desirability* of each neighbor node (that is, to the value $r^i_{n,d}(t)$) and to the local queues status.
- The identifier of every visited node i and the time elapsed from its launching time to its arrival at this ith node are pushed onto a memory stack $S_{s\to d}$ carried by the forward ant.

[9]Backward ants use high-priority queues which, on real networks, are usually available for control packets.

- If a cycle is detected, that is, if an ant is forced to return to an already visited node because none of the neighbor nodes has an "unvisited" status, the cycle's nodes are popped from the ant's stack and all the memory about them is destroyed.
- When the ant $F_{s \to d}$ reaches the destination node d, it generates a backward ant*backward ants* $B_{d \to s}$, transfers to it all of its memory, and then dies.
- The backward ant makes the same path as that of its corresponding forward ant, but in the opposite direction. At each node i along the path it pops its stack $S_{s \to d}$ to know the next hop node. Arriving in a node i coming from a neighbor node $i-1$, the backward ant updates Γ_i and the routing table R_i as follows:
  - The model Γ_i is updated with the values stored in the stack memory $S_{s \to d}(i)$; the times elapsed to arrive in every node f on the path $i \to d$, that is the path followed by ant $F_{i \to d}$ starting from the current node i, are used to update the corresponding sample means and variances. To be exact, only those subpaths that have a smaller trip time than the average trip time memorized in the model are used to update the model. Poor subpaths' trip times are not used because they do not give a correct idea about the time to go toward the subdestination node. In fact, all the forward ant routing decisions were made only as a function of the destination node. In this perspective, subpaths are side effects, and they are intrinsically suboptimal because of the local variations in the traffic load. As already noted when discussing Guérin's work [159] on smart ants in the previous section, the dynamic programming optimality principle must be applied with care because of the nonstationarity of the problem representation. Obviously, in case of a good subpath we can use it: the ant has discovered, at zero-cost, an additional good route.
  - The routing table R_i is changed by incrementing the probability $r^i_{i-1,d}(t)$ associated with the neighbor node $i-1$ and the destination node d, and decreasing (by normalization) the probabilities $r^i_{n,d}(t)$ associated with the other neighbor nodes $n, n \neq i-1$, for the same destination. The trip time $T_{i \to d}$ experienced by the forward ant $F_{s \to d}$ is used to assign the probability increments. $T_{i \to d}$ is the only explicit feedback signal available: it gives an indication about the goodness r of the followed route because it is proportional to its length from a physical point of view (number of hops, transmission capacity of the used links, processing speed of the crossed nodes) and from a traffic congestion point of view.[10] A problem in using the value $T_{i \to d}$ is that it cannot be associated with an exact error measure, given that the optimal trip times are not known since they depend on the net load status. It can, therefore, only be used as a reinforcement signal. The values stored in the model Γ_i are used to score the trip times so that they can be transformed in a reinforcement signal

[10]Forward ants share the same queues as data packets (backward ants do not, they have priority over data to propagate the accumulated information faster), so if they cross a congested area, they will be delayed. This has a double effect (which is very similar to the one artificially induced by delaying ants in congested nodes as implemented in Schoonderwoerd et al.'s [283] ABC algorithm): (i) the trip time will grow and therefore back-propagated probability increments will be small, and (ii) at the same time these increments will be assigned a larger delay.

$r \equiv r(T_{i\to d}, \Gamma_i), r \in [0,1]$ (r is such that the smaller $T_{i\to d}$, the higher r). This dimensionless value takes into account an average of the observed values and their dispersion: $r \propto (1 - W_{i\to d}/T_{i\to d}) + \Delta(\sigma, W)$, where $W_{i\to d}$ is the best trip time experienced over a time window, and $\Delta(\sigma, W)$ is a correcting term [97]; r is used by the current node i as a positive reinforcement for the node $i-1$ the backward ant $B_{d\to s}$ comes from. The probability $r^i_{i-1,d}(t)$ is increased by the reinforcement value as follows:

$$r^i_{i-1,d}(t+1) = r^i_{i-1,d}(t) + r \cdot (1 - r^i_{i-1,d}(t)) = r^i_{i-1,d}(t) \cdot (1-r) + r\,. \quad (2.35)$$

In this way, the probability $r^i_{n,d}(t)$ will be increased by a value proportional to the reinforcement received and to the previous value of the node probability (that is, given a same reinforcement, small probability values are increased proportionally more than big probability values).

Probabilities $r^i_{n,d}(t)$ for destination d of the other neighboring nodes n implicitly receive a negative reinforcement by normalization. That is, their values are reduced so that the sum of probabilities will still be 1:

$$r^i_{n,d}(t+1) = r^i_{n,d}(t) \cdot (1-r), n \neq i-1\,. \quad (2.36)$$

The transformation from the raw value $T_{i\to d}$ to the definition of the more refined reinforcement r is similar to what happens in Actor-Critic systems [8]: the raw reinforcement signal ($T_{i\to d}$, in our case) is processed by a critic module which is learning a model of the underlying process, and then is fed to the learning system.

Every discovered path receives a positive reinforcement in its selection probability. In this way, not only the (explicit) assigned value r plays a role, but also the (implicit) ant's arrival rate. This strategy is based on trusting paths that receive either high, independent of their frequency, reinforcements, or low and frequent reinforcements. In fact, for any traffic load condition, a path receives one or more high reinforcements only if it is much better than previously explored paths. On the other hand, during a transient phase after a sudden increase in network load all paths will likely have high traversing times with respect to those learned by the model Γ in the preceding, low congestion, situation. Therefore, in this case good paths can only be differentiated by the frequency of ants' arrivals. Assigning always a positive, but low, reinforcement value in the case of paths with high traversal time allows the implementation of the above mechanism based on the frequency of the reinforcements, while, at the same time, avoids giving excessive credit to paths with high traversal time due to their poor quality.

Algorithm 2.8 reports a high-level description of the AntNet algorithm [97]. Before AntNet starts to control the network, an initialization phase is run for $t = 500$ sec. During this initialization phase the algorithm is exactly the same as during regular operation, but operates on the unloaded network. The goal of the initialization phase is to build initial routing tables which match the network topology,

Algorithm 2.8 High-level description of AntNet

```
/* Initialization */
/* The initialization phase is the same as the main loop, but without data traffic */
t = current time
Δt = time interval between ants generations
/* Main loop */
For each node do /* concurrently over all network nodes */
  While t ≤ t_max do
    If ((t mod Δt) = 0) do
      Select destination node
      Launch forward ant
    End If
    For each forward ant do /* concurrently for all forward ants */
      While (current node ≠ destination node) do
        Select link using routing table and link queues
        Put ant on link queue
        Wait on data link queue
        Cross the link
        Push the elapsed time information on the stack
      End While
      Launch backward ant
      Die
    End For
    For each backward ant do /* concurrently for all backward ants */
      While (current node ≠ destination node) do
        Choose next node by popping the stack
        Wait on high priority link queue
        Cross the link
        Update the traffic model
        Update the routing table
      End While
    End For
  End While
End For

/* Values of algorithm parameters used in experiments */
δ = 1.2, Δt = 0.3 seconds
```

so that when data traffic is added AntNet can concentrate its efforts on adapting routing tables to traffic load.

In order to fully characterize the system, one has to define how packets are routed within a node. Each node holds a buffer in which incoming and outgoing packets are stored; a different buffer is used for each outgoing link. Packets at the same priority level are served on a first-in-first-out basis; high-priority packets (backward ants) are served first. A packet which has not yet reached its destination reads the relevant information from the routing table and selects its next node. It is important to note that not only ants, but also data packets use routing tables in a probabilistic manner. In fact, the use of probabilistic entries is very specific to AntNet. It has been observed to improve AntNet performance, which means that the way the routing tables are built in AntNet is well matched with a probabilistic distribution of the data packets over all the good paths. Data packets are prevented from choosing links with very low probability by remapping the T's entries by means of a power function $f(p) = p^\delta, \delta > 1$ which emphasizes high-probability values and reduces lower ones (in their experiments Di Caro and Dorigo set $\delta = 1.2$).

If link resources are available, the transfer is made, and arrival of the packet at the next node is scheduled with a certain transmission delay characteristic of the link and packet size. If there are not enough resources, the packet waits in the node's buffer. If, upon arrival at a given node, there is not enough buffer space, the packet is discarded.

Di Caro and Dorigo [97, 99, 100, 101] have tested their algorithm on a set of model networks, among which are the US NSFNET-T1, and the Japanese NTTnet. NSFNET is composed of 14 nodes and 21 bidirectional links (Figure 2.22(a)) with a bandwidth of 1.5 Mbits/s and propagation delay which range from 4 to 20 ms. NSFNET is the old T1 US backbone, and is a well balanced network: the distance between a pair of nodes in terms of hops ranges from 1 to 4, while the ratio between the mean connectivity degree and the number of nodes is about 0.2. NTTnet is the NTT (Nippon Telephone and Telegraph company) corporate backbone, and is the major Japanese network (Figure 2.22(b)). NTTnet has 57 nodes and 162 bidirectional links. Links have a 6 Mbit/sec bandwidth, while their propagation delays range from 1 to 5 msec. NTTnet is not well balanced: the distance between a pair of nodes in term of hops ranges from 1 to 20, while the ratio between the mean connectivity degree and the number of nodes is about 0.05.

A number of different traffic patterns, both in terms of spatial and temporal characteristics, have been considered. Traffic is defined in terms of open sessions between pairs of different nodes. Each single session is characterized by the number of transmitted packets, and by their size and interarrival time distributions. Sessions over the network are characterized by their interarrival time distribution and by their geographical distribution. The latter is controlled by the probability assigned to each node to be selected as a session start or end-point. They considered three basic patterns for the temporal distribution of the sessions, Poisson (P), fixed (F), and temporary (TMP), and three for their spatial distribution, uniform (U), random (R), and hot spots (HS). General traffic patterns are then obtained by combining the above temporal and spatial characteristics. For example, UP traffic is obtained by letting an identical Poisson process regulate the arrival of new sessions for all nodes, while in RP traffic the Poisson process is different for each node, and in the UP-TMPHS case a temporary hot spots traffic is superimposed to a UP traffic.

The performance of the AntNet algorithm and of several other algorithms is expressed in throughput (delivered bits/s) and delivery time from source to desti-

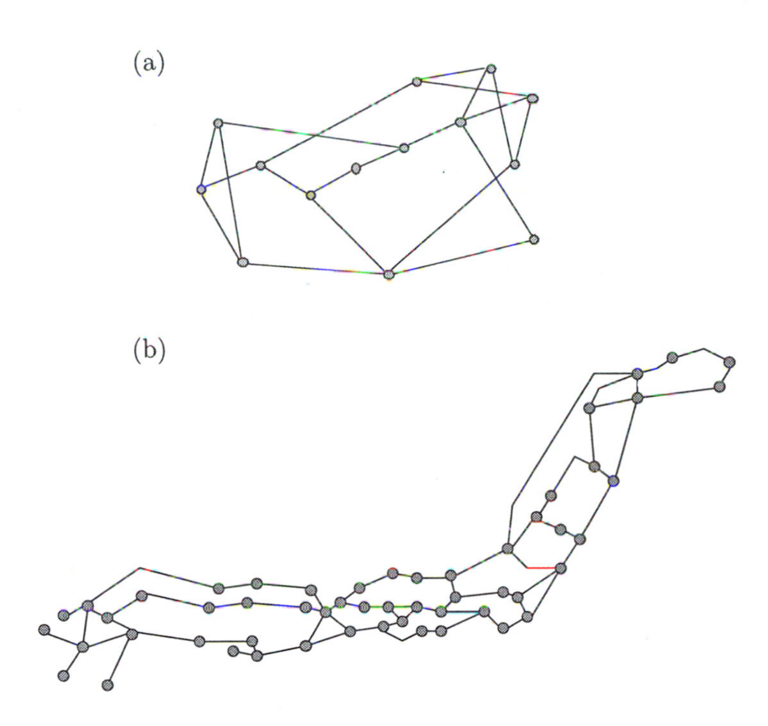

FIGURE 2.22 Two of the network topologies used by Di Caro and Dorigo: (a) NSFNET, with 14 nodes and 21 bidirectional links, and (b) NTTnet, with 57 nodes and 162 bidirectional links.

nation (s) during a total simulated time of 1000 s. The competing algorithms with which AntNet is compared are:

1. A simplified version of open shortest path first (OSPF), the current official Internet routing algorithm [244].
2. A sophisticated version of the asynchronous distributed Bellman-Ford algorithm (BF) with a dynamic cost metric [23].
3. The shortest path first (SPF) algorithm with a dynamic cost metric [197].
4. The Q-routing algorithm as proposed by Boyan and Littman [39].
5. The predictive Q-routing algorithm [71], an extension of Q-routing.

In order to evaluate the "absolute" performance of the AntNet algorithm, Di Caro and Dorigo [96, 97] introduced an approximation of an "ideal" algorithm, called *daemon* algorithm, which provides an empirical bound to achievable performance. In this algorithm, which is an abstraction and cannot be implemented in a real network, a daemon is capable of reading at every instant the states of the queues of all nodes and of computing on the basis of this global knowledge instantaneous costs for all the links. This algorithm can, therefore, compute before each packet hop new network-wide shortest paths. The cost of a given link is computed on the basis of the characteristics of the link and of its queue status. For the BF and SPF algorithms, where the cost metric is dynamic, several cost metrics were tested (a detailed description can be found in Di Caro and Dorigo [97]). The parameters of all competing algorithms were carefully optimized to make the comparison as fair as possible.

A wide range of experiments were run on NSFNET and NTTnet, as well as on other networks, using UP, RP, UP-HS and TMPHS-UP traffic patterns. For each traffic pattern considered Di Caro and Dorigo studied the behavior of the algorithms when moving the traffic load toward a saturation region. To this aim, they ran five distinct groups of ten trial experiments gradually increasing the generated workload. In all the cases considered, differences in throughput were of minor importance with respect to those shown by packet delays. Because the distribution of packet delays presents a very high variance, they measured the performance by the 90th percentile statistic, instead of the more usual average packet delay. In this case, in fact, the 90th percentile gives a much better representation of the algorithm's performance.

Figures 2.23 and 2.24 show some results regarding throughput and packet delays for RP traffic on NSFNET and NTTnet, respectively. These results are exemplar of the behavior of AntNet: results obtained on other traffic pattern and network topology combinations are qualitatively equivalent. It is interesting to note that AntNet throughput was at least as good as that of the competing algorithms, while its 90th percentile of packet delays was much better than that of the others. Moreover, it was rather close to that generated by the ideal daemon algorithm. Di Caro and Dorigo also investigated the response of AntNet to a sudden increase of traffic which temporarily took the network to a saturation condition.

Figures 2.25 and 2.26 show the results of an experiment in which a temporary hot-spot load was superimposed to a light Poisson traffic. In both the NSFNET (Figure 2.25) and NTTnet (Figure 2.26) case AntNet was the best performing algorithm with the exception of the daemon which in the NSFNET case did slightly better.

Di Caro and Dorigo's AntNet is an interesting contribution to both the fields of swarm-based algorithms and ant-based routing algorithms. Although it is close in spirit to Schoonderwoerd et al.'s [283] algorithm, Di Caro and Dorigo's [97, 98, 99, 100, 101, 102, 103] work goes further in several respects: it uses a network model which is more complex, it deals with general packet-switching networks avoiding the "symmetric path costs" constraint of Schoonderwoerd et al.'s model, it introduces additional information at the level of nodes about the reliability of agent-based information in the form of estimated trip times and their variances, it compares the algorithm's performance with the performance of (simplified versions of) classical routing schemes, and, in doing so, it introduces a "benchmark" algorithm (the daemon algorithm) the performance of which provides an empirical bound to achievable performance.

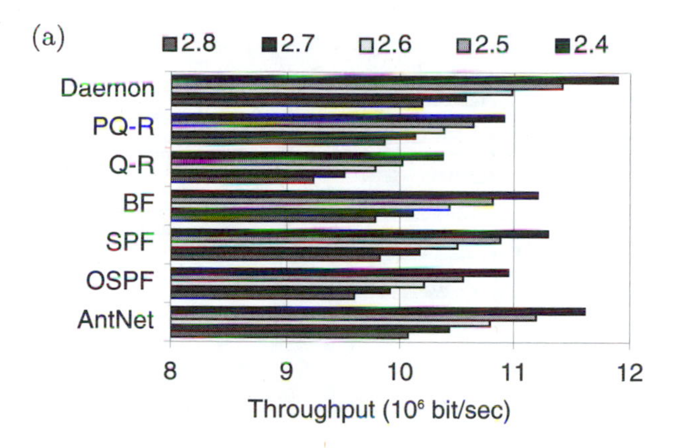

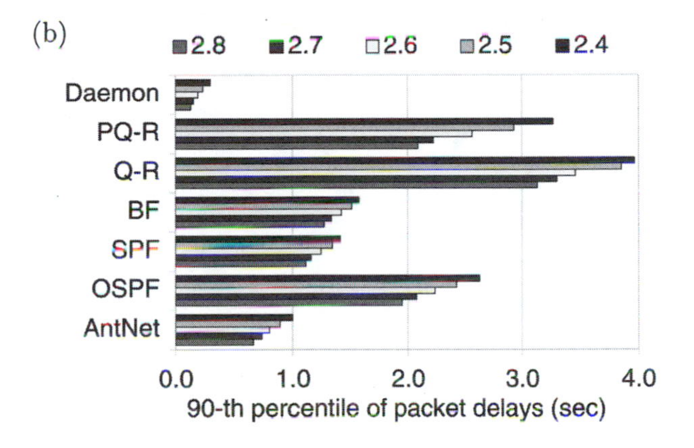

FIGURE 2.23 Comparison of algorithms for increasing load for RP traffic on NSFNET. The load is increased reducing the mean session inter arrival value from 2.8 to 2.4 seconds. (a) Throughput, and (b) 90th percentile of the packet delays empirical distribution.

2.7.3 AN ANT ROUTING ALGORITHM BASED ON ANT SYSTEM

White et al. [325] have proposed a scheme for routing in circuit-switched networks which was also inspired by ant colonies' foraging behavior, and in particular by Ant System (AS) (section 2.3.1). Although they had no significant results at the time this book was being written, we report a brief description of their method for completeness. In their system, ants are launched each time there is a connection request. A connection request can be a point-to-point (PTP) or a point to multi-point (PMP) request. In the PTP case, in a network of N nodes, N ants are launched to look for the best path to the destination, given a cost function associated with every node. For a PMP request with m destinations, $N \cdot m$ ants are launched.

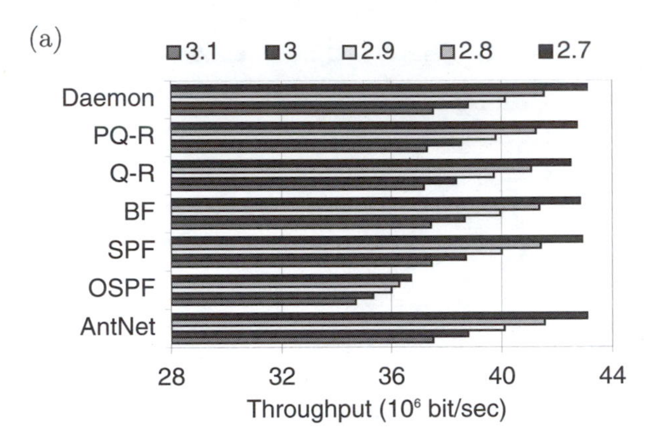

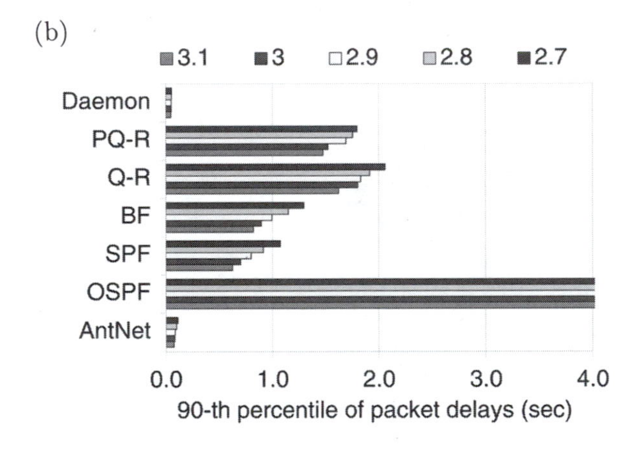

FIGURE 2.24 Comparison of algorithms for increasing load for RP traffic on NTTnet. The load is increased reducing the mean session inter arrival value from 3.1 to 2.7 seconds. (a) Throughput, and (b) 90th percentile of the packet delays empirical distribution. The OSPF value for packet delays is out of scale.

Contrary to the previous algorithms, but similar to what happens in Ant System this one relies on pheromone concentrations on links. Let $\tau_{ij}(t)$ be the amount of pheromone trail on the link connecting node i and node j, $C_{ij}(t)$ the cost associated with this link ($C_{ij}(t)$ is a dynamic variable that depends on the link's load), $C^k(t)$ the cost of the kth ant's route, $\Delta\tau^k$ the amount of pheromone deposited by ant k ($\Delta\tau^k$ is independent of the ant's performance $C^k(t)$; it could be modified to be, for example, proportional to $1/C^k(t)$), $p_{ij}^k(t)$ the probability that ant k chooses to hop from node i to node j, and C_{max} the maximum allowed cost of a route. In a way

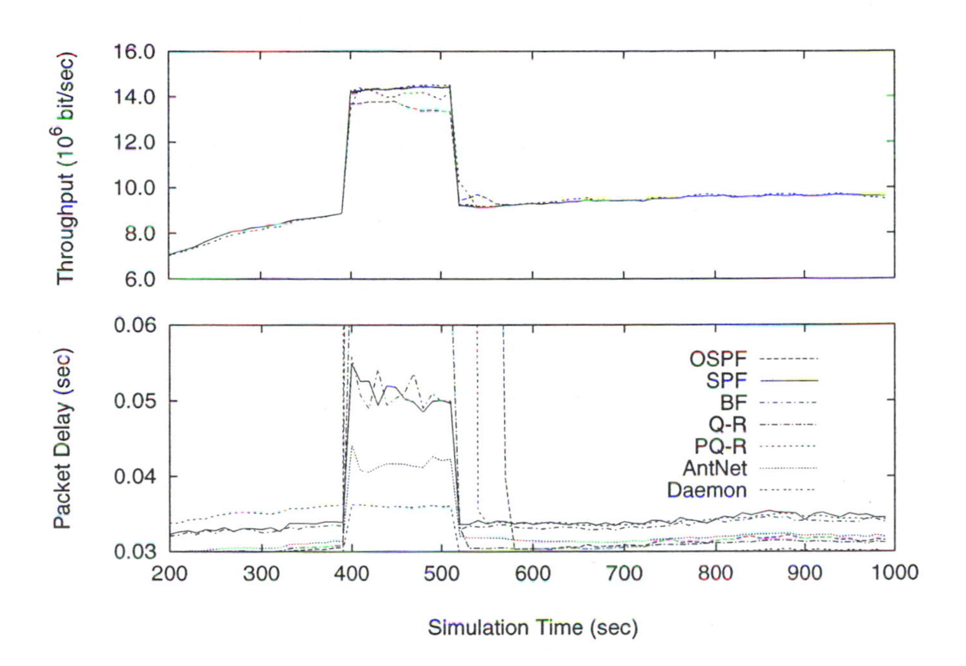

FIGURE 2.25 NSFNET: Comparison between algorithms for a temporary hot-spot load superimposed to a Poisson traffic. Packet delays are averaged over a 10 seconds moving window.

similar to what is done in the AS-TSP algorithm, $p_{ij}^k(t)$ is defined by:

$$p_{ij}^k(t) = \frac{\left[\tau_{ij}(t)\right]^\alpha \cdot \left[1/C_{ij}(t)\right]^\beta}{\sum_{l \in J_i^k} \left[\tau_{il}(t)\right]^\alpha \cdot \left[1/C_{il}(t)\right]^\beta}, \tag{2.37}$$

where J_i^k is the set of nodes connected to node i (that is, the neighbors of node i) which have not been traversed yet by ant k, and α and β are two tunable parameters that determine the respective influences of pheromone concentration and link cost. After the initial generation of N (or $N \cdot m$) ants, ants are created at a frequency f_e during the time of the session, so that it is possible to adapt to changing network conditions within a session. Ant k hops from node to node to reach its destination node and memorizes the visited nodes. Ant k selects the next node to hop to according to probability $p_{ij}^k(t)$ and updates its route cost each time it traverses a link: $C^k(t+1) = C^k(t) + C_{ij}(t)$. If $C^k(t+1) > C_{\max}$, ant k is deleted from the network. Once it has reached its destination, ant k goes all the way back to its source node through all the nodes visited during the forward path, and updates the pheromone concentration of every edge (i,j) traversed: $\tau_{ij}(t+1) = \tau_{ij}(t) + \Delta\tau^k$. In

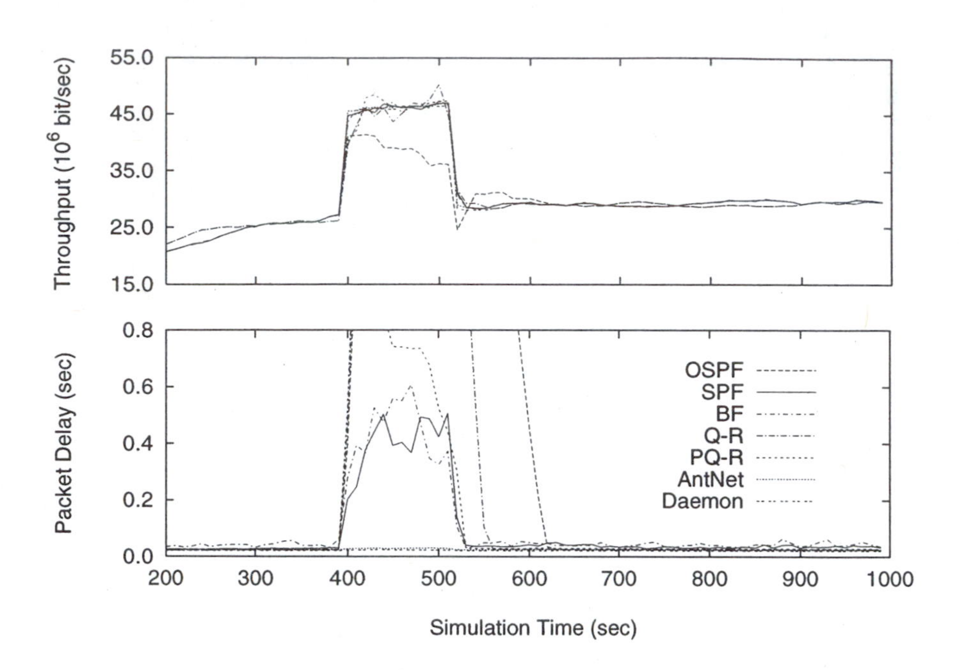

FIGURE 2.26 NTTnet: Comparison between algorithms for a temporary hot-spot load superimposed to a Poisson traffic. Packet delays are averaged over a 10 seconds moving window. Since AntNet and daemon have very similar performance their packet delay graphs are superimposed.

addition to this pheromone reinforcement, there is evaporation, characterized by an evaporation rate ρ (evaporation per unit time). The source node keeps in memory all paths that have been found by ants. If the path found by ant k is the same as $p\%$ of all paths in the node's memory, then the path is reserved and a connection is eventually set up (in the case of a PMP request, it is the spanning trees found by ants to the multiple destination nodes which are compared).

2.7.4 LIMITATIONS AND DISCUSSION

In the previous sections we have seen how ant-based algorithms can be used to find good, adaptive routing tables in both connection-oriented and connectionless types of networks. Although ABC and AntNet differ for a number of important details, they share some important ideas:

- Both approaches view ants as a mechanism to explore the network and to collect information about its status.

- In both systems the implicit information represented by the arrival rate of ants, one of the most original aspects of the approach, plays an important role.
- They both use probabilistic routing tables instead of routing tables based on cost estimates (e.g., of expected times required to reach destination nodes).

These aspects make ant colony routing algorithms a novel, and very promising, approach to distributed routing. Before the ant colony routing methods presented in this chapter can actually be implemented in real communications networks, some limitations have to be overcome:

- The model network used in the examples is obviously a simplification of reality: the "ant routing" methods have to be tested on realistic, and therefore more complex, network models. Not only should more realistic traffic conditions be simulated, specific constraints should also be taken into account. In the considered examples, routing is limited to finding the shortest path from one point to another under given network conditions. On the contrary, in the context of modern high-speed networks like ATM (Asynchronous Transfer Mode), routing is often complicated by the notion of guaranteed Quality of Service (QoS), which can either be related to time, packet loss or bandwidth requirements: constraints related to various types of QoS make some routing solutions unacceptable. Currently, AntNet is being extended, see [98], to allow best effort QoS [229]. The idea is to use faster ants (not waiting in data queues) which use the values maintained in the probabilistic routing tables to concurrently build a set of possible routes; the best one will be chosen and used to set up a virtual channel with loosely defined characteristics.
- Flow and congestion control has been mostly neglected in the systems described in the previous sections[11]: flow control (see chapter 6 in Bertsekas and Gallager [23]) is a mechanism that prevents some of the offered traffic from entering the network to avoid congestion when the network is heavily loaded. Obviously, flow control works in synergy with routing, influences throughput and average delay, and therefore affects routing algorithms.

Designing more complex simulation environments, however, points to the problem of analyzing the routing's behavior. Routing algorithms are generally difficult to analyze mathematically, especially when the underlying network is complex and/or not fully connected, and it is not possible to know a priori the characteristics of the traffic flow. For example, it would be crucial for "self-organizing" algorithms such as those presented here, where control by humans can only be limited, to be able to prove that they (i) converge to a solution asymptotically close to the optimal one for stationary traffic loads, (ii) adapt reasonably quickly to any new traffic situation, and (iii) do not enter oscillatory behavior. One also needs to have

[11] Although in AntNet a very simple form of flow control based on a limited production window is implemented.

a clear understanding of the limits and constraints of communications networks: for example, if there are sufficient computational power and spare capacity in the network to launch a large number of complex agents without affecting traffic, why bother to design simple agents?

2.7.5 FINDING INFORMATION IN A NETWORKED ENVIRONMENT

As a final example of the use of virtual pheromones in telecommunications networks, although in an application different from routing, a group of researchers at the School of Communication, Information, and Library Studies at Rutgers University is developing a new approach to finding information in networked environments based on digital information pheromones (DIP) (see Holden [170]). The idea is to use other people's experience in prior searches: for example, net surfers that are looking for interesting links or sites that deal with particular subjects can, if they belong to a "colony" of users, access information that has been left in the form of DIP (ratings) in previous searches by other colony members: as links are turned up by a web search, an "ant icon" appears next to every hyperlink for which the "ant server" had data from previous similar searches; clicking on the icon would produce a chart showing how many previous searches went through a particular hyperlink, and how the hyperlink was rated [170]. Documents that are found interesting could also get reinforced, while the DIP associated with those that are not found interesting could evaporate. Trail decay also allows obsolete information about documents to disappear naturally: if no user has visited a given site for a certain time, the DIP will have disappeared, so that only information which is regularly verified and reinforced is conserved. This could be implemented by discounting ratings by a factor that decreases with time. Although this project is only beginning, it looks promising and relies on principles that we have explored throughout this chapter.

2.8 POINTS TO REMEMBER

- In several species of ants, cooperative foraging through trail laying–trail following allows decentralized selection of the food source that is closest to the nest. Despite its ability to select the closest food source, a colony may not be able to track a changing environment because trail reinforcement is too strong. Pheromone evaporation could, in principle, solve this problem, but real trail pheromones seem to be long-lived. Virtual trail pheromones in simulations of collective foraging can be made as volatile as necessary if biological plausibility is not an issue.
- In army ants, foraging patterns are adapted to the spatial distribution of prey. A single self-organizing algorithm, in which both exploration and prey retrieval are based on trail laying–trail following, can account for different raid patterns in different species of army ants. One conjecture that remains to be tested is

that the energy expense is optimized in the raid patterns: if that is the case, the self-organizing algorithm provides an optimal response to different environments.

- The foraging behavior of ants has inspired a novel approach to distributed optimization, Ant Colony Optimization (ACO). Most of current ACO applications are either in combinatorial optimization or in communications networks routing. Ant System (AS), the first ACO algorithm, was applied to the traveling salesman problem (TSP). Although it had limited success from a performance standpoint, it opened up the road to improvements and new applications. Ant Colony System (ACS), a modified version of AS that extends AS and includes local search routines, exhibits top performance on asymmetric TSPs.
- A whole family of ACO algorithms is now applied to many different combinatorial optimization problems ranging from the quadratic assignment and the sequential ordering problems to vehicle routing and graph coloring problems. ACO algorithms for routing in communications networks, also called ant routing algorithms, have been particularly successful. This success is probably due to the dynamic nature of the routing problem which is well matched to the distributed and adaptive characteristics of ACO algorithms.
- AntNet, an ACO algorithm tested on several packet-switching communications networks, outperformed some of the best-known routing algorithms. These results, although obtained using a simulator which only approximates real networks, are very promising.
- The ACO area of research is now very active and the interested reader can find up-to-date information about publications as well as workshops and conferences on the Web at: http://iridia.ulb.ac.be/dorigo/ACO/ACO.html.

CHAPTER 3
Division of Labor and Task Allocation

3.1 OVERVIEW

Many species of social insects have a division of labor. The resilience of task allocation exhibited at the colony level is connected to the elasticity of individual workers. The behavioral repertoire of workers can be stretched back and forth in response to perturbations.

A model based on response thresholds connects individual-level plasticity with colony-level resiliency and can account for some important experimental results. Response thresholds refer to likelihood of reacting to task-associated stimuli. Low-threshold individuals perform tasks at a lower level of stimulus than high-threshold individuals.

An extension of this model includes a simple form of learning. Within individual workers, performing a given task induces a decrease of the corresponding threshold, and not performing the task induces an increase of the threshold. This double reinforcement process leads to the emergence of specialized workers, that is, workers that are more responsive to stimuli associated with particular task requirements, from a group of initially identical individuals. The fixed response threshold model can be used to allocate tasks in a multiagent system, in a way that is similar to market-based models, where agents bid to get resources or perform tasks.

The response threshold model with learning can be used to generate differentiation in task performance in a multiagent system composed of initially identical entities. Task allocation in this case is emergent and more robust with respect to perturbations of the system than when response thresholds are fixed. An example application to distributed mail retrieval is presented.

3.2 DIVISION OF LABOR IN SOCIAL INSECTS

In social insects, different activities are often performed simultaneously by specialized individuals. This phenomenon is called division of labor [253, 272]. Simultaneous task performance by specialized workers is believed to be more efficient than sequential task performance by unspecialized workers [188, 253]. Parallelism avoids task switching, which costs energy and time. Specialization allows greater efficiency of individuals in task performance because they "know" the task or are better equipped for it.

All social insects exhibit reproductive division of labor: only a small fraction of the colony, often limited to a single individual, reproduces. Beyond this primary form of division of labor between reproductive and worker castes, there most often exists a further division of labor among workers, which may take three, possibly coexisting, basic forms:

1. *Temporal polyethism.* With temporal polyethism, individuals of the same age tend to perform identical sets of tasks. Individuals in the same age class form an *age caste.* It is not clear at the moment whether or not absolute aging is the main cause of temporal polyethism. Social and external environment, as well as genetic characteristics, seem to influence the rate of behavioral ontogeny.
2. *Worker polymorphism.* In species that exhibit worker polymorphism, workers have different morphologies. Workers that differ by their morphologies are said to belong to different *morphological or physical castes.* Workers in different morphological castes tend to perform different tasks. An example of a worker caste is the soldier or major caste which is observed in several species of ants.
3. *Individual variability.* Even within an age or morphological caste, differences among individuals in the frequency and sequence of task performance may exist. One speaks of *behavioral castes* to describe groups of individuals that perform the same set of tasks within a given period.

A key feature of division of labor is its plasticity [272]. Division of labor is rarely rigid. The ratios of workers performing the different tasks that maintain the colony's viability and reproductive success can vary in response to internal perturbations or external challenges. Such factors as food availability, predation, climatic conditions, phase of colony development, or time of year influence the size and structure of a colony's worker population in natural conditions. The worker force must be allocated to tasks so as to adjust to changing conditions. Changes in the pattern of task allocation can be induced experimentally by altering colony size, structure, or demography, or by increasing the need for nest maintenance, nest repair, defense, etc.

Wilson [330] experimentally altered the structure of colonies of polymorphic ant species from the *Pheidole* genus. In most species of this genus, the worker population is divided into two morphological subcastes: the minor and major subcastes. Minors, which take care of most of the quotidian tasks of the colony, are smaller than the large-headed majors (often called soldiers), which are specialized either for seed milling, abdominal food storage, defense, or some combination of these functions. Wilson [330] showed that majors in the species he studied exhibit elasticity: their normally limited behavioral repertoire can be stretched back and forth in a predictable manner in response to perturbations. He artificially tuned the ratio of majors to minors and observed a change in the rate of activity within one hour of the ratio change. When the fraction of minors in the colony becomes small, majors engage in tasks usually performed by minors and efficiently replace the missing minors. Figure 3.1 shows the number of behavioral acts per major per hour as a function of the fraction of majors in the colony for two tasks that are normally performed by minors, for three species of the genus *Pheidole*. The usual proportion of majors in these species varies between 5% and 30%. It can be seen that majors become progressively more and more involved in performing minors' tasks as the fraction of majors increases. These experiments suggest that the *resilience* of a colony as a whole, that is, the degree to which the colony responds to alterations, is determined by the *elasticity* observed at the individual level.

In the next section, we introduce a model that explicitly connects the two levels, that is, elasticity at the worker level and resilience at the colony level. The model applies equally well to division of labor based on physical, temporal or behavioral castes.

3.3 RESPONSE THRESHOLDS

3.3.1 INTRODUCTION

In order to explain Wilson's [330] observations, Bonabeau et al. [29] have developed a simple model which relies on response thresholds [270, 272]. In this model, every individual has a response threshold for every task. Individuals engage in task performance when the level of the task-associated stimuli exceeds their thresholds. When individuals performing a given task are withdrawn from the colony—they have low response thresholds with respect to stimuli related to the task—the associated demand increases and so does the intensity of the stimulus, until it eventually reaches the higher response thresholds of the remaining individuals. The increase of stimulus intensity beyond threshold has the effect of stimulating these individuals into performing the task.

Let us give two simple examples. If the task is larval feeding, the associated stimulus may be larval demand, which is expressed, for example, through the emission of pheromones. Feeding the larvae reduces larval demand. Another example is

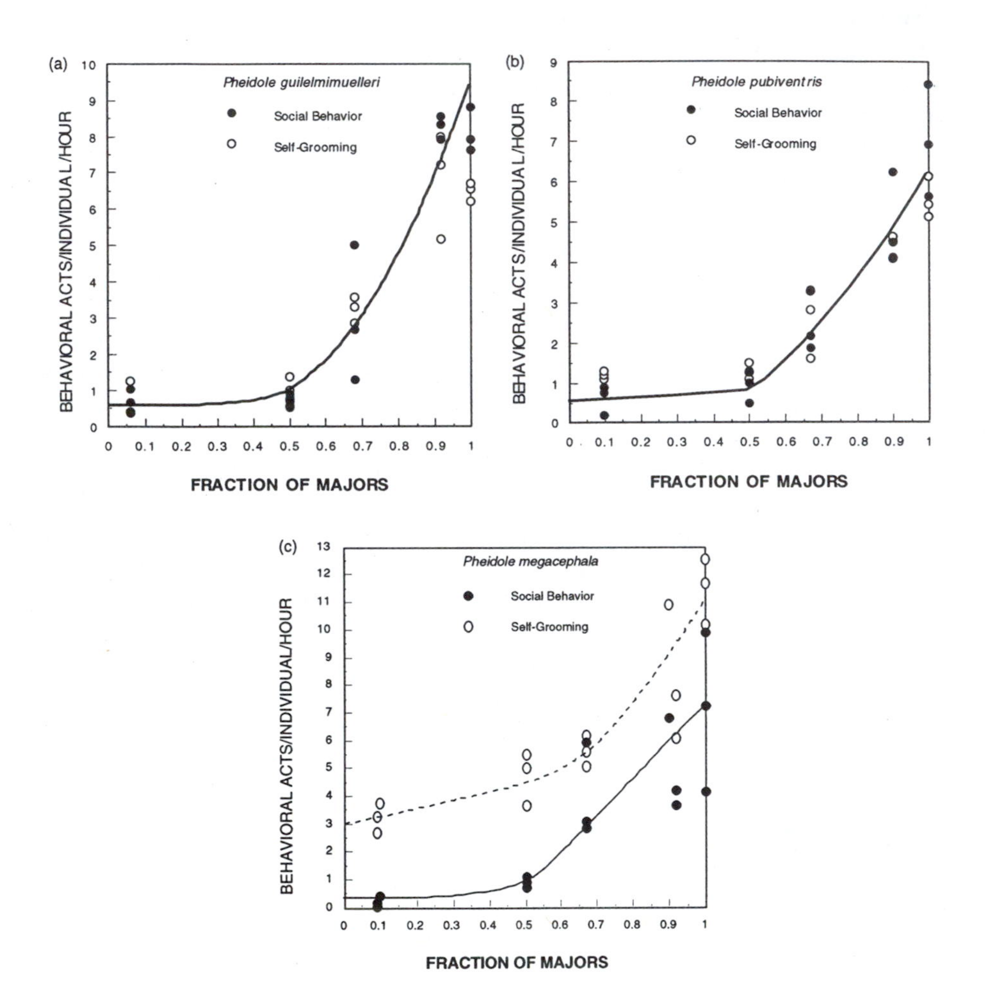

FIGURE 3.1 (a) Number of behavioral acts (social behavior and self-grooming) per major per hour as a function of the fraction of majors in the colony for *Pheidole guilelmimuelleri*. (b) Number of behavioral acts (social behavior and self-grooming) per major per hour as a function of the fraction of majors in the colony for *Pheidole pubiventris*. (c) Number of behavioral acts (social behavior and self-grooming) per major per hour as a function of the fraction of majors in the colony for *Pheidole megacephala*. "Fitting lines" are only visual aides. After Wilson [330]. Reprinted by permission © *Springer-Verlag*.

the removal of corpses from the nest in honey bees [256]: "undertaker" honey bees drag the corpses of their nestmates, fly away with them, and drop them at some distance from the nest. Dead nestmates within the nest provide the stimuli that trigger the "undertaker" behavior. The removal of corpses has the effect of actually reducing or eliminating the stimuli. In these two examples, task performance by an individual reduces the intensity of task-associated stimuli and, therefore, modifies the stimulatory field of its nestmates: information about the necessity of performing such or such task is transmitted through a combination of direct and indirect channels. Indirect information transfer through modifications of the environment is stigmergy [157, 158].

The nature of task-related stimuli may vary greatly from one task to another and so can information sampling techniques, which may involve direct interactions among workers (trophallaxis, antennation, etc.) [153], nest "patrolling" [223], or more or less random exposure to task-related stimuli.

3.3.2 RESPONSE THRESHOLDS: INTRODUCTION AND EXPERIMENTAL EVIDENCE

The first question that we have to answer is: how do we formally define a response threshold? Let s be the intensity of a stimulus associated with a particular task. s can be a number of encounters, a chemical concentration, or any quantitative cue sensed by individuals. A response threshold θ, expressed in units of stimulus intensity, is an internal variable that determines the tendency of an individual to respond to the stimulus s and perform the associated task. More precisely, θ is such that the probability of response is low for $s \ll \theta$ and high for $s \gg \theta$. One family of response functions $T_\theta(s)$ (the probability of performing the task as a function of stimulus intensity s) that satisfy this requirement is given by

$$T_\theta(s) = \frac{s^n}{s^n + \theta^n}, \tag{3.1}$$

where $n > 1$ determines the steepness of the threshold. In the rest of the chapter, we will be concerned with the case $n = 2$, but similar results can be obtained with other values of $n > 1$. Figure 3.2 shows several such response curves, with $n = 2$, for different values of θ. The meaning of θ is clear: for $s \ll \theta$, the probability of engaging task performance is close to 0, and for $s \gg \theta$, this probability is close to 1. At $s = \theta$, this probability is exactly 1/2. Therefore, individuals with a lower value of θ are likely to respond to a lower level of stimulus. Figure 3.3 shows a schematic representation of how the threshold model can be used to understand the experiments of Wilson [330].

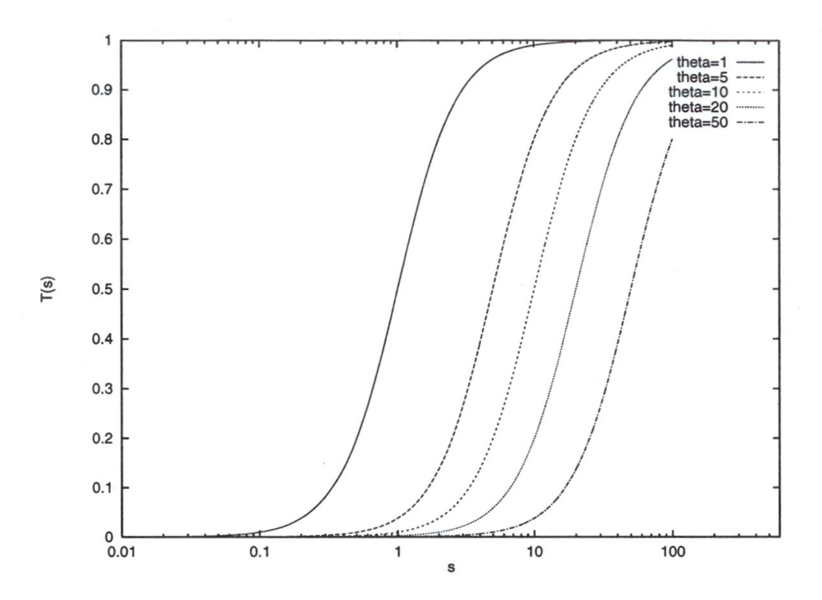

FIGURE 3.2 Semi-log plot of threshold response curves ($n = 2$) with different thresholds ($\theta = 1, 5, 10, 20, 50$).

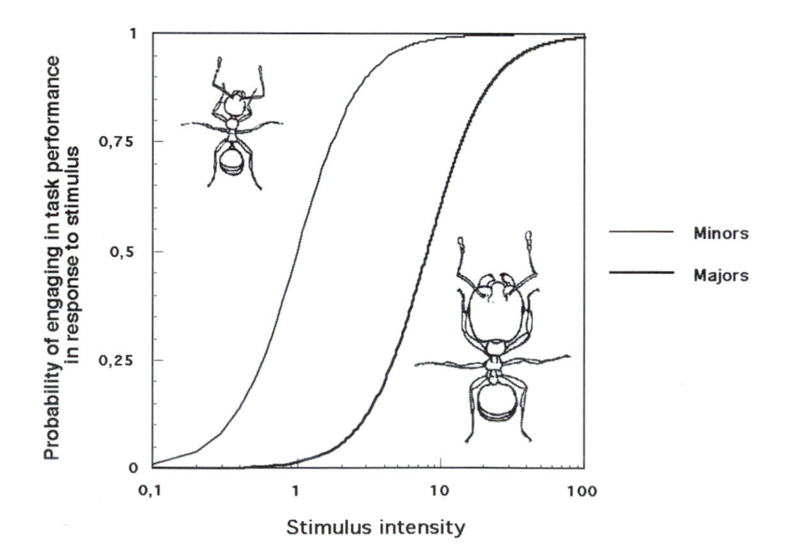

FIGURE 3.3 Schematic representation of hypothetical response curves for minors and majors in the polymorphic species of ants studied by Wilson [330]. In this figure, majors have a higher response threshold than minors to task-associated stimulus intensity.

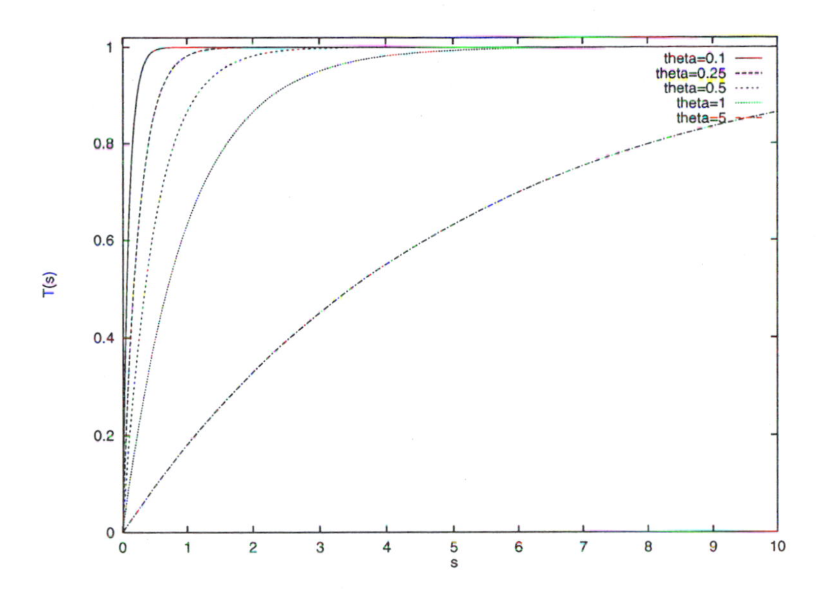

FIGURE 3.4 Exponential response curves with different thresholds ($\theta = 0.1, 0.25, 0.5, 1, 5$).

Another important example of threshold response function is given by[1]

$$T_\theta(s) = 1 - e^{-s/\theta} . \qquad (3.2)$$

Figure 3.4 shows $T_\theta(s)$ given by Eq. (3.2) for different values of θ. We see here again that the probability of engaging task performance is small for $s \ll \theta$, and is close to 1 for $s \gg \theta$. Although there is no change in concavity in the curve, this response function produces behaviors which are comparable to those produced by response functions based on Eq. (3.1) [34]. Figure 3.5 also shows that a semi-logarithmic plot of $T_\theta(s)$ given by Eq. (3.2) exhibits a change in concavity, and Figure 3.5 is actually very similar to Figure 3.2. In most of this chapter, we make use of Eq. (3.1) with $n = 2$ rather than Eq. (3.2), simply because analytical results are possible for Eq. (3.1) with $n = 2$.

It is important to emphasize that threshold models encompass exponential response functions: the main ingredient of a threshold model is the existence, within each individual, of a value θ of the stimulus intensity s such that task performance is likely for $s \gg \theta$ and unlikely for $s \ll \theta$. Exponential response functions are particularly important because they are likely to be common. For example, imagine a stimulus that consists of a series of encounters with items that have to be processed, such as grains to be carried to the nest or corpses to be removed

[1]Plowright and Plowright [261]: use this type of response function in their model of the emergence of specialization.

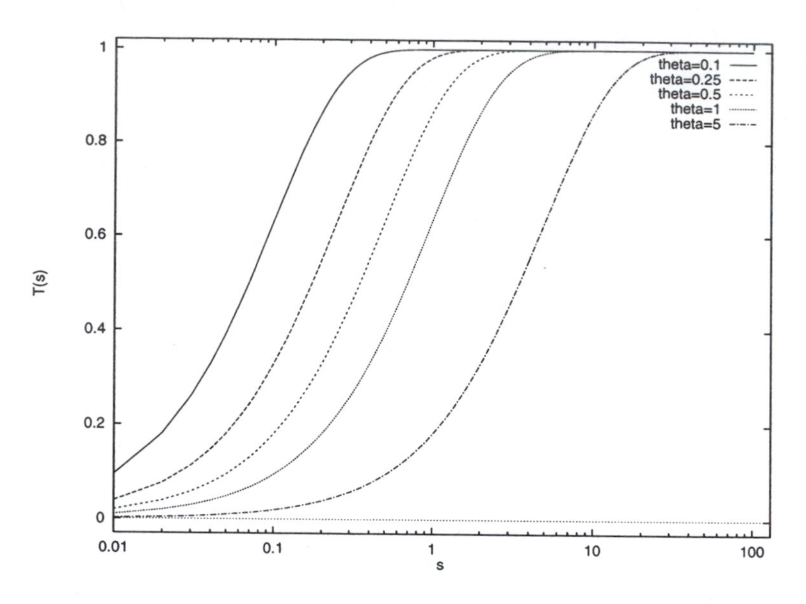

FIGURE 3.5 Semi-log plot of exponential response curves with different thresholds ($\theta = 0.1, 0.25, 0.5, 1, 5$).

from the nest. If, at each encounter with an item, an individual has a fixed probability ρ of processing the item, then the probability that the individual will not respond to the first N encountered items is given by $(1-\rho)^N$. Therefore, the probability $P(N)$ that there will be a response within the N encounters is given by $P(N) = 1 - (1-\rho)^N = 1 - e^{N \ln(1-\rho)}$, which is exactly Eq. (3.2) with $s = N$ and $\theta = -1/\ln(1-\rho)$. The organization of cemeteries by ants provides a good illustration of this process (see also Chapter 4). The probability of dropping a corpse (or a dead item, i.e., a thorax or an abdomen) has been studied experimentally by Chrétien [72] in the ant *Lasius niger*: the probability that a laden ant drops an item next to a N-cluster can be approximated by $P(N) = 1 - (1-p)^N = 1 - e^{N \ln(1-p)}$, for N up to 30, where $p \approx 0.2$ (Figure 3.6). Here, the intensity of the stimulus is the number of encountered corpses, and the associated response is dropping an item.

Another situation in which exponential response functions may be observed is when there are waiting times involved, although it may not always be the case. Let us assume that tasks A and B are causally related in the sense that a worker performing task A has to wait for a worker performing task B to, say, unload nectar or pulp, or any kind of material. If a task-A worker has a fixed probability p per unit waiting time of giving up task-A performance, the probability that this worker will still be waiting after t time units is given by $P(t) = 1 - (1-p)^t = 1 - e^{t \ln(1-p)}$.

In summary, response functions which exhibit an explicit threshold-like shape, such as the ones given by Eq. (3.1), or exponential response functions, such as the

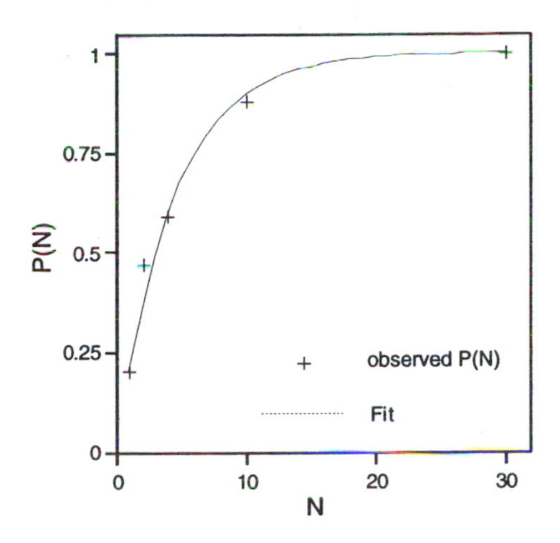

FIGURE 3.6 Probability $P(N)$ of dropping a corpse next to a N-cluster as a function of N. Fit $P(N) = 1(1-p)^N$ with $p = 0.2$ is shown. After Chrétien [72]. Reprinted by permission.

ones given by Eq. (3.2), can be encountered in various situations and yield similar results. Both types of response functions can be used in response threshold models.

Viewed from the perspective of response thresholds, castes may correspond to possible physical differences, but also to innate differences in response thresholds without any visible physical difference.

Robinson and Page [255, 271] have shown that honey bee workers belonging to different patrilines[2] may have different response thresholds. For example, assume for simplicity that workers of patriline A engage in nest guarding as soon as there are less than 20 guards, whereas workers of patriline B start performing this task when there are less than 10 workers guarding the hive's entrance: workers of patriline B have a higher response threshold to perform this task. Here, the stimulus intensity is the number of honey bees that are already guarding the nest.

Another example of response threshold has been found in honey bees. If it takes a honey bee forager too long to unload her nectar to a storer bee, she gives up foraging with a probability that depends on her search time in the unloading area. She will then start a "tremble dance" [286] to recruit storer bees (the tremble dance also inhibits waggle dancing). If, on the other hand, her in-hive waiting or search time is very small, she starts recruiting other foragers with a waggle dance. If her in-hive waiting or search time lies within a given window, she is likely not to dance at all and return to the food source. If one plots the probability of either

[2]Two bees belong to the same patriline if they have the same mother and the same father. They are also called supersisters.

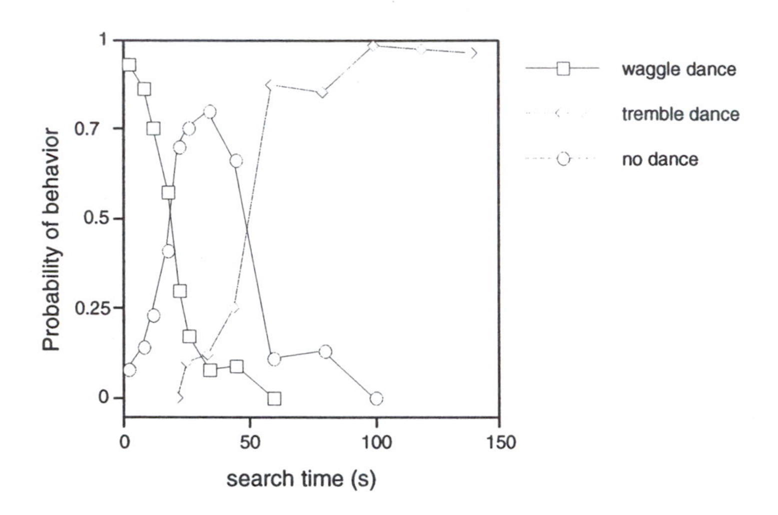

FIGURE 3.7 Probabilities of performing a waggle dance and a tremble dance as a function of in-hive search time for foragers visiting a rich nectar source. After Seeley [286]. Reprinted by permission © *Springer-Verlag*.

waggle or tremble dancing as a function of search time, a clear threshold function can be observed (Figure 3.7).

Other experimental results show the existence of response thresholds in two species of ants, *Pheidole pallidula* and *Ectatomma ruidum*:

- A series of experiments by Detrain et al. [90, 91, 92] clearly indicate the existence of differential response thresholds in the ant *Pheidole pallidula* in at least two activities, foraging and nest defense. The intensity of behavioral stimuli (measured by trail concentration and the number of tactile invitations in the case of foraging, supplemented by the number of intruders in the case of defense) required to induce the effective recruitment of majors is greater than for minors for both tasks, indicating that majors have higher response thresholds. An interesting discussion of the adaptive significance of these findings is given by Detrain and Pasteels [91, 92].
- Schatz [281] presents convincing evidence of response thresholds in the ant *Ectatomma ruidum*: when presented with an increasing number of prey, specialized "stinger ants" (or killer ants) start to become involved in the retrieval process (in addition to transporters, to which dead prey are transferred), the number of such stinger ants being dependent on the number of prey in a characteristic sigmoid-like manner (Figure 3.8). This suggests that within-caste specialization among hunters is indeed based on response thresholds.

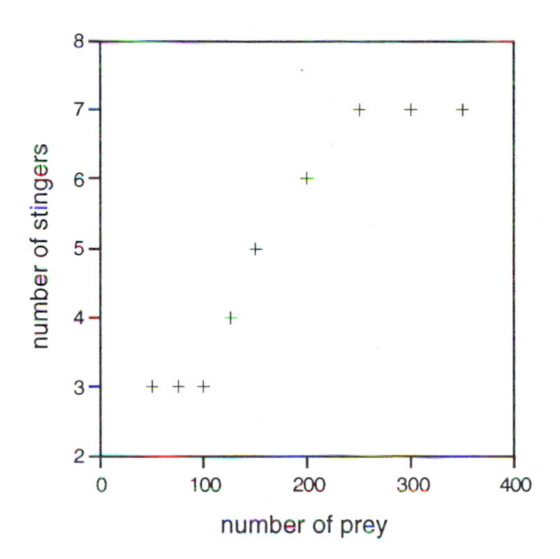

FIGURE 3.8 Number of stingers involved in prey retrieval as a function of the number of prey in the ant *Ectatomma ruidum*, for a nest composed of 130 workers. After Schatz [281]. Reprinted by permission.

3.3.3 MODEL WITH ONE TASK

Assume that only one task needs to be performed. This task is associated with a stimulus or demand, the intensity s of which increases if it is not satisfied—because either the task is not performed by enough individuals or it is not performed with enough efficiency. Let X_i be the state of an individual i: $X_i = 0$ corresponds to inactivity and $X_i = 1$ corresponds to performing the task. Let θ_i be the response threshold of i. An inactive individual starts performing the task with a probability P per unit time:

$$P(X_i = 0 \to X_i = 1) = T_{\theta_i}(s) = \frac{s^2}{s^2 + \theta_i^2} \,. \tag{3.3}$$

The probability that individual i will perform the task depends on s, the magnitude of the task-associated stimulus (which also affects the probability of being exposed to it) and on θ_i, individual i's response threshold [272]. Again, the choice of the threshold function in Eq. (3.3) is dictated by the availability of some exact analytical results. Any threshold function is expected to generate similar results.

An active individual gives up task performance and becomes inactive with probability p per unit time:

$$P(X_i = 1 \to X_i = 0) = p \,. \tag{3.4}$$

Parameter p is assumed to be identical for all individuals. $1/p$ is the average time spent by an individual in task performance before giving up the task. It is assumed

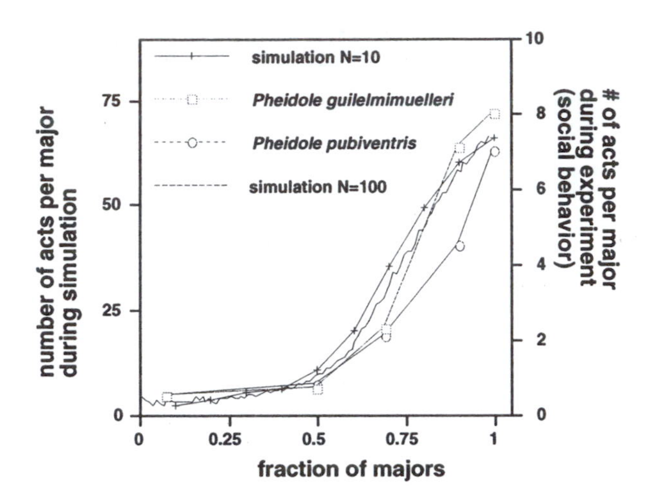

FIGURE 3.9 Fraction of active majors (type-1 workers) found in Monte Carlo simulations as a function of the proportion f of majors in the colony, for $\theta_1 = 8$, $\theta_2 = 1$, $\alpha = 3$, $\delta = 1$, $p = 0.2$. Comparison with Wilson's [330] results (scaled so that curves of model and experiments lie within the same range): number of acts of social behavior per major within time of experiments in *Pheidole guilelmimuelleri and Pheidole pubiventris*. After Bonabeau et al. [29]. Reprinted by permission © *Elsevier Science.*

that p is fixed and independent of stimulus. Individuals give up task performance after $1/p$ time units, but may become engaged again immediately if stimulus is still large.

Variations in stimulus intensity result from task performance, which reduces stimulus intensity, and from the natural increase of demand irrespective of whether or not the task is performed. The resulting equation for the discrete-time dynamics of stimulus intensity s is given by:

$$s(t+1) = s(t) + \delta - \frac{\alpha N_{\text{act}}}{N}, \tag{3.5}$$

where N_{act} is the number of active individuals, N is the total number of potentially active individuals in the colony, δ is the increase in stimulus intensity per unit time, and α is a scale factor measuring the efficiency of task performance. The amount of work performed by active individuals is scaled by N to reflect the intuitive idea that the demand is an increasing function of N, that we take linear here. This corresponds to the idea that brood should be divided by two when colony size is divided by two: in other words, colony requirements scale (more or less) linearly with colony size.

In Monte Carlo simulations [29], this simple fixed-threshold model shows remarkable agreement with experimental results in the case where there are two castes characterized by two different values of the response threshold: when "minors," with low response thresholds, are removed from the simulated colony, "majors," with higher response thresholds, start to perform tasks usually performed by minors. Figure 3.9 shows the fraction of majors engaged in task peformance as a function of the fraction of majors in the colony. This curve is very similar to the one observed by Wilson [330]. This simple model with one task can be easily extended to the case where there are two or more tasks to perform. Such an extension is examined in section 3.3.5.

3.3.4 FORMAL ANALYSIS OF THE MODEL: DETERMINISTIC EQUATIONS

In order to further analyze the behavior of the system described in section 3.3.3, we may write down "mean-field" differential equations based on the transition probabilities given by Eqs. (3.3) and (3.4). Assuming the existence of two types of workers, such as minors and majors, characterized by two thresholds θ_1 and θ_2, let n_1 and n_2 be the respective numbers of workers of type 1 and type 2 in the colony, N the total number of workers in the colony ($n_1 + n_2 = N$), $f = n_1/N$ the fraction of workers of type 1 in the colony, N_1 and N_2 the respective numbers of workers of type 1 and type 2 engaged in task performance, and x_1 and x_2 the corresponding fractions $x_i = N_i/n_i$). The variable x_i can be viewed as the fraction of type-i workers performing the task or, equivalently, as the fraction of time spent by a typical type-i worker in task performance, or also as the probability of finding an arbitrary type-i worker performing the task within a time unit. The average deterministic differential equations describing the dynamics of x_1 and x_2 are given by:

$$\partial_t x_1 = T_{\theta_1}(s)(1 - x_1) - p x_1 \,, \tag{3.6}$$

$$\partial_t x_2 = T_{0_2}(s)(1 - x_2) - p x_2 \,, \tag{3.7}$$

where θ_i is the response threshold of type i workers, and s the integrated intensity of task-associated stimuli. The first terms on the right-hand side of Eqs. (3.6) and (3.7) describe how the $(1 - x_i)$ fraction of inactive type i workers responds to the stimulus intensity or demand s, with a threshold function $T_{\theta_i}(s)$:

$$T_{\theta_i}(s) = \frac{s^2}{s^2 + \theta_i^2} \,. \tag{3.8}$$

The continuous-time dynamics of the stimulus intensity is given by:

$$\partial_t s = \delta - \frac{\alpha}{N}(N_1 + N_2) \,, \tag{3.9}$$

that is, since $(N_1 + N_2)/N = f x_1 + (1 - f) x_2$,

$$\partial_t s = \delta - \alpha f x_1 - \alpha(1 - f) x_2 \,, \tag{3.10}$$

where, as for Eq. (3.5), δ is the increase in stimulus intensity per unit time, and α is the scale factor that measures the efficiency of task performance. Let us introduce for convenience $z = \theta_1^2/\theta_2^2$, and let f be the fraction of majors in the colony. Assuming that type-1 workers are majors ($\theta_1 > \theta_2$), it can be shown analytically [34] that the average fraction of time x_1^s spent by majors in task performance (which is also the average frequency of task performance per major, or the fraction of majors involved in task performance per unit time, or the probability per unit time of finding a major performing the task), for the model described by Eqs. (3.6), (3.7) and (3.10), is given by

$$x_1^s = \frac{\chi + (\chi^2 + 4f(p+1)(z-1)(\delta/\alpha))^{1/2}}{2f(p+1)(z-1)}, \tag{3.11}$$

where

$$\chi = (z-1)\left(f + (p+1)\frac{\delta}{\alpha}\right) - z. \tag{3.12}$$

This exact result is in very good agreement with Monte Carlo simulations. Figure 3.10 shows how x_1^s varies as a function of f, for $z = 64$, $p = 0.2$, $\delta = 1$, and $\alpha = 3$, and a comparison with Wilson's [330] results. Figure 3.11 illustrates the fact that it is possible to find appropriate parameters with an exponential response function ($T_\theta(s) = 1 - e^{-s/\theta}$) such that it reproduces the same results as the threshold response function.

Only three parameters influence the shape of $x_1^s(f)$: δ/α, z, and p. Figures 3.12, 3.13, and 3.14 show how the $x_1^s(f)$ relationship varies with these parameters. When individuals are very efficient at performing the task (δ/α small), the value of f above which an important fraction of majors is performing the task is larger and the crossover becomes smoother; conversely, a decrease in efficiency leads to an earlier and more abrupt modification of the number of majors engaged in task performance (Figure 3.12: $\delta = 1$, α varying). This result is relatively natural, as more efficient task performance by individuals which have a low response threshold prevents task-related stimuli from growing large, and therefore from eliciting task performance by individuals that have larger response thresholds. The crossover becomes more abrupt as z increases, and the point at which the crossover is observed decreases; when z is close to 1, the proportion of majors engaged in task performance starts from a larger value (Figure 3.13). When the probability of giving up task performance becomes small, the involvement of majors in task performance becomes less progressive, and starts at a larger value of f (Figure 3.14): this is due to the fact that task performers, mostly minors for low values of f, spend more time on average in task performance, so that fewer majors are required. But when majors have to engage in task performance, they must do so more massively because missing minors were performing a lot of work. Finally, it is important to note that in Figures 3.12, 3.13, and 3.14, there is always a fraction of active majors as f becomes close to 0 (this property may be difficult to see directly on Eq. (3.11)).

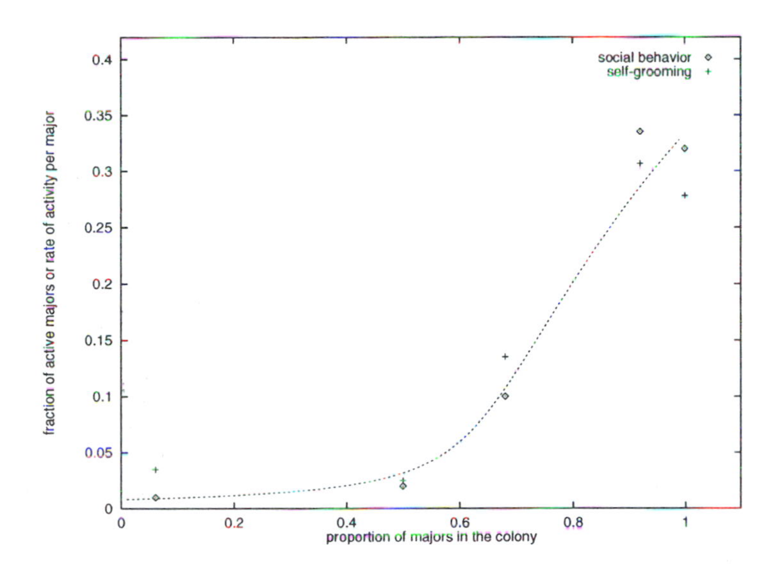

FIGURE 3.10 Fraction of active majors given by Eq. (3.11) as a function of the proportion f of majors in the colony, for $\theta_1 = 8$, $\theta_2 = 1$, $\alpha = 3$, $\delta = 1$, $p = 0.2$. Comparison with Wilson's [330] results (scaled so that model's curve and experimental data lie within the same range): number of acts of social behavior and self-grooming per major within time of experiments in *Pheidole guilelmimuelleri*. After Bonabeau et al. [34]. Reprinted by permission © *Elsevier Science*.

3.3.5 MODEL WITH SEVERAL TASKS

Let us now proceed to the case of m tasks. By analogy with the previous case, let us define N_{ij}, the number of workers of type i engaged in task j performance, x_{ij}, the corresponding fraction ($x_{ij} = N_{ij}/n_i$), and θ_{ij}, the associated response threshold. The average deterministic differential equations describing the dynamics of the x_{ij}'s are given by:

$$\partial_t x_{ij} = \frac{s_j^2}{s_j^2 + \theta_{ij}^2}\left(1 - \sum_{k=1}^{m} x_{ik}\right) - p x_{ij}\,, \tag{3.13}$$

where s_j is the intensity of task-j-associated stimuli. Let us assume for simplicity that $m = 2$ and $i = 1, 2$. Results for other cases can readily be inferred from those obtained with these parameters. The dynamics of the demand s_j associated with task j is given by:

$$\partial_t s_j = \delta - \alpha f x_{1j} - \alpha(1 - f) x_{2j}\,. \tag{3.14}$$

Here again, we assume that efficiency in task performance, mesured by α, is identical for workers of both types. Furthermore, α is the same for all tasks. Let us now distinguish two cases. In case (1), type-2 workers are more responsive to

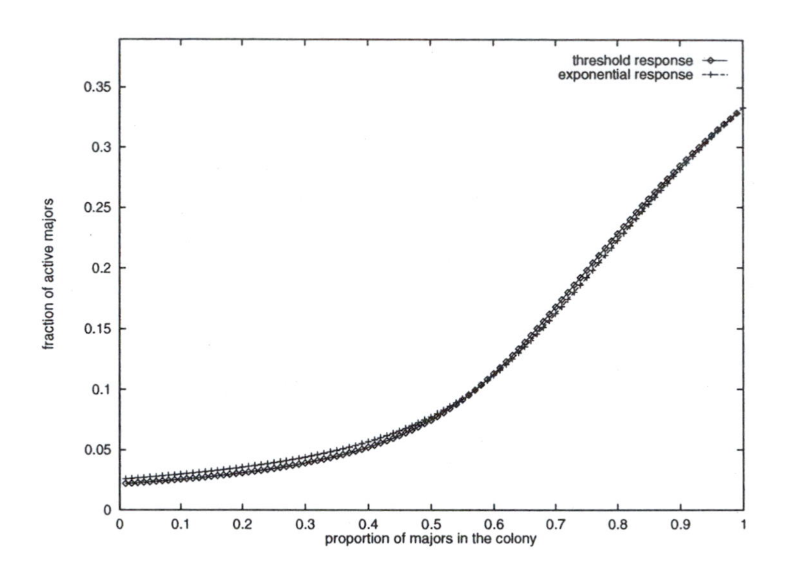

FIGURE 3.11 Comparison between the fraction of active majors as a function of the proportion of majors in the colony obtained with an exponential response function ($\theta_1 = 0.1$, $\theta_2 = 1$, $\alpha = 3$, $\delta = 1$, $p = 0.2$) and a threshold response function ($\theta_1 = 8$, $\theta_2 = 1$, $\alpha = 3$, $\delta = 1$, $p = 0.2$). After Bonabeau et al. [34]. Reprinted by permission © *Elsevier Science*.

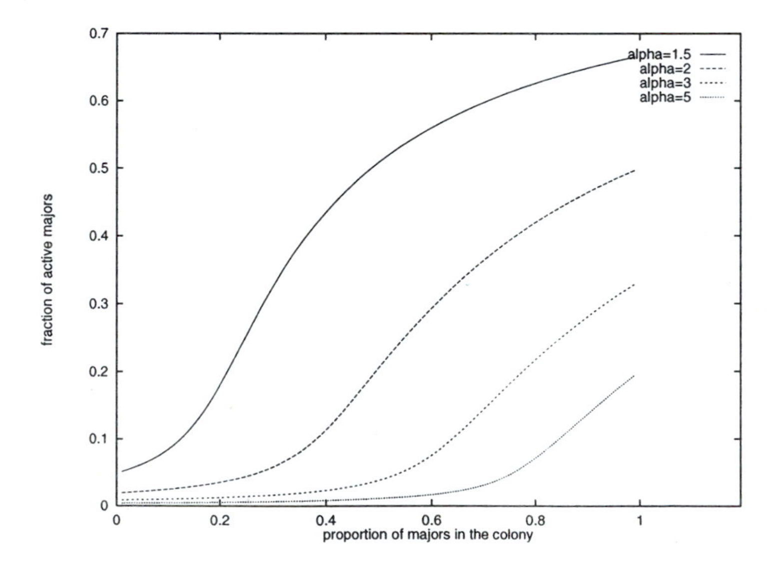

FIGURE 3.12 Fraction of active majors given by Eq. (3.11) as a function of the proportion f of majors in the colony, for $z = 64$, $\delta = 1$, $p = 0.2$, and $\alpha = 1.5, 2, 3, 5$. After Bonabeau et al. [34]. Reprinted by permission © *Elsevier Science*.

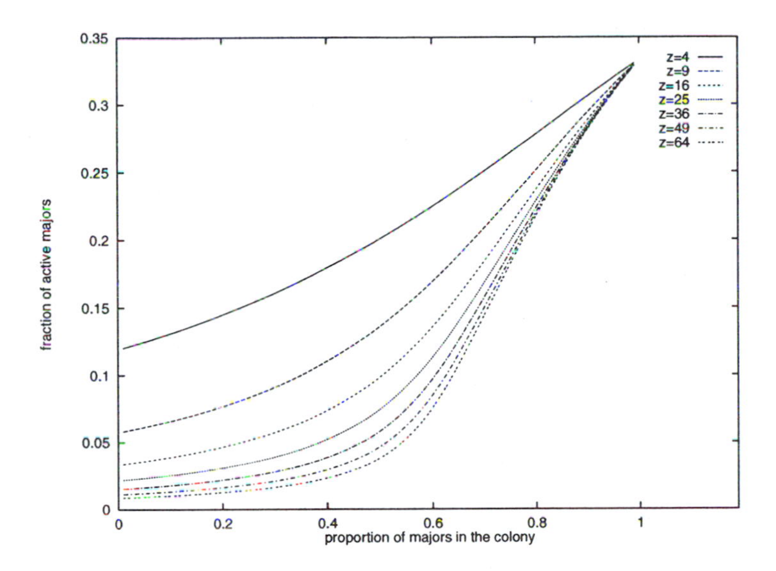

FIGURE 3.13 Fraction of active majors given by Eq. (3.11) as a function of the proportion f of majors in the colony, for $\alpha = 3$, $\delta = 1$, $p = 0.$, $z = 4, 9, 16, 25, 36, 49$, and 64. After Bonabeau et al. [34]. Reprinted by permission © *Elsevier Science.*

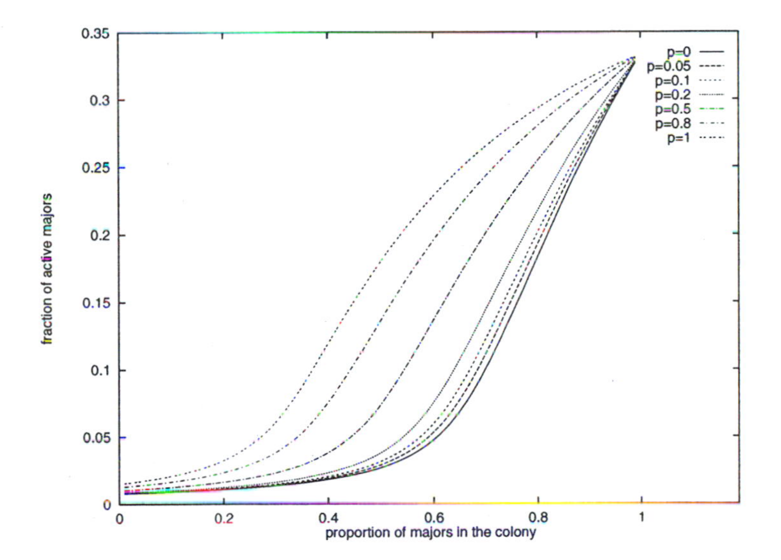

FIGURE 3.14 Fraction of active majors given by Eq. (3.11) as a function of the proportion f of majors in the colony, for $\alpha = 3$, $\delta = 1$, $z = 64$, and $p = 0, 0.05, 0.1, 0.2, 0.5, 0.8$ and 1. After Bonabeau et al. [34]. Reprinted by permission © *Elsevier Science.*

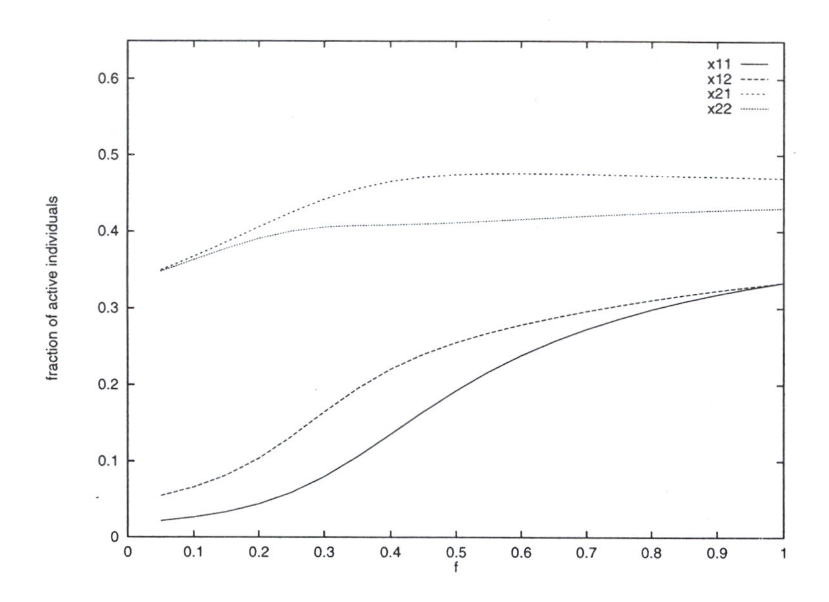

FIGURE 3.15 Fractions of type-1 and type-2 workers performing tasks 1 and 2 as a function of the proportion f of majors in the colony, obtained by numerical integration of Eqs. (3.13) and (3.14). $\theta_{11} = 8$, $\theta_{12} = 5$, $\theta_{21} = 1$, $\theta_{22} = 1$, $\alpha = 3$, $\delta = 1$, $p = 0.2$ After Bonabeau et al. [34]. Reprinted by permission © *Elsevier Science.*

task-associated stimuli for both tasks ($\theta_{11} > \theta_{21}$ and $\theta_{12} > \theta_{22}$). In case (2), type-1 workers are more responsive than type-2 workers to stimuli associated with task 1, but type-2 workers are more responsive than type-1 workers to stimuli associated with task 2; in other words, type-i workers are specialists of task i ($\theta_{11} < \theta_{21}$ and $\theta_{12} > \theta_{22}$).

Numerical integration is necessary to find the stationary values of x_{ij} as a function of the fraction f of type-1 workers. Let us introduce $z_j = \theta_{1j}^2/\theta_{2j}^2$. Figures 3.10 show x_{ij} as a function of f in case (1) ($z_1 = 64$ and $z_2 = 25$) and case in (2) ($z_1 = 64 = 1/z_2$) for $p = 0.2$, $\delta = 1$, and $\alpha = 3$. In case (1), when type-1 workers have lower response thresholds to both tasks, the obtained curves are qualitatively similar to those observed in Figure 3.12–3.14. The curves are, however, quantitatively different because z_1 and z_2 are different. Wilson [330] measured the number of acts per major for two tasks, social behavior and self-grooming, in *Pheidole megacephala*. There is a reasonably good agreement between Wilson's observations and the model's curves. In case (2), when a caste is specialized in one of the two tasks and the other caste in the other task, behavioral flexibility is observed on both sides: workers of type 1 can replace workers of type 2, and vice versa (Figure 3.16).

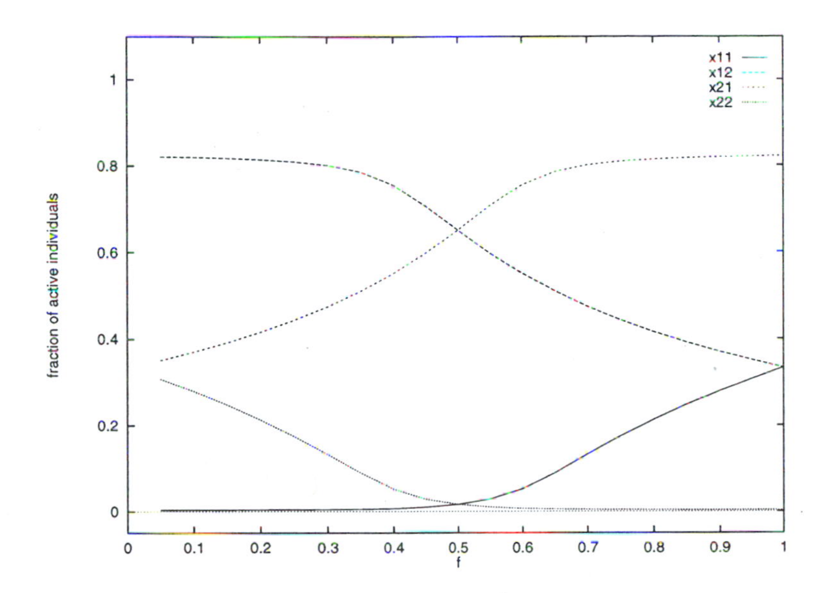

FIGURE 3.16 Fractions of type-1 and type-2 workers performing tasks 1 and 2 as a function of the proportion f of majors in the colony, obtained by numerical integration of Eqs. (3.13) and (3.14). $q_{ij} = 1/2$, $\theta_{11} = 8$, $\theta_{12} = 1$, $\theta_{21} = 1$, $\theta_{22} = 8$, $\alpha = 3$, $\delta = 1$, $p = 0.2$. After Bonabeau et al. [34]. Reprinted by permission © *Elsevier Science.*

Although the results of the model have been compared only with Wilson's [330] data on species with physical castes, the model can equally well describe the behavioral flexibility observed within physical, age, or behavioral castes.

3.3.6 MODELS OF "EMERGENT" TASK SUCCESSION

Before proceeding to the case of variable response thresholds, we give here two simple examples where the fixed-threshold model can be used to generate "emergent" task succession. These examples are of particular interest in that they show that coordination (whereby individuals tend to perform the same task before switching in relative synchrony to another task) does not have to be directly coded into individual behavior. Such a result may have direct application to the design of distributed problem-solving systems. In addition, the models presented below lend themselves to a robotic implementation. Puck clustering, detailed in chapter 4, is an important proof of concept but remains extremely primitive in terms of functionality. In chapter 4 we describe how a distributed robotic system capable of clustering could easily be modified to implement task succession.

It is possible, with fixed response thresholds, to observe a simulated colony perform some tasks in succession in situations that are not uncommon. There are two possible models. Model 1 assumes that individuals all have identical response thresholds, but these thresholds are different for the various tasks to be performed, and, moreover, success rate in task performance also varies with the task. This model can describe the dynamics of brood sorting in *Leptothorax* ants [130] or seed piling in harvester ants. Model 2 assumes that performing a given task increases the demand for another task: for example, creating a refuse pile just at the entrance of the nest generates a need for cleaning [72]. Both models will give the impression that individuals have decided to perform the tasks in sequence.

Model 1 Let us assume that m different types of items need to be processed. Let L_i be the number of workers loaded with item type i, U the number of unloaded workers, s_i the number of items that still need to be processed, r_i the number of items that have been successfully processed, and τ_i the time it takes to process an item of type i (for example, to carry a larva or seed to the appropriate location), and $f_i = e^{-\tau_i/\tau}$ (where τ is a characteristic time) the probability of success of the task (the longer it takes to carry a larva or seed to the appropriate location, the less likely the individual is to succeed). The idea behind this model is that, for example, heavier items or items that are harder to work with will be processed after lighter or easier items have been processed. Heavier items are naturally associated with a lower probability of success. The dynamics of U and L_i are given by

$$\partial_t U = \sum_{i=1}^{m} \left(\frac{L_i}{\tau_i} - \alpha_i U s_i \right) \tag{3.15}$$

$$\partial_t L_i = \alpha_i U s_i - \frac{L_i}{\tau_i} , \tag{3.16}$$

where:

$$\alpha_i = p_i \frac{s_i^2}{\theta_i^2 + s_i^2} . \tag{3.17}$$

In Eq. (3.17), p_i is the probability of finding an item of type i, that we approximate simply by

$$p_i = \frac{s_i}{\sum_{j=1}^{m} s_j} . \tag{3.18}$$

Equation (3.16) expresses the fact that laden individuals deposit their loads every τ_i (either successfully or not) and that unloaded workers pick-up items with a probability that combines the probability of encountering an item of type i and that of responding to such an item (Eq. (3.17)). Equation (3.16) also expresses how laden workers become unladen, and vice versa, but with signs opposite to those of Eq. (3.15), as it describes the dynamics of the number of laden workers. The threshold θ_i in Eq. (3.17) is such that if $s_i \ll \theta_i$, few workers will be stimulated to carry an item of type i, and, if $s_i \gg \theta_i$, workers will, on the contrary, be stimulated to process items of type i. The dynamics of r_i and s_i are given by

$$\partial_t s_i = -\alpha_i U s_i + (1 - f_i) \frac{L_i}{\tau_i} , \tag{3.19}$$

$$\partial_t r_i = f_i \frac{L_i}{\tau_i} \,. \tag{3.20}$$

Equation (3.19) expresses that the number of items that can be processed decreases when an item is picked up, but increases when a laden worker deposits an item at a wrong location (unsuccessful deposition), which happens with probability $1 - f_i$. Eqs. (3.15) to (3.20) have been integrated numerically for three types of items. Figure 3.17(a) shows the respective numbers of workers performing tasks 1, 2, and 3 as a function of time. Figure 3.17(b) shows the fraction of processed items of types 1, 2, and 3 as a function of time. Clearly, workers tend to process items of type 1, then items of type 2, and eventually items of type 3.

Model 2 Let us now assume that there are m potential tasks to be performed by the workers. Let x_i define the fraction of workers engaged in performing task i, s_i the demand (stimulus intensity) associated with task i, θ_i the threshold associated with task i (similar to model 1), p_i the probability of encountering stimuli associated with task i, p the probability of stopping performing task i (the average time spent performing a task before task switching or before becoming inactive is given by $1/p$), and α the efficiency of task performance, which we also take to be the rate of stimulus production per unit working time for the next task.

An example in which one task provides the stimulus for the next task is the polyethism among hunters in the ant *Ectatomma ruidum* [281] that was introduced in Section 3.3.2. When live *drosophila* are presented to the colony, stinger behavior is activated; stingers start killing the prey, which leads to the presence of an increasing number of corpses to be transported to the nest; the presence of these corpses stimulate transporter behavior. Figure 3.18(a) shows how the fractions of ants that are, respectively, stingers and transporters, and how the stimuli associated with, respectively, killing prey and transporting corpses, vary over time. The number of live prey in the experimental arena decreases rapidly, whereas the number of corpses that are to be transported as a result of the activity of stingers increases and then decreases.

The dynamics of x_i and s_i are described by

$$\partial_t x_i = p_i \frac{s_i^2}{\theta_i^2 + s_i^2} \left(1 - \sum_{k=1}^{m} x_k\right) - p x_i \,, \tag{3.21}$$

$$\partial_t s_i = \alpha(x_{i-1} - x_i)(\text{with } x_{i-1} = 0 \text{ if } i = 1) \,. \tag{3.22}$$

Equation (3.21) is similar to Eq. (3.13), and Eq. (3.22) expresses the fact that performing task $i - 1$ increases s_i, which in turn decreases when task i is being performed. Numerical integration of Eqs. (3.21) and (3.22) has been performed for three tasks ($i = 1, 2, 3$). Figure 3.18(b) shows the respective numbers of workers performing tasks 1, 2, and 3 as a function of time. Figure 3.18(c) shows the dynamics of stimulus intensities associated with tasks 1, 2, and 3, respectively. It can be seen that workers first tend to perform task 1, then task 2 and finally task 3. If performing

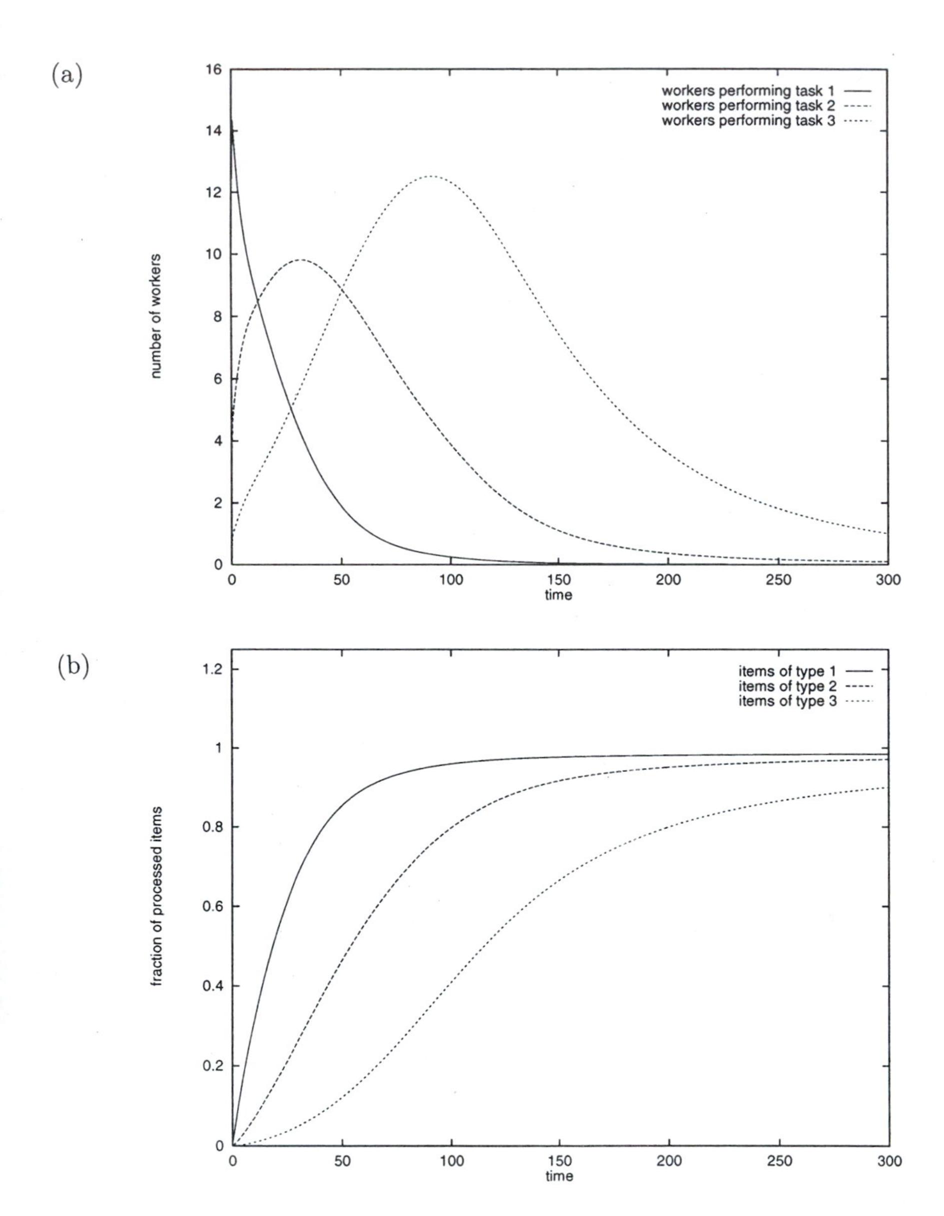

FIGURE 3.17 (a) Respective numbers of workers performing tasks 1, 2, and 3 as a function of time. Dynamics given by Eqs. (3.15) to (3.20), with three types of items. $\theta_1 = 10, \theta_2 = 40, \theta_3 = 80, \tau = 5, \tau_1 = 2, \tau_2 = 5, \tau_3 = 8, s_1(t = 0) = 120, s_2(t = 0) = 70, s_3(t = 0) = 50$. After Bonabeau et al. [34]. (b) Respective fractions of processed items of types 1, 2, and 3 as a function of time for the same simulation as in Figure 3.17(a). Time unit: $\tau/5$. After Bonabeau et al. [34]. Reprinted by permission © *Elsevier Science*.

(a)

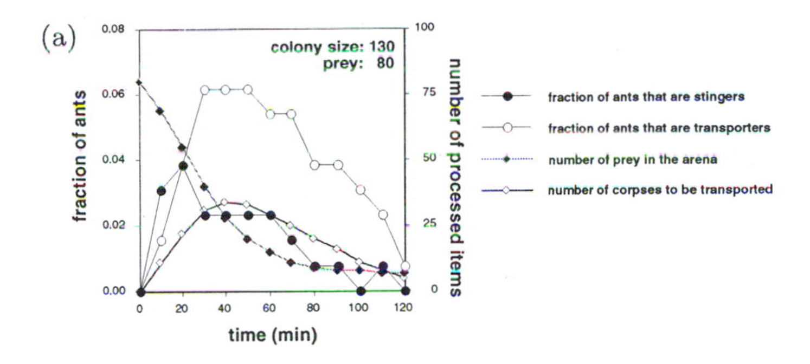

(b)

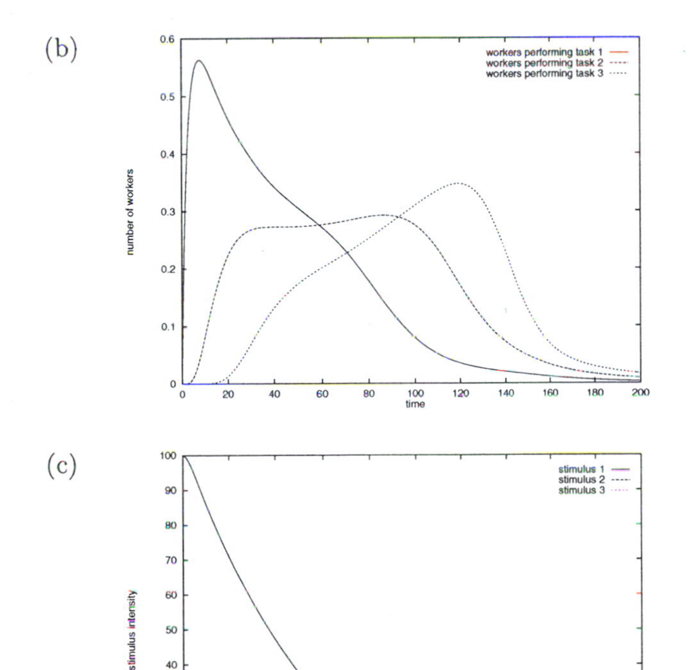

(c)

FIGURE 3.18 (a) Fraction of the workers that are stingers, fraction of the workers that are transporters, number of live prey in the arena, and number of corpses to be transported as a function of time. The total number of workers is 130, and the initial number of live prey is 80. The experiment's duration is 2 hours. After Schatz [281]. (b) Respective numbers of workers performing tasks 1, 2, and 3 as a function of time. Dynamics given by Eqs. (3.21) and (3.22) with three tasks. $s_1(t=0) = 100, s_2(t=0) = s_3(t=0) = 0, s_1 \rightarrow s_2 \rightarrow s_3, \alpha = 3, \theta_1 = \theta_2 = \theta_3 = 20, p_1 = p_2 = p_3 = 0.3, p = 0.2$. Time unit: $0.2p$. After Bonabeau et al. [34]. *continued.*

FIGURE 3.18 *continued* (c) Respective stimulus intensities associated with tasks 1, 2, and 3 as a function of time for the same simulation as in Figure 3.18(a). After Bonabeau et al. [34]. Reprinted by permission © *Elsevier Science*.

task 3 increased the demand associated with task 1, a cyclical activity would be observed.

3.4 SPECIALIZATION

The simple response threshold model introduced in the previous section, which assumes that each worker responds to a given stimulus when stimulus intensity exceeds the worker's threshold, can explain how flexibility at the colony level results from the workers' behavioral flexibility [29]. Unfortunately, this model is of limited applicability because it assumes that workers' thresholds are fixed over the studied time scale:

1. It cannot account for the genesis of task allocation, including temporal polyethism, for it assumes that individuals are differentiated and roles preassigned.
2. It cannot account for *strong* task specialization within (physical or temporal) castes.
3. It is valid only over sufficiently short time scales, where thresholds can be considered constant.
4. It is not consistent with recent experiments on honey bees [56, 273], showing that aging and/or learning play a role in task allocation.

In order to overcome these limitations, Theraulaz et al. [308, 312] have extended the fixed-threshold model by allowing thresholds to vary in time, following a simple reinforcement process: a threshold decreases when the corresponding task is performed and increases when the corresponding task is not performed. The original model of Theraulaz et al. [308] is slightly more complicated than a model of division of labor, since it is coupled to a model of self-organizing hierarchies. A simplified version of this model is described in this section. This idea had been previously introduced by Oster [252], Deneubourg et al. [84], and Plowright and Plowright [261]. The main competing hypothesis for overcoming limitations 1 to 4 (above) is absolute aging, or, within the present context, age-based changes in response thresholds.

Several experiments suggest the existence of a reinforcement process or support the reinforcement hypothesis. Deneubourg et al. [84] proposed that such a hypothesis could be consistent with experimental observations of foraging in ants. Sendova-Franks and Franks [288] suggest that reinforcement learning plays a role in the ability of *Leptothorax* ant colonies to quickly re-assemble after dissociation.

Withers et al. [331] observed that important changes in some regions of the brain are associated with aging in honey bees: the brain of a forager (> 20 days old) is significantly different from that of a 1-day-old bee. While their observations could result from absolute aging, further experiments, where worker bees were forced to start foraging early, have shown that precocious foragers were similar in brain organization to normal older foragers. This suggests that behavior influences brain organization, which in turn certainly influences the conditions under which tasks are performed. Other experiments, which have been interpreted as suggesting aging as the main factor, can be reinterpreted along the same lines. For example, the probability of behavioral reversion from foraging to nursing in honey bees [285] is a decreasing function of the time spent foraging, which suggests that some kind of learning may occur. Another example is the recent study by Calderone and Page [56], who have shown that deprived bees (raised in isolation in wire cages with a queen) exhibit precocious foraging, suggesting that the lack of certain stimuli may influence the rate of behavioral ontogeny (a possibility that Calderone and Page [56] did not rule out; see also Huang and Robinson [181]). More generally, relative age (that is, age relative to the rest of the colony and to the individual's own experience) is often a more relevant parameter than absolute age [54, 186, 217, 288, 320], which means that stimuli provided by the environment and other colony members, as well as individual history, are likely to play an important role in shaping behavioral ontogeny. These studies suggest that individual experience shapes behavioral ontogeny, and that response thresholds may be dynamic, rather than static.

Using the same notations as in sections 3.3.4 and 3.3.5, in the fixed-threshold model [29] individual i engages in task-j performance with probability

$$T_{\theta_{ij}}(s_j) = \frac{s_j^2}{s_j^2 + \theta_{ij}^2} \,. \tag{3.23}$$

In addition to the fixed-threshold model, θ_{ij} is now updated in a self-reinforcing way [308, 312]. The more individual i performs task j, the lower θ_{ij}, and vice versa. Let ξ and φ be the coefficients that describe learning and forgetting, respectively. In this time-incremental model, individual i becomes more (respectively less) sensitive by an amount $\xi\Delta t$ (respectively $\varphi\Delta t$) to task-j-associated stimuli when performing (respectively not performing) task j during a time period of duration Δt:

$$\theta_{ij} \leftarrow \theta_{ij} - \xi\Delta t \tag{3.24}$$

if i performs task j within Δt, and

$$\theta_{ij} \leftarrow \theta_{ij} + \varphi\Delta t \tag{3.25}$$

if i does not perform the task within Δt. Let x_{ij} be the fraction of time spent by individual i in task j performance: within Δt, individual i performs task j during $x_{ij}\Delta t$, and other tasks during $(1 - x_{ij})\Delta t$. The resulting change in θ_{ij} within Δt is therefore given by

$$\theta_{ij} \leftarrow \theta_{ij} - x_{ij}\xi\Delta t + (1 - x_{ij})\varphi\Delta t \,, \tag{3.26}$$

which combines Eqs. (3.24) and (3.25). ξ and φ are assumed to be identical for all tasks, and the dynamics of θ_{ij} is restricted to an interval $[\theta_{\min}, \theta_{\max}]$. The decision to perform task j is still given by $T_{\theta_{ij}}(s)$ in Eq. (3.23), but θ_{ij} now varies in time according to Eq. (3.26). The continuous-time formulation of Eq. (3.26) is given by:

$$\partial_t \theta_{ij} = [(1 - x_{ij})\varphi - x_{ij}\xi]\Theta(\theta_{ij} - \theta_{\min})\Theta(\theta_{\max} - \theta_{ij}), \tag{3.27}$$

where ∂_t denotes the time derivative and $\Theta(\cdot)$ is a step function ($\Theta(y) = 0$ if $y \leq 0$, $\Theta(y) = 1$ if $y > 0$). $\Theta(\cdot)$ is used to maintain θ_{ij} within bounds. The average temporal dynamics of x_{ij} is given by the following equation:

$$\partial_t x_{ij} = T_{\theta_{ij}}(s_j)\left(1 - \sum_{k=1}^{m} x_{ik}\right) - p x_{ij} + \psi(i,j,t). \tag{3.28}$$

The first term on the right-hand side of Eq. (3.28) describes how the $1 - \sum_{k=1}^{m} x_{ik}$ fraction of time potentially available for task performance is actually allocated to task-j performance. The second term on the right-hand side of Eq. (3.28) expresses the assumption that an active individual gives up task performance and becomes inactive with probability p per time unit. The average time spent by an individual in task j performance before giving up that task is $1/p$. It is assumed that p is fixed, identical for all tasks and individuals, and independent of stimulus. Individuals give up task performance after $1/p$, but may engage again immediatly if stimulus intensity is still large. $\psi(i,j,t)$ is a centered gaussian stochastic process of variance σ^2, uncorrelated in time and uncorrelated among individuals and among tasks:

$$\forall i,j,t \quad \langle \psi(i,j,t) \rangle = 0, \tag{3.29}$$

and

$$\forall i,j,h,k,t,t' \quad \langle \psi(i,j,t)\psi(h,k,t') \rangle = \sigma^2 \delta_0(i-h)\delta_0(j-k)\delta_0(t-t'), \tag{3.30}$$

where δ_0 is the Dirac distribution. $\psi(i,j,t)$ is a stochastic term that simulates the fact that individuals encounter slightly different local conditions. Assuming for simplicity that the demand for each task increases at a fixed rate per unit time, the dynamics of s_j, the intensity of task-j-associated stimuli, is described by

$$\partial_t s_j = \delta - \frac{\alpha_j}{N}\left(\sum_{i=1}^{N} x_{ij}\right), (j = 1, 2), \tag{3.31}$$

where δ is the increase in stimulus intensity per unit time, and α_j is a scale factor measuring the efficiency of task j performance. It is assumed that both factors are identical for all tasks ($\forall j$, $\alpha_j = \alpha$), and that α is fixed over time and identical for all individuals. In reality, however, α can vary as a result of specialization [116, 190] or genotypic differences. The amount of work performed by active individuals is again scaled by the number of individuals N.

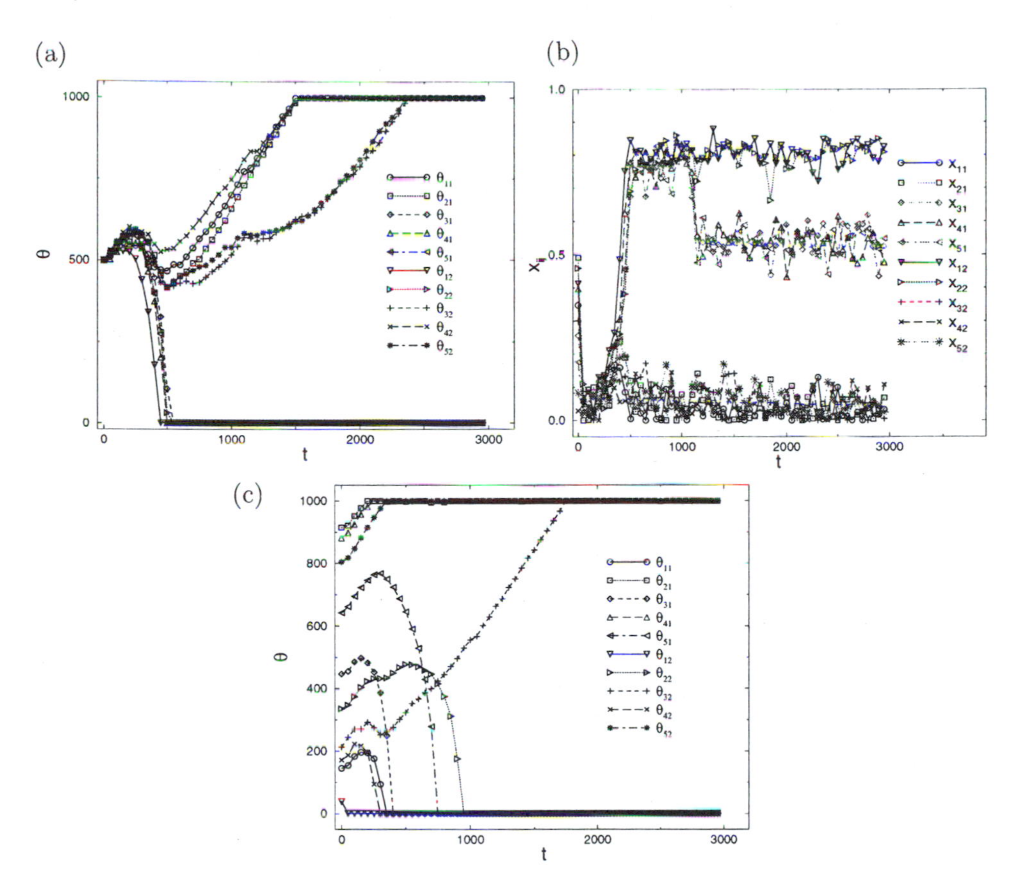

FIGURE 3.19 (a) Dynamics of response thresholds θ_{ij} for $N = 5$ individuals and $m = 2$ tasks. $\alpha = 3$, $\delta = 1$ $p = 0.2$, $\xi = 10$, $\varphi = 1$, $\sigma = 0.1$. Initial conditions: $\forall i \in \{1, \ldots, N\}$, $\forall j \in \{1, \ldots, m\}$, $\theta_{ij}(t = 0) = 500$, and $x_{ij}(t = 0) = 0.1$. θ_{ij} is the response threshold of individual i to task j: for example, θ_{21} is the response threshold of individual 2 to task 1. A low value of θ_{ij} indicates that individual i is highly sensitive to task-j-associated stimuli and is therefore a specialist of task j. Individuals 3, 4, and 5 are task-1 specialists, and individuals 1 and 2 are task-2 specialists. After Theraulaz et al., [312]. Time unit $0.2/p$. (b) Dynamics of the fractions of time spent in task performance x_{ij} for the same parameter values as in Figure 3.19(a). x_{ij} is the fraction of time spent by individual i in task j performance: for example, x_{42} is the fraction of time spent by individual 4 performing task 2. When x_{ij} is close to 1, individual i spends most of its time performing task j. Individuals 3, 4, and 5, which all perform mostly task 1, are less active $x_{i1} \approx 0.55$, $x_{i2} \approx 0.05$) than individuals 1 and 2, which perform mostly task 2 ($x_{i1} \approx 0.05$, $x_{i2} \approx 0.8$). After Theraulaz et al. [312]. (c) The same as in Figure 3.19(a), except that the initial distribution of thresholds is uniform over $[\theta_{\min} = 1, \theta_{\max} = 1000]$. Individuals 1, 3 and 5 are task-1 specialists, and individuals 1, 2, and 4 are task-2 specialists (individual 1 is a specialist of both tasks). After Theraulaz et al. [312]. Reprinted by permission © *Elsevier Science.*

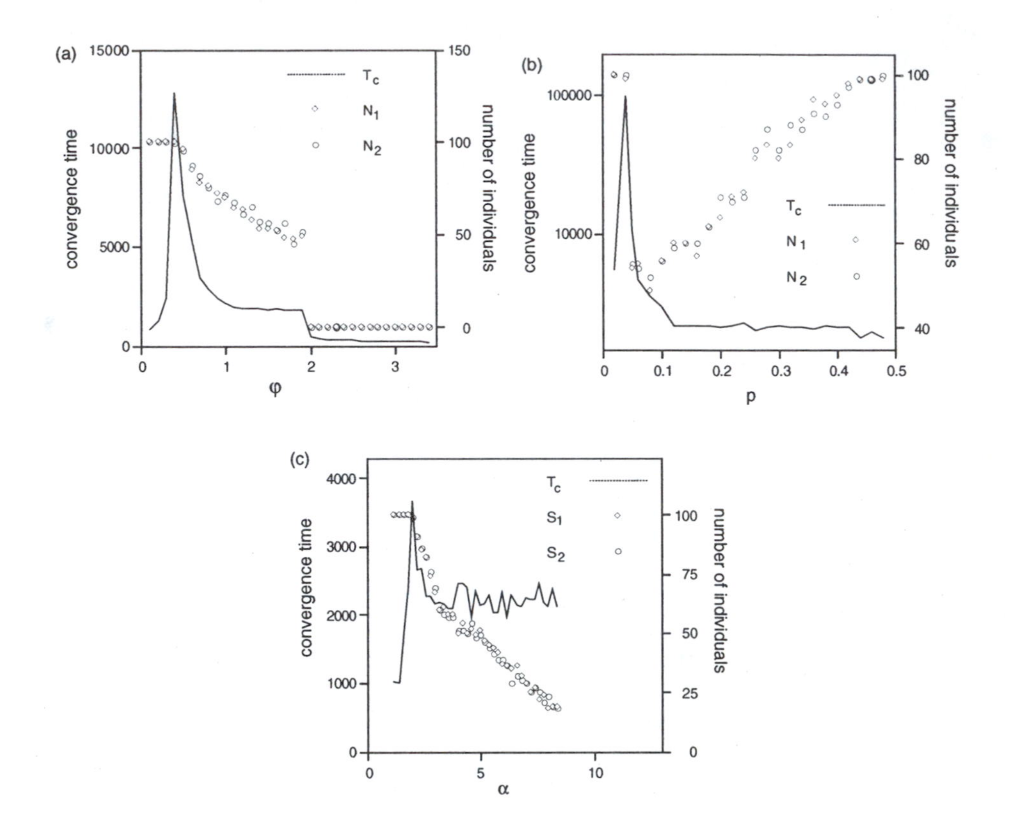

FIGURE 3.20 (a) Convergence time T_c (time it takes for the following condition to be satisfied: $\forall i \in \{1, \ldots, N\}, \forall j \in \{1, \ldots, m\}, \theta_{ij} > 900$ or $\theta_{ij} < 100$), and N_1 and N_2 (the number of specialists of tasks 1 and 2, respectively) as a function of φ (forgetting rate) for $N = 100$ and $m = 2$, assuming that $\varphi + \xi = 11$ (where ξ is the learning rate), $\alpha = 3, \delta = 1, p = 0.2, \sigma = 0.1$. $\forall i \in \{1, \ldots, N\}, \forall j \in \{1, \ldots, m\}, \theta_{ij}(t = 0) = 500$, and $x_{ij}(t = 0) = 0.1$. After Theraulaz et al. [312]. (b) Convergence time T_c, N_1 and N_2 as a function of p (probability per unit time of becoming inactive) for $N = 100, m = 2, \alpha = 3, \delta = 1, \xi = 10, \varphi = 1, \sigma = 0.1$. $\forall i \in \{1, \ldots, N\}, \forall j \in \{1, \ldots, m\}, \theta_{ij}(t = 0) = 500$, and $x_{ij}(t = 0) = 0.1$. After Theraulaz et al. [312]. (c) Convergence time T_c, N_1, and N_2 as a function of α (efficiency of task performance) for $N = 100, m = 2, p = 0.2, \delta = 1, \xi = 10, \varphi = 1, \sigma = 0.1$. $\forall i \in \{1, \ldots, N\}, \forall j \in \{1, \ldots, m\}, \theta_{ij}(t = 0) = 500$, and $x_{ij}(t = 0) = 0.1$.

The dynamics of the model described by Eqs. (3.27) to (3.31) can lead to specialization, or differentiation out of an initially homogeneous population. Individual i is considered a specialist of task j if θ_{ij} is small. Figure 3.19(a) shows an example of specialization: thresholds are represented as a function of time for five individuals and two tasks ($N = 5, m = 2$) (initially, $\forall i \in \{1, \ldots, 5\}$, $\forall j \in \{1, 2\}$, $\theta_{ij} = 500$, $x_{ij} = 0.1$, $\alpha = 3$, $\delta = 1$, $p = 0.2$, $\xi = 10$, $\varphi = 1$, $\sigma = 0.1$). Some of these individuals become highly responsive to stimuli associated with task 1 but not to those associated with task 2, and others exhibit the opposite behavior. Individuals 1 and 2 are task-2 specialists, while individuals 3, 4, and 5 are task-1 specialists. Figure 3.19(b) shows the frequencies x_{i1} and x_{i2} of task performance for all individuals as a function of time. Individuals with small θ_{ij} (respectively large) tend to have large (respectively small) x_{ij}. Individuals adjust their levels of activity to maintain stimulus intensity below threshold, so that the three individuals 3, 4, and 5, which perform mostly task 1, are less active $x_{i1} \approx 0.55, x_{i2} \approx 0.05$) than individuals 1 and 2, which perform mostly task 2 ($x_{i1} \approx 0.05, x_{i2} \approx 0.8$). This specialization seems to well describe what is observed in the primitively eusocial wasp *Polistes dominulus*, where a clear division of labor, including reproductive division of labor, emerges after a few days [309]) despite the lack of initial individual differentiation.

When, instead of initially being identical, thresholds are initially randomly distributed in $[\theta_{\min}, \theta_{\max}]$, individuals with an initially small θ_{ij} are more likely to become task-j specialists (Figure 3.19(c)). In a genetically diverse colony, individuals with close genotypic characteristics (for example, belonging to the same patriline) may have similar response thresholds and are, therefore, predisposed to perform the same tasks [42, 55, 135, 255, 271] (genes might also influence α, p, ξ, φ, $\theta_{\min}$, and $\theta_{\max}$). On Figure 3.19(c), individuals 1, 3, and 5 are task-1 specialists, and individuals 1, 2, and 4 are task-2 specialists; individual 1 is a specialist of both tasks.

Specialization can be observed over a certain portion of parameter space. Let T_c be the convergence time, that is, the time it takes until all individuals become either specialists or nonspecialists of both tasks (the criterion is: $\forall i \in \{1, \ldots, N\}$, $\forall j \in \{1, \ldots, m\}$, $\theta_{ij} > 900$ or $\theta_{ij} < 100$), and let N_1 and N_2 be the number of specialists after convergence of tasks 1 and 2, respectively. Figure 3.20(a) shows T_c, N_1, and N_2 as a function of φ (forgetting rate) for $N = 100$ and $m = 2$, assuming that $\varphi + \xi = 11$ (where ξ is the learning rate). When $\varphi < 0.4$, all individuals are specialists because the forgetting rate is small. T_c becomes large as φ approaches 0.4: s_j always being very low thanks to the large number of specialists; individuals are not easily stimulated to engage in task performance, so that their thresholds fluctuate a lot. For $0.4 < \varphi < 2$, the number of specialists for each task decreases, and so does T_c: differentiation is observed in this region. For $\varphi > 2$, forgetting is quick and no specialization is observed. A different pattern is observed when T_c, N_1, and N_2 are plotted as a function of p (probability per unit time of becoming inactive) (Figure 3.20(b)). All individuals are specialists for $p < 0.04$ because they spend a lot of time in task performance each time they engage. For the same reason as above, T_c becomes large as p approaches 0.04. The number of specialists for each

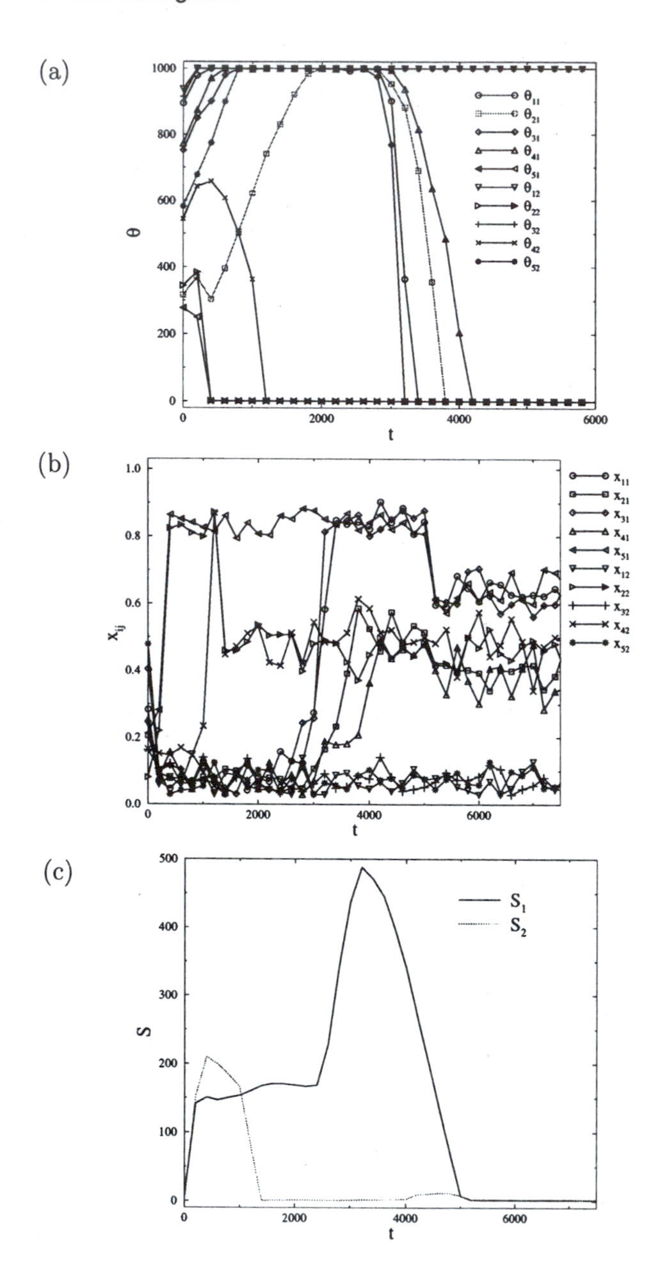

FIGURE 3.21 (a) Dynamics of response thresholds θ_{ij} for $N = 5$ individuals and $m = 2$ tasks. Before $t = 2500$: $\alpha_1 = \alpha = 2 = 5, \delta = 1, p = 0.2, \xi = 10, \varphi = 1, \sigma = 0.1$. Initial conditions: $\forall i \in \{1, \ldots, N\}, \forall j \in \{1, \ldots, m\}, x_{ij}(t = 0) = 0.1$, and initial distribution of thresholds is uniform over $[\theta_{\min} = 1, \theta_{\max} = 1,000]$. Individual 5 is a task-1 specialist, and individuals 2 and 4 are task-2 specialists. After $t = 2,500$: $\alpha_1 = 2$. All individuals become task-1 specialists. Individuals 2 and 4 remain task-2 specialists. Time unit: $0.2/p$. *continued.*

FIGURE 3.21 *continued.* (b) Dynamics of the fractions of time spent in task performance x_{ij} for the same parameter values as in Figure 3.21(a). The same remarks apply as to Figure 3.21(a). (c) Dynamics of the demands s_1 and s_2. s_1 increases at $t = 2,500$ as a response to the perturbation and decreases again after $t = 3,200$ because of the engagement in task 1 performance of four new workers.

task drops dramatically just after $p = 0.04$. For $0.04 < p < 0.42$, T_c decreases and the number of specialists increases with p, until all individuals become specialists, for $p > 0.42$, because they spend so little time in task performance that it requires a lot of specialists to maintain stimulus intensity at a low level. When (efficiency of task performance) is varied (Figure 3.20(c)), N_1 and N_2 decrease with increasing α because a smaller number of specialists is required as the efficiency or success rate of task performance increases. A transition is observed around $\alpha = 1.8$, where T_c becomes large before stabilizing.

One important advantage of this strategy over the fixed-threshold strategy is that in the case where individuals with low thresholds are removed, the demand is quickly taken back to a much lower level because individuals with previously high thresholds lower their thresholds and become more responsive to task-associated stimuli. Flexibility and robustness can also be observed when task performance becomes less efficient (lower value of α). A lower value of α may be due to competition with neighboring colonies, changing environmental conditions (temperature, humidity, rain, wind, etc.), exhaustion of food sources, or many other events. Figure 3.21(a) shows how thresholds change when the value of task 1 efficiency suddenly decreases from 5 to 2 at $t = 2,500$. Before $t = 2,500$, there is only one specialist of task 1 (individual 5), and two specialists of task 2 (individuals 2 and 4). After $t = 2,500$, all other individuals become task-1 specialists by lowering their thresholds to the minimum value. Individuals 2 and 4 remain task-2 specialists. Therefore, the system has generated task-1 specialists in response to the perturbation. The same behavior can be seen in Figure 3.21(b), where frequencies of task performance are plotted as a function of time. The system's ability to respond to the perturbation also reflects on the dynamics of the demand, as can be seen in Figure 3.21(c): demand s_1 (which is higher than s_2 before the perturbation because there is only one individual performing task 1) exhibits a sudden increase as soon as the perturbation is applied. This increase is rapidly corrected for, and starts to decrease around $t = 3,200$, until it reaches s_2's low level around $t = 5,000$. By contrast, a strategy based on fixed-response thresholds would be unable to influence the demand in response to the perturbation.

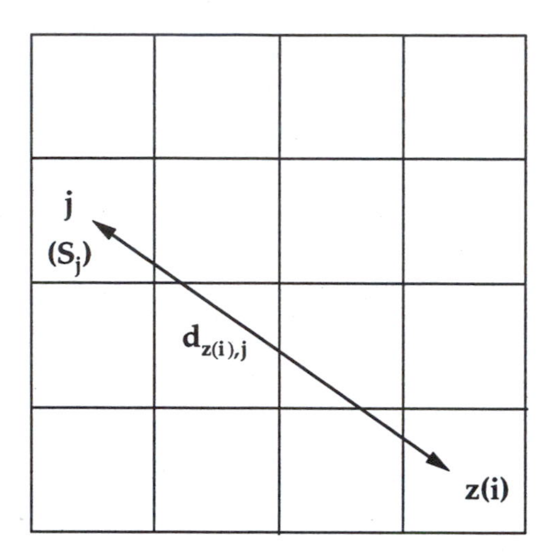

FIGURE 3.22 Schematic representation of the problem's formulation: individual i, located in zone $z(i)$, responds to a stimulus S_j from zone j at a distance $d_{z(i),j}$ from zone $z(i)$, with probability $P_{ij} = S_j^2/(S_j^2 + \alpha\theta_{i,j}^2 + \beta d_{z(i),j}^2)$.

3.5 DIFFERENTIATION IN A MULTIAGENT SYSTEM AND ADAPTIVE TASK ALLOCATION

3.5.1 DIFFERENTIATION IN A MULTIAGENT SYSTEM OR A GROUP OF ROBOTS

Simple identical robots or agents are easy to design and produce: they do not require a "predifferentiation," neither at the hardware level nor at the software level. If several tasks need to be performed simultaneously by the group of robots or agents, the undifferentiated entities need to be differentiated online. The response threshold model combined with the reinforcement procedure described in section 3.4 can be used to differentiate initially identical agents in a multiagent or multirobot system. It provides a widely applicable scheme for organizing task allocation in a homogeneous group of agents or robots. For example, Murciano et al.[246] have used a simple reinforcement learning procedure to achieve specialization in an artificial multiagent system composed of homogeneous agents, that is, agents with identical capabilities, which do not communicate directly. Agents specialize and eventually cooperate to collectively perform a complex gathering task. Each agent has an "affinity" variable for each type of item to be processed. The affinity of an agent for an object type is equivalent to the inverse of a response threshold. The distribution of affinities in the group converges to an optimum as affinities are reinforced as a function of the successes and failures of each agent.

But even the fixed-threshold model can be used to organize a group of robots. Krieger [203] designed a group of Khepera robots (miniature mobile robots aimed at "desktop" experiments [241] to collectively perform a puck-foraging task. In that remarkable work, pucks spread in the environment are taken back to the "nest" where they are dropped in a basket. The available "energy" of the group increases when pucks are dropped into the basket, and decreases regularly with time. More energy is consumed during foraging trips than when robots are immobile at the nest. Each robot has a foraging threshold: when the energy of the colony goes below the foraging threshold of a robot, the robot leaves the nest to look for pucks in the environment (in addition, a robot can detect a patch of pucks, and goes back directly to a patch that was detected in its previous foraging trip). Different robots have different thresholds. But even a robot that is not very responsive to the energy level can be recruited by another, more responsive, robot.

3.5.2 ADAPTIVE TASK ALLOCATION

A simple toy application illustrates how the threshold-based model of specialization presented in section 3.4 can be used to allocate tasks in a multiagent system in a flexible and robust way [32]. Imagine a group of mailmen (hereafter called agents), belonging to some express mail company, that have to pick up letters in a large city. Customers should not have to wait more than a given amount of time: the aim of the mail company is, therefore, to allocate the agents to the various demands that appear in the course of the day, keeping the global demand as low as possible. The probability that individual i, located in zone $z(i)$, responds to a demand S_j in zone j is given by

$$P_{ij} = \frac{S_j^2}{S_j^2 + \alpha\theta_{i,j}^2 + \beta d_{z(i),j}^2}\,, \tag{3.32}$$

where $\theta_{i,j}(\in [\theta_{\min}, \theta_{\max}])$ is the response threshold of agent i to the demand from zone j, $d_{z(i),j}$ is the distance between $z(i)$ and j (this distance can either be euclidian or include factors such as one-way thoroughfares, lights, usual traffic jams, etc.), and α and β are two positive coefficients that modulate the respective influences of θ and d. Each time an agent i allocates itself to zone j to retrieve mail, its response thresholds are updated in the following way:

$$\theta_{i,j} \leftarrow \theta_{i,j} - \xi_0\,, \tag{3.33}$$

$$\theta_{i,n(j)} \leftarrow \theta_{i,n(j)} - \xi_1,\ \forall n(j)\,, \tag{3.34}$$

$$\theta_{i,k} \leftarrow \theta_{i,k} + \varphi \text{ for } k \neq j, k \notin \{n(j)\}\,, \tag{3.35}$$

where $\{n(j)\}$ is the set of zones surrounding j, ξ_0, and ξ_1 are two learning coefficients corresponding to zone j and its neighboring zones respectively ($\xi_0 > \xi_1$), and φ is the forgetting coefficient applied to response thresholds associated with other zones. This procedure allows an increase in spatial fidelity not only purely locally but also with respect to neighboring zones.

Algorithm 3.1 High-level description of zone allocation algorithm

```
/* Initialization */
For j = 1 to L × L do
  S_j = 0 /* initial demand in zone j is equal to 0 */
End For
For i = 1 to m do
  For j = 1 to L × L do
    θ_ij = 500 /* initialization of thresholds at neutral values */
  End For
  Place agent i in randomly selected zone z(i)
End For
/* Main loop */
For t = 1 to t_max do
  /* increase demand in randomly selected zone and assign agent if possible */
  For k = 1 to RS do
    Draw random integer number n between 1 and L × L /* select zone */
    If (zone n not covered) then
      S_n ← S_n + δ /* increase demand by δ in selected zone */
      Reponse = 0 /* no agent has responded to demand from zone n */
      Sweep = 1 /* first sweeping of all agents */
      Repeat
      For i = 1 to m do
        If (agent i available) /* agent i is not traveling */ then
          Draw real number r between 0 and 1
          Compute P_in /* Eq. (3.32) */
          If (r < P_in) then
            Zone z(i) is no longer covered
            Zone n is covered
            S_n ← 0
            Agent i unavailable for d_z(i)n time units
            z(i) ← n
          End If
        End If
      End For
      Sweep ← Sweep +1
      Until ((Response=1) or (Sweep = SW)
    End If
  End For
  /* threshold updates */
  For i = 1 to m do
    For j = 1 to L × L do
      If (j = z(i)) then
        θ_ij ← θ_ij − ξ_0 /* agent i "learns" zone j */
      Else If (j is in the neighborhood of z(i)) then
        θ_ij ← θ_ij − ξ_1 /* agent i "learns" zone j */
      Else
        θ_ij ← θ_ij + φ /* agent i "forgets" zone j */
      End If
    End For
  End For
End For
/* Values of parameters used in simulations */
α = 0.5, β = 500, θ_min = 0, θ_max = 1000, ξ_0 = 150, ξ_1 = 70, δ = 50, φ = 10, m = 5, RS = 5,
SW = 5, L = 5
```

Simulations have been performed with a grid of $L \times L = 5 \times 5$ zones (we consider four neighbors for the update of Eq. (3.35), with periodic boundary conditions, see Figure 3.22), and $m = 5$ agents; at every iteration, the demand increases in $RS = 5$ randomly selected zones by an amount $\delta = 50$. Other parameter values are: $\alpha = 0.5$, $\beta = 500$, $\theta_{\min} = 0$, $\theta_{\max} = 1{,}000$, $\xi_0 = 150$, $\xi_1 = 70$, $\varphi = 10$. Agents are swept in a random order and decide to respond to the demand from a particular zone according to Eq. (3.32). If no agent responds after $SW = 5$ sweepings, the next iteration starts. If an agent responds, this agent will be unavailable for an amount of time that we take to be equal to the distance separating its current location from the zone where the demand originates. Once the agent decides to allocate itself to that zone, the associated demand in that zone is maintained at 0 (since any demand emerging between the time of the agent's response and its arrival in the zone will be satisfied by the same agent). Algorithm 3.1 shows a high-level algorithmic description of the Monte Carlo simulations. Figure 3.23(a) shows how the demand increases but is still kept under control when one individual fails to perform its task. Figure 3.23(b) shows how the threshold of an individual with respect to a single zone can vary as a function of time. A special behavior can be observed after the removal of another individual. The removed individual was a specialist of a given zone: another individual lowers its threshold with respect to that zone and becomes in turn a new specialist of that zone. This is observed in Figure 3.23(b). However, because the workload may be too high to allow agents to settle into a given specialization, response thresholds may oscillate in time. All these features point to the flexibility and robustness of this algorithm.

Although we have presented the performance of the algorithm based on one specific (toy) example, it can certainly be modified to apply to virtually any kind of task allocation: the demand S_j can be the abstract demand associated with some task j, $\theta_{i,j}$ is a response threshold of agent i with respect to the task-associated stimulus S_j. The distance $d_{z(i),j}$ is an abstract distance between i and task j which can, for example, represent the ability or lack of ability of i to deal with task j: if i is not the most efficient agent to perform task j, it will not respond preferentially to S_j, but if no other agent is in a position to respond, it will eventually perform the task. It is certainly possible to design a scheme in which d can vary depending on the efficiency of i in performing task j. Moreover, tasks may not have the same priority, which can be reflected in the magnitude of their associated stimuli and in the rate of increase of the stimulus intensities.

The task allocation scheme advocated in this section is easy to implement, provides a flexible method for allocating tasks, and allows adjustment to unpredictable changes in demand, disturbances to individual agents, or perturbations due to other factors. It constitutes a new programming methodology to control task or resource allocation in complex systems. More work is needed to explore its behavior and performance.

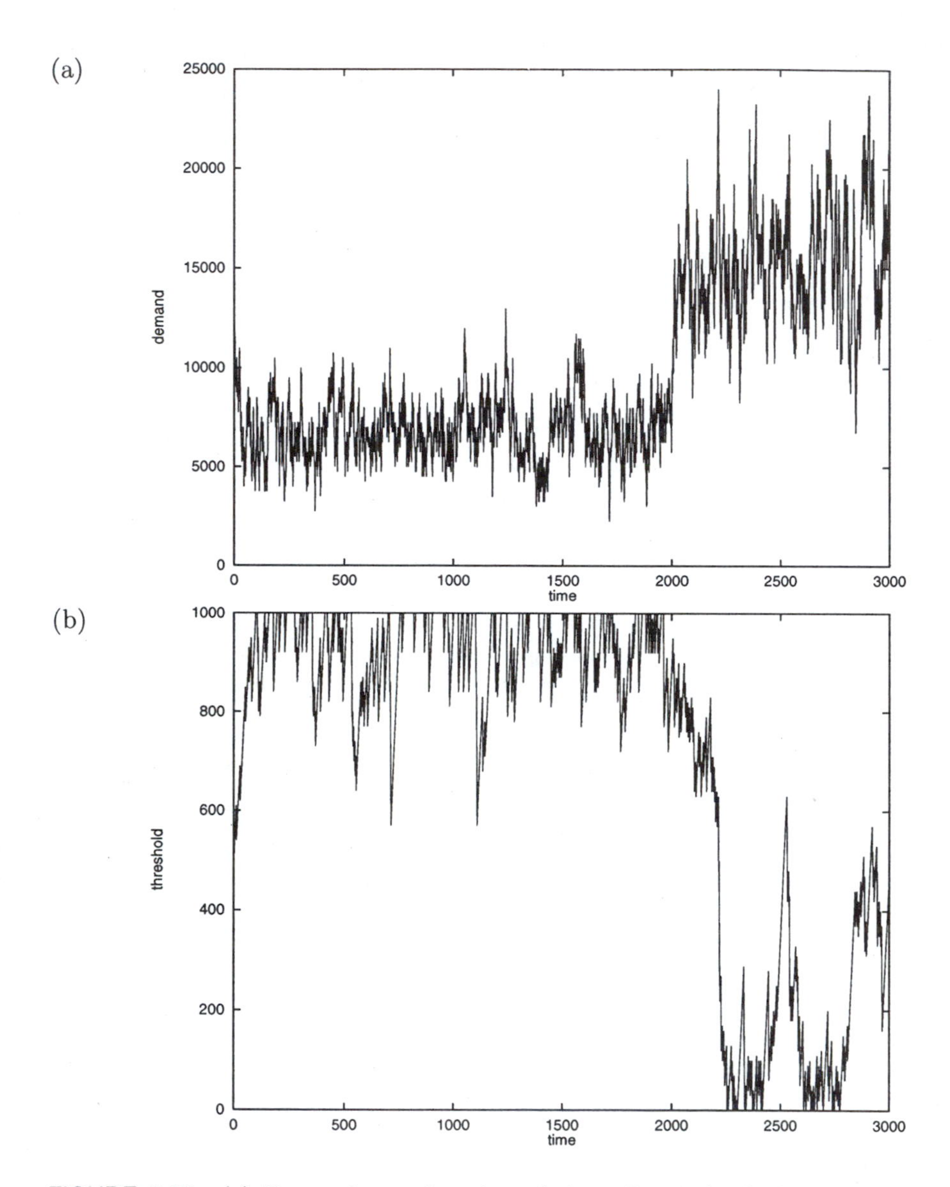

FIGURE 3.23 (a) Demand as a function of time. One individual is removed at time $t = 2,000$. After Bonabeau et al. [32]. (b) Threshold dynamics of a particular individual with respect to the zone from which a specialist is removed at $t = 2,000$. After Bonabeau et al. [32]. Reprinted by permission © *World Scientific.*

3.5.3 CONNECTION WITH "BIDDING" ALGORITHMS

Algorithms inspired by the division of labor in social insects share fundamental features with multiagent "market-based" algorithms where the allocation of tasks or resources is regulated by a bidding system [73, 74, 184, 212, 321]. The agent that makes the highest bid gets the resource or gets to perform the task. A market-clearing mechanism ensures that all tasks or resources have been allocated. For instance, Waldspurger et al. [321] have developed Spawn, a computational system where each task, starting with a certain amount of "money" that corresponds to its relative priority, bids for the use of machines on a network. Examples of tasks include searching through a database for an item, printing a paper, etc. Through bidding for the use of machines, each task can allocate its budget to those resources that are most important for it [184, 321]. When prices are low enough, some tasks may use several machines in parallel; the number of machines used by, and the computation time devoted to, each task are adjusted to the demand from other tasks. In another example [73], the problem of thermal resource distribution in a building is solved using market-based control: computational agents that represent individual temperature controllers bid to buy or sell cool or warm air via a double-blind computerized auction moderated by a central computer auctioneer. Clearwater et al. [73] have shown that this system results in an equitable temperature distribution throughout the building at a low cost. Market-based control is a paradigm for controlling complex systems. A social insect colony is a complex system in which, as we have argued in this chapter, task allocation is controlled by response thresholds. Tasks that need to be performed correspond to a job market and insects are individuals (more or less) actively seeking a job (Tofts and Franks [314] speak of "foraging-for-work"). Insects with a low response threshold get the job.

A high bid may, therefore, be similar to a low threshold: if the agent has a low response threshold with respect to the task or the resource, it makes a high bid. Morley [242] and Morley and Ekberg [243] describe a situation in which the bidding system is equivalent to an adaptive task allocation system inspired by social insects. The problem to be solved is to assign trucks to paint booths in a truck facility: trucks get out of the assembly line at a quick pace (one per minute) to be painted. The color that a truck should have is predetermined directly by a customer order. The paint fit-and-finish operations are a traditional bottleneck that can significantly reduce production throughput. It takes three minutes to paint a truck, but it takes more time to change the color of a paint booth: if, for example, a given paint booth is applying black paint and the next truck to be processed by that paint booth requires white paint, the booth will be down for several minutes. The cost of paint changeover is also high. Therefore, the number of such changeovers should be minimized online. On the other hand, if the paint booth which is applying black is the only one that is available or that has a small waiting queue, and if painting the white truck is urgent, the white truck should be allocated to that paint booth. In other words, a black paint booth does not accept white trucks unless no other better-positioned paint booth is available; in a polymorphic ant colony, majors do not perform brood care unless minors are not doing it. Morley [242, 243] developed a system in which each paint booth is an agent that follows four simple rules:

1. Try to take another truck the same color as the current color.
2. Take particularly important jobs.
3. Take any job to stay busy.
4. Do not take another job if paint booth is down or queue is large.

These rules were implemented through a bidding system and a market-clearing mechanism (every truck must be painted). Booths bid based on their ability to do the job efficiently—low cost, minimal delay. A paint booth that is down or the queue of which is full does not participate in the bidding. The optimal parameters of the bidding system were determined with evolutionary techniques. The bidding system, implemented in a real truck factory,[3] resulted in a 10% reduction in paint usage, half as many paint changeovers, a higher global throughput (booths are busier), and a significantly shorter software code than with the previous method, which was based on a global preoptimization of the paint schedule. Most of these nice results come from the fact that the system is more robust and more flexible: a booth that breaks down has a limited impact on the throughput because the schedule is generated online.

In this example, the parallel with threshold-based task allocation is striking. Imagine that each paint booth has a response threshold with respect to an arriving truck—the stimulus: the response threshold is low if the truck is the same color as the paint booth's current color, if the job is urgent, or if the queue is empty or small; the response threshold is high if the truck's color is different from the paint booth's current color, if the queue is large, or if the paint booth is down (in which case the threshold is infinite). It is therefore easy to transform this market-based task-allocation algorithm into a threshold-based algorithm.

3.6 POINTS TO REMEMBER

- Division of labor in social insect colonies is not rigid. Workers switch tasks. The behavioral elasticity of individual workers leads to colony-level resilience. For example, major workers in polymorphic species of the ant genus *Pheidole* start performing tasks that are normally performed by minor workers when minor workers are removed from the colony.
- A simple model based on response thresholds allows a formal connection between the level of individual workers and the colony level. Workers with low response thresholds respond to lower levels of stimuli than workers with high response

[3] The system has recently been dismantled to replace hydraulic paint robots with electric ones that require new software. Interestingly, despite the efficiency of the system, there always was some resistance against using it because of the autonomy that was given to the paint booths.

thresholds. Task performance reduces the intensity of stimuli. If workers with low thresholds perform their normal tasks, the task-associated stimuli never reach the thresholds of the high-threshold workers. But if workers with low thresholds are removed, or if for any reason the intensity of task-associated stimuli increases, high-threshold workers start task performance. There is a good agreement between the model and experimental observations of colony-level resiliency in ants of the genus *Pheidole*.

- An extension of the model, where a threshold is lowered when the associated task is performed and increased when the associated task is not performed, generates differential task allocation and specialization in a population of initially identical workers.
- Threshold models can serve as a basis to generate differential task allocation and specialization in a multiagent system. They are very similar to market-based algorithms used to allocate resources and tasks, in which agents bid for a resource or a task. The agent that makes the highest bid gets the resource or performs the task. A toy application example, distributed mail retrieval, shows how a task-allocation system based on a response-threshold model with learning responds to perturbations.

CHAPTER 4

Cemetery Organization, Brood Sorting, Data Analysis, and Graph Partitioning

4.1 OVERVIEW

In the previous two chapters, foraging and division of labor were shown to be useful metaphors to design optimization and resource allocation algrithms. In this chapter, we will see that the clustering and sorting behavior of ants has stimulated researchers to design new algorithms for data analysis and graph partitioning.

Several species of ants cluster corpses to form a "cemetery," or sort their larvae into several piles. This behavior is still not fully understood, but a simple model, in which agents move randomly in space and pick up and deposit items on the basis of local information, may account for some of the characteristic features of clustering and sorting in ants.

The model can also be applied to data analysis and graph partitioning: objects with different attributes or the nodes of a graph can be considered items to be sorted. Objects placed next to each other by the sorting algorithm have similar attributes, and nodes placed next each other by the sorting algorithm are tightly connected in the graph. The sorting algorithm takes place in a two-dimensional space, thereby offering a low-dimensional representation of the objects or of the graph.

Distributed clustering, and more recently sorting, by a swarm of robots have served as benchmarks for swarm-based robotics. In all cases, the robots exhibit extremely simple behavior, act on the basis of purely local information, and communicate indirectly except for collision avoidance.

4.2 CEMETERY ORGANIZATION AND LARVAL SORTING

In several species of ants, workers have been reported to form piles of corpses—literally cemeteries—to clean up their nests. Chrétien [72] has performed experiments with the ant *Lasius niger* to study the organization of cemeteries. Other experiments on the ant *Pheidole pallidula* are also reported in Deneubourg et al. [88], and many species actually organize a cemetery. Figure 4.1 shows the dynamics of cemetery organization in another ant, *Messor sancta*. If corpses, or, more precisely, sufficiently large parts of corposes are randomly distributed in space at the beginning of the experiment, the workers form cemetery clusters within a few hours. If the experimental arena is not sufficiently large, or if it contains spatial heterogeneities, the clusters will be formed along the edges of the arena or, more generally, following the heterogeneities.

The basic mechanism underlying this type of aggregation phenomenon is an attraction between dead items mediated by the ant workers: small clusters of items grow by attracting workers to deposit more items. It is this positive feedback that leads to the formation of larger and larger clusters. The exact individual behavior that actually implements this positive feedback is, however, still unclear, as is the adaptive significance of corpse clustering. The formation of cemeteries may be the (possibly nonadaptive) side-effect of clustering behavior, which can be applied to all kinds of items, including brood items: clustering would then be adaptive in a different context, although cleaning the nest by getting rid of corpses is certainly an important task by itself.

Brood sorting is observed in the ant *Leptothorax unifasciatus* (this phenomenon is certainly widespread, but has been clearly studied only in this species [130]). Workers of this species gather the larvae according to their size, that is, all larvae tend to be aggregated with smaller larvae located in the center and larger larvae in the periphery (Figure 4.2). Franks and Sendova-Franks [130] have intensively analyzed the distribution of brood within the brood cluster. The brood items are sorted, and different amounts of space are allocated to different types of brood. Eggs and microlarvae are clustered compactly at the center of the brood area, with each item being given little individual space. Individual space tends to increase from the center of the cluster outward. The largest larvae, which are located at the periphery of the brood cluster, receive more individual space. When pupae and prepupae are present, they are located between peripheral large larvae and the more central larvae of medium size. This suggests that space and location are not determined purely on the basis of size, since pupae and prepupae are at least as large as the largest larvae. In fact, pupae and prepupae do not require feeding but only grooming: Franks and Sendova-Franks have shown that there is a positive correlation between individual space and metabolic rate.

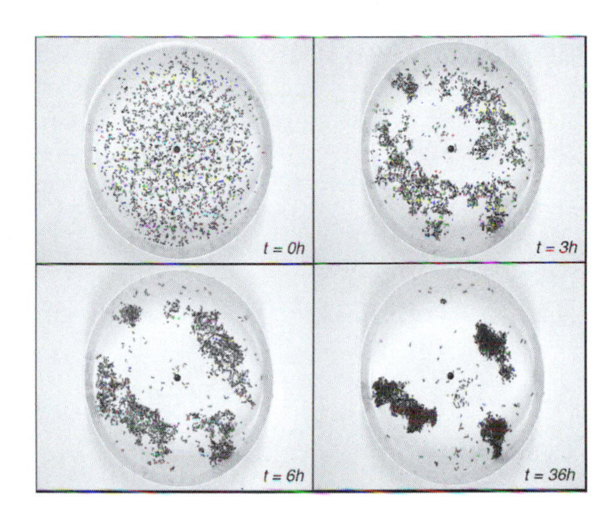

FIGURE 4.1 1500 corposes are randomly located in a circular arena ($\phi = 25$ cm), where *Messor sancta* workers are present. The figure shows four successive pictures of the arena: the initial state, 2 hours, 6 hours, and 26 hours after the beginning of the experiment.

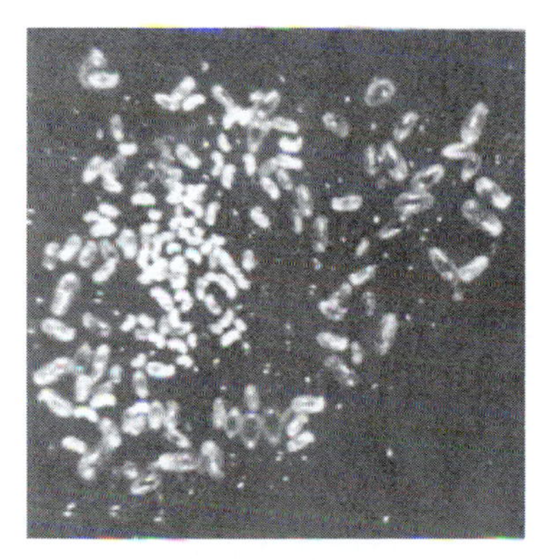

FIGURE 4.2 Piles of larvae in *Leptothorax unifasciatus*. Smaller larvae can be seen at the center, and larger larvae at the periphery. This picture was taken 72 hours after all larvae were randomly placed in the experimental arena. After Deneubourg et al. [88]. Reprinted by permission © *MIT Press*.

A possible adaptive explanation for the phenomenon of brood sorting is that, because items with similar needs are located nearby, care can be provided with more efficiency. For the sake of simplicity, the model presented in the next section assumes that items are sorted on the basis of their perceived characteristics, which may include size, metabolic rate, as well as other features. Recent robotic "modeling" by Melhuish et al. [237] (described in section 4.6.3) suggests that it is sufficient to assume that one item needs more space to actually segregate items (see also Barker and Grimson [7]).

4.3 A MODEL OF CORPSE CLUSTERING AND BROOD SORTING

4.3.1 CLUSTERING

Deneubourg et al. [88] have proposed two closely related models to account for the two above-mentioned phenomena of corpse clustering and larval sorting in ants. Although the model of clustering reproduces experimental observations more faithfully, the second one gives rise to more applications. Both models rely on the same principle, and, in fact, the clustering model is merely a special case of the sorting model.

The general idea is that isolated items should be picked up and dropped at some other location where more items of that type are present. Let us assume that there is only one type of item in the environment. The probability p_p for a randomly moving, unladen agent (representing an ant in the model) to pick up an item is given by

$$p_p = \left(\frac{k_1}{k_1 + f}\right)^2 \tag{4.1}$$

where f is the perceived fraction of items in the neighborhood of the agent, and k_1 is a threshold constant. When $f \ll k_1$, p_p is close to 1, that is, the probability of picking up an item is high when there are not many items in the neighborhood. p_p is close to 0 when $f \gg k_1$, that is, items are unlikely to be removed from dense clusters. The probability p_d for a randomly moving loaded agent to deposit an item is given by:

$$p_d = \left(\frac{f}{k_2 + f}\right)^2 \tag{4.2}$$

where k_2 is another threshold constant: for $f \ll k_2$, p_d is close to 0, whereas for $f \gg k_2$, p_d is close to 1. As expected, the depositing behavior obeys roughly opposite rules.

How is f evaluated? For the sake of robotic implementation, Deneubourg et al. [88] have assumed that f is computed through a short-term memory that each

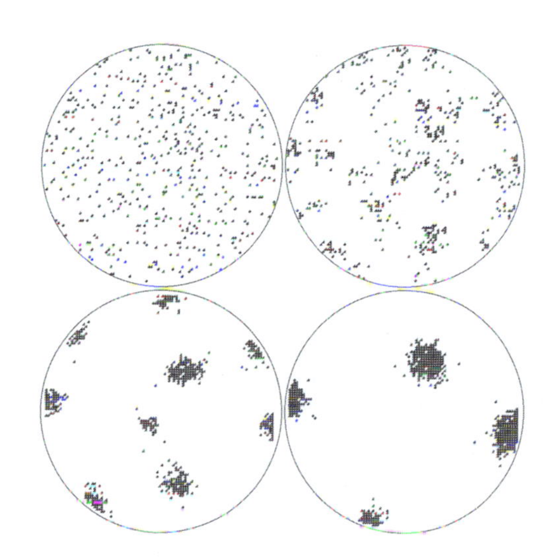

FIGURE 4.3 Simulation of the clustering model. The figure shows four successive pictures of the simulated circular arena (diameter = 200 grid sites; total area: 31,416 sites): the initial state, with 5,000 items placed randomly in the arena, the arena at $t = 0$, $t = 50,000$, $t = 1,000,000$, and $t = 5,000,000$. $T = 50$, $k_1 = 0.1$, $k_2 = 0.3$, 10 agents.

agent possesses: an agent keeps track of the last T time units, and f is simply the number N of items encountered during these last T time units, divided by the largest possible number of items that can be encountered during T time units. If one assumes that only 0 or 1 object can be found within a time unit, then $f = N/T$.

As mentioned above, this procedure lends itself more easily to a robotic implementation. Real ants are likely to use chemical or tactile cues to orient their behavior. The algorithms described in sections 4.4 and 4.6, which are inspired by this idea, rely on a more direct evaluation of f. This procedure should, therefore, be taken as an example among many possible procedures, and changing the detail of the procedure does not drastically alter the results [30]. Figure 4.3 shows a simulation of this model with $T = 50$, $k_1 = 0.1$, $k_2 = 0.3$, and 10 agents clustering 5,000 items in a circular arena, the diameter of which is 200 sites, at $t = 0$, $t = 50,000$, $t = 1,000,000$, and $t = 5,000,000$. Small evenly spaced clusters emerge within a relatively short time and then merge, more slowly, into fewer larger clusters.

4.3.2 SORTING

Let us now assume that there are two types, A and B, of items in the environment. The principle is the same as before, but now f is replaced by f_A and f_B, the respective fractions of items of types A and B encountered during the last T time units. Figures 4.4(a), 4.4(b), and 4.4(c) show a simulation of this model with two items, $T = 50$, $k_1 = 0.1$, $k_2 = 0.3$, and 10 agents sorting 200 items of each type on a 100×100 square grid, at $t = 0$, $t = 500,000$, and $t = 5,000,000$. Notice that this model is unable to reproduce exactly the brood sorting patterns observed in *Leptothorax* ants, where brood is organized into concentric areas of different brood types. In order to fully reproduce such patterns, one would have to include some kind of differential adhesion [293, 294, 295].

4.3.3 SPATIAL ENTROPY AND VARIANTS

Gutowitz [160] has suggested the use of spatial entropy to track the dynamics of clustering. Spatial entropy measures how well items are clustered at different spatial scales. Let s be a certain spatial scale (for example, $s = 8$ means that we are considering patches of 8×8, hereafter called s-patches): the spatial entropy E_s at scale s is defined by

$$E_s = \sum_{I \in \{\text{s-patches}\}} P_I \log P_I \,, \tag{4.3}$$

where P_I is the fraction of all objects on the lattice that are found in s-patch I. By definition, $\sum_{I \in \{\text{s-patches}\}} P_I = 1$. E_s, which decreases as clustering proceeds, is useful to compare various clustering strategies, and how fast they lead to clusters at different scales.

A variant of Deneubourg et al.'s [88] model (hereafter called BM, for basic model) was proposed by Oprisan et al. [251], in which the influence of previously encountered objects is discounted by a time factor: recently encountered objects have more weight than objects encountered long ago. Different patterns can also be obtained with other weighting functions. Bonabeau [30] also explored the influence of various weighting functions, especially those with short-term activation and long-term inhibition, and found that it is possible to generate reaction-diffusion-like patterns.

4.4 EXPLORATORY DATA ANALYSIS

Lumer and Faieta [225] have generalized Deneubourg et al's [88] BM to apply it to exploratory data analysis. The idea is to define a distance or dissimilarity d between objects in the space of object attributes. For instance, in the BM, two objects o_i and o_j can only be either similar or different, so that a binary distance can be defined, where, for example, $d(o_i, o_j) = 0$ if o_i and o_j are identical objects, and

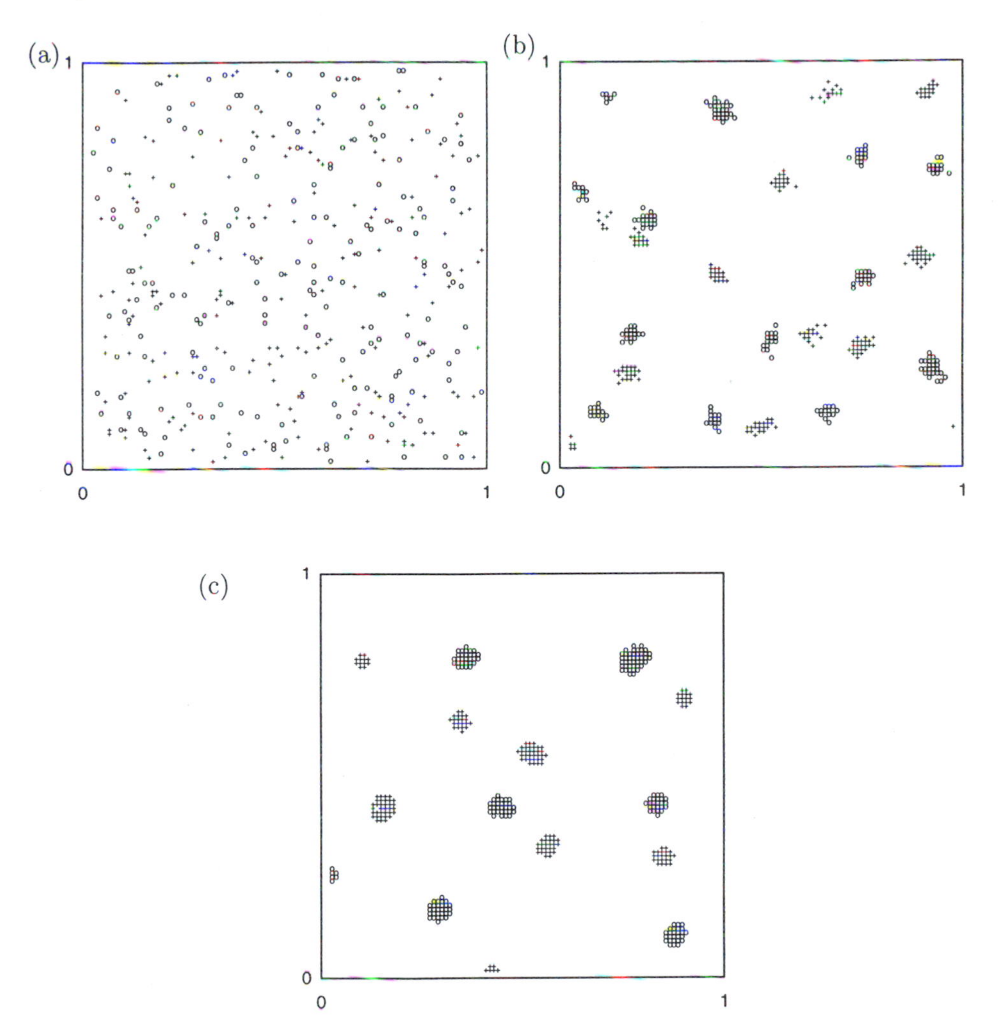

FIGURE 4.4 Simulation of the sorting model. (a) Initial spatial distribution of 400 items of two types, denoted by o and +, on a 100×100 grid. (b) Spatial distribution of items at $t = 500,000$. $T = 50$, $k_1 = 0.1$, $k_2 = 0.3$, 10 agents. (c) Same as (b) at $t = 5,000,000$.

$d(o_i, o_j) = 1$ if o_i and o_j are not identical objects. Obviously, the very same idea can be extended to include more complicated objects, that is, objects with more attributes, and/or more complicated distances. It is classical in data analysis to have to deal with objects that can be described by a finite number n of real-valued attributes, so that objects can be seen as points in $\mathcal{R}^n$, and $d(o_i, o_j)$ is the euclidian norm, or any other usual metric, such as the infinite norm$|| \ldots ||_\infty$.

The algorithm introduced by Lumer and Faieta [225] (herafter LF) consists of projecting the space of attributes onto some lower dimensional space, typically of dimension $z = 2$, so as to make clusters appear with the following property: intra-cluster distances (i.e., attribute distances between objects within clusters) should be small with respect to inter-cluster distances, that is, attribute distances between objects that belong to different clusters. Such a mapping should, therefore, keep some of the neighborhood relationships present in the higher-dimensional space (which is relatively easy since, for instance, any continuous mapping will do the job) without creating too many new neighbors in z-dimensions that would be false neighbors in n-dimensions (which is much less trivial since projections tend to compress information and may map several well-separated points in the n-dimensional space onto one single point in the z-dimensional space).

The LF algorithm works as follows. Let us assume that $z = 2$. Instead of embedding the set of objects into $\mathcal{R}^2$ or a subspace of $\mathcal{R}^2$, they approximate this embedding by considering a grid, that is, a subspace of Z^2, which can also be considered a discretization of a real space. Agents that are moving in this discrete space can directly perceive a surrounding region of area s^2 (a square $\text{Neigh}_{(s \times s)}$ of $s \times s$ sites surrounding site r). Direct perception allows a more efficient evaluation of the state of the neighborhood than the memory-based procedure used in the BM: while the BM was aimed to a robotic implementation, the LF algorithm is to be implemented in a computer, with significantly fewer material constraints. Let $d(o_i, o_j)$ be the distance between two objects o_i and o_j in the space of attributes. Let us also assume that an agent is located at site r at time t, and finds an object o_i at that site. The "local density" $f(o_i)$ with respect to object o_i at site r is given by

$$f(o_i) = \begin{cases} \frac{1}{s^2} \sum_{o_j \in \text{Neigh}_{(s \times s)}(r)} \left[1 - \frac{d(o_i, o_j)}{\alpha}\right] & \text{if } f > 0\,, \\ 0 & \text{otherwise}\,. \end{cases} \tag{4.4}$$

$f(o_i)$ is a measure of the average similarity of object o_i with the other objects o_j present in the neighborhood of o_i. $f(o_i)$ replaces the fraction f of similar objects in the basic model (BM). α is a factor that defines the scale for dissimilarity: it is important for it determines when two items should or should not be located next to each other. For example, if is too large, there is not enough discrimination between different items, leading to the formation of clusters composed of items which should not belong to the same cluster. If, on the other hand, α is too small, distances between items in attribute space are amplified to the point where items which are relatively close in attribute space cannot be clustered together because discrimination is too high.

Let us briefly examine the behavior of f by looking at the extreme cases. If, for example, all s^2 sites around r are occupied by objects that are similar to o_i ($\forall o_j \in \text{Neigh}_{(s \times s)}(r), d(o_i, o_j) = 0$), then $f(o_i) = 1$, and the object should be picked up with low probability. If all s^2 sites around r are occupied by objects that are maximally dissimilar to o_i ($\forall o_j \in \text{Neigh}_{(s \times s)}(r), d(o_i, o_j) = d_{\text{max}}$), then $f(o_i)$ is small, and the object should be picked up with high probability. Finally, if all sites around r are empty, then, obviously, $f(o_i) = 0$, and the object should be picked up with a high probability. An easy generalization shows that o_i should be picked up with high (respectively, low) probability when f is close to 0 (respectively, close to 1). Lumer and Faieta [225] define picking up and dropping probabilities as follows:

$$p_p(o_i) = \left(\frac{k_1}{k_1 + f(o_i)}\right)^2 , \tag{4.5}$$

$$p_d(o_i) = \begin{cases} 2f(o_i), & \text{if } f(o_i) < k_2 \\ 1, & \text{if } f(o_i) \geq k_2 s \end{cases} \tag{4.6}$$

where k_1 and k_2 are two constants that play a role similar to k_1 and k_2 in the BM.

Algorithm 4.1 High-level description of the Lumer Faieta algorithm

```
/* Initialization */
For every item o_i do
  place o_i randomly on grid
End For
For all agents do
  place agent at randomly selected site
End For
/* Main loop */
For t = 1 to t_max do
  For all agents do
    If ((agent unladen) and (site occupied by item o_i)) then
      Compute f(o_i) and p_p(o_i)
      Draw random real number R between 0 and 1
      If (R ≤ p_p(o_i)) then
        Pick up item o_i
      End If
    Else If ((agent carrying item o_i) and (site empty)) then
      Compute f(o_i) and p_d(o_i)
      Draw random real number R between 0 and 1
      If (R ≤ p_d(o_i)) then
        Drop item
      End If
    End If
    Move to randomly selected neighboring site not occupied by other agent
  End For
End For
Print location of items
/* Values of parameters used in experiments */
k_1 = 0.1, k_2 = 0.15, α = 0.5, s^2 = 9, t_max = 10^6 steps.
```

To illustrate the functioning of their algorithm, Lumer and Faieta [225] use a simple example in which the attribute space is $\mathcal{R}^2$, and the values of the two attributes for each object correspond to its coordinates (x, y) in $\mathcal{R}^2$. Four clusters of 200 points are generated in attribute space, with x and y distributed according to normal (or gaussian) distributions $N(\mu, \sigma)$ of average μ and variance σ^2:

$$
\begin{array}{lll}
(1) & x \propto N(0.2, 0.1), & y \propto N(0.2, 0.1), \\
(2) & x \propto N(0.8, 0.1), & y \propto N(0.2, 0.1), \\
(3) & x \propto N(0.8, 0.1), & y \propto N(0.8, 0.1), \\
(4) & x \propto N(0.2, 0.1), & y \propto N(0.8, 0.1),
\end{array}
$$

for clusters 1, 2, 3, and 4, respectively (see Figure 4.5). The data points were then assigned to random locations on a 100×100 grid, and the clustering algorithm was run with 10 agents. Figures 4.6(a), 4.6(b), and 4.6(c) show the system at $t = 0$, $t = 500,000$, and $t = 1,000,000$. At each time step, all agents have made a random move and possibly performed an action. Grid coordinates have been scaled to $[0, 1] \times [0, 1]$. Items that belong to different clusters are represented by different symbols: o, $+$, $*$, and $\times$. Objects that are clustered together belong to the same initial distribution, and objects that do not belong to the same initial distribution are found in different clusters. But there are generally more clusters in the projected system than in the initial distribution.

In order to correct this tendency to create more clusters than desired, Lumer and Faieta [225] have added three features:

1. Agents with different moving speeds. Let v be the speed of an agent (v is the number of grid units walked per time unit by an agent along a given grid axis). v is distributed uniformly in $[1, v_{\max}]$. The simulations use $v_{\max} = 6$. v also influences, through the function f, the tendency of an agent to either pick up or drop an object:

$$
f(o_i) = \begin{cases} \frac{1}{s^2} \sum_{o_j \in \text{Neigh}_{(s \times s)}(r)} \left[1 - \frac{d(o_i, o_j)}{\alpha (1 + \frac{v-1}{v_{\max}})} \right] & \text{if } f > 0\,, \\ 0 & \text{otherwise} \end{cases} \qquad (4.7)
$$

   Therefore, fast moving ants are not as selective as slow ants in their estimation of the average similarity of an object to its neighbors. The diversity of agents allows the formation of clusters over various scales simultaneously: fast agents form coarse clusters on large scales, that is, drop items approximately in the right coarse-grained region, while slow agents take over at smaller scales by placing objects with more accuracy. Since v serves as a sort of temperature, the presence of agents with different values of v corresponds to a system operating at different temperatures at the same time.
2. A short-term memory. Agents can remember the last m items they have dropped along with their locations. Each time an item is picked up, the agent compares the properties of the item with those of the m memorized items and goes toward the location of the most similar instead of moving randomly. This behavior leads to a reduction in the number of equivalent clusters, since similar items have a low probability of initiating independent clusters.

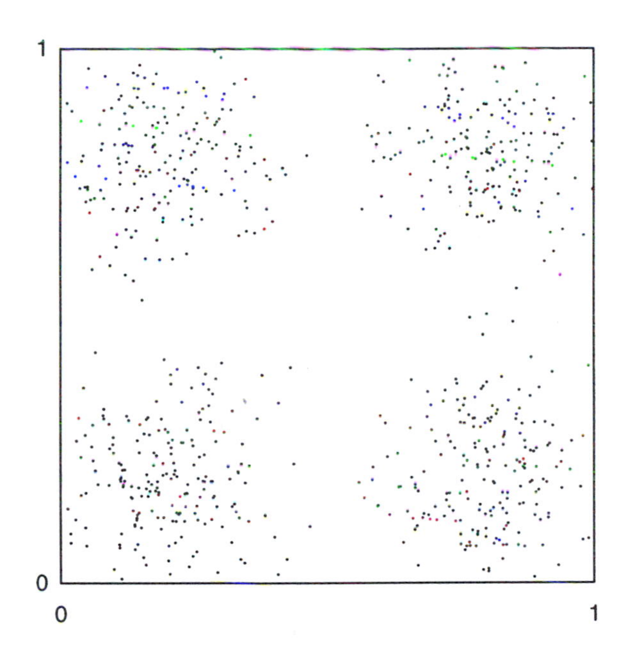

FIGURE 4.5 Distribution of points in "attribute space": 4 clusters of 200 points each are generated in attribute space, with x and y distributed according to normal (or gaussian) distributions $N(\mu, \sigma)$: $[x \propto N(0.2, 0.1),\ y \propto N(0.2, 0.1)]$, $[x \propto N(0.8, 0.1),\ y \propto N(0.2, 0.1)]$, $[x \propto N(0.8, 0.1),\ y \propto N(0.8, 0.1)]$, $[x \propto N(0.2, 0.1),\ y \propto N(0.8, 0.1)]$, for clusters 1, 2, 3, and 4, respectively.

3. Behavioral switches. The system exhibits some kind of self-annealing since items are less and less likely to be manipulated as clusters of similar objects form. Lumer and Faieta [225] have added the possibility for agents to start destroying clusters if they haven't performed an action for a given number of time steps. This procedure allows a "heating up" of the system to escape local nonoptimal configurations.

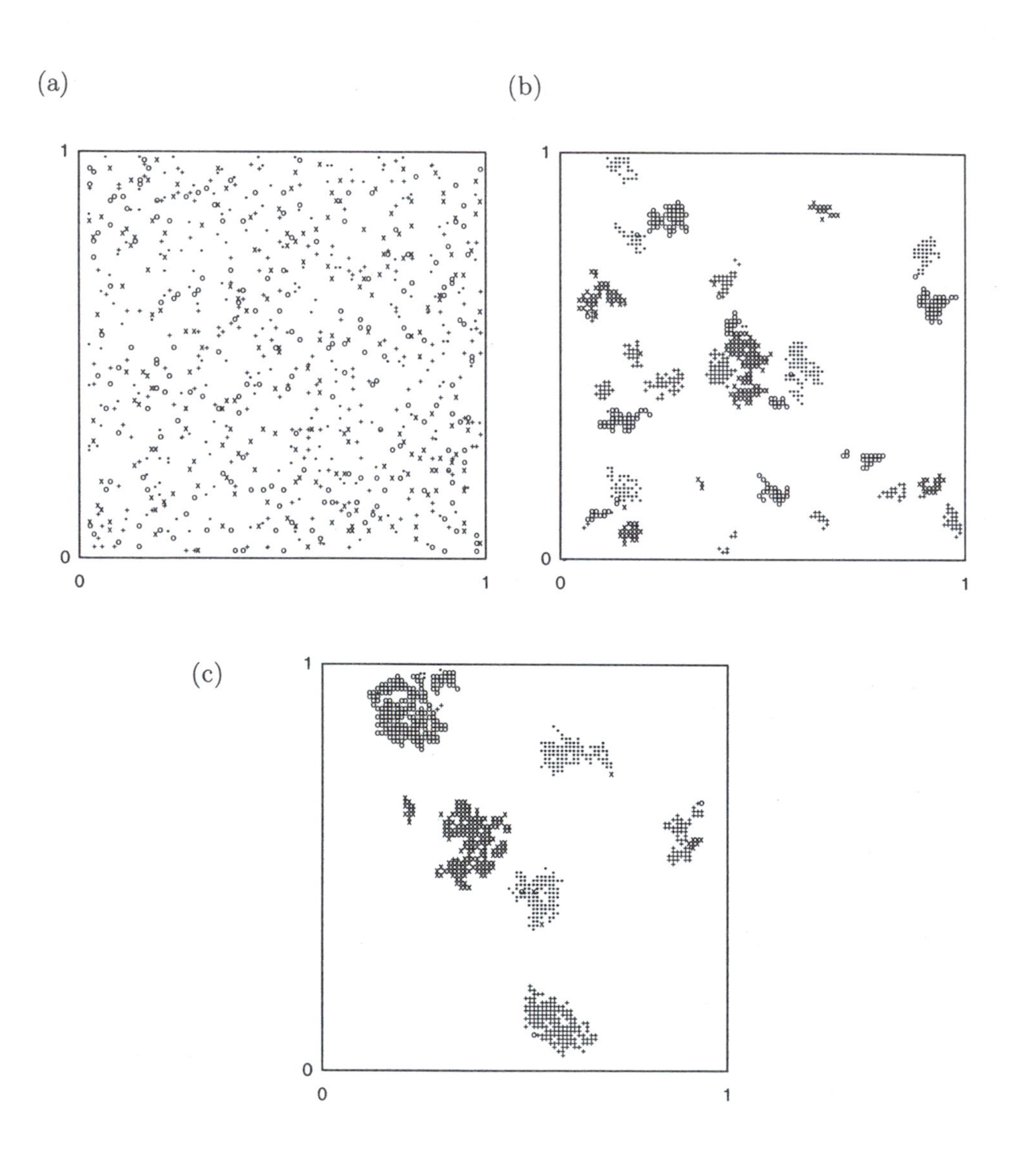

FIGURE 4.6 (a) Initial spatial distribution of 800 items on a 100×100 grid. Grid coordinates have been scaled to $[0, 1] \times [0, 1]$. Items that belong to different clusters are represented by different symbols: o, $+$, $*$, and $\times$. (b) Distribution of items at $t = 500,000$. $k_1 = 0.1$, $k_2 = 0.15$, $\alpha = 0.5$, $s^2 = 9$. (c) Same as (b) at $t = 1,000,000$.

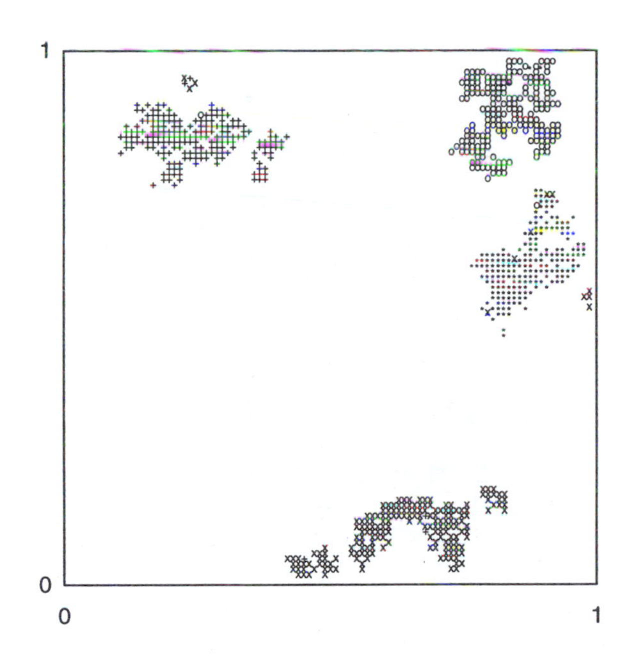

FIGURE 4.7 Spatial distribution of 800 items on a 100×100 grid at $t = 1,000,000$. Grid coordinates have been scaled to $[0, 1] \times [0, 1]$. Items that belong to different clusters are represented by different symbols: o, $+$, $*$, and $\times$. $k_1 = 0.1$, $k_2 = 0.15$, $\alpha = 0.5$, $s^2 = 9$, $m = 8$, $v_{\max} = 6$.

Figure 4.7 shows the system at $t = 1,000,000$ with ants characterized by different speeds and a short-term memory ($m = 8$). The effects of behavioral switches, not included here, can be found in Lumer and Faieta [225].

Simulations of this algorithm with appropriate parameter values show that the spatial distribution of objects becomes strongly correlated with the distance in attribute space. Let $w_i = (x_i, y_i)$ be the position in the plane of item o_i. The correlation can be seen clearly on Figure 4.8, which shows $||w_i - w_j||$ as a function of $d(o_i, o_j)$ in all possible pairs (o_i, o_j) for the spatial distribution of items obtained at $t = 1,000,000$ in the previous simulation (Figure 4.7). However, because the positions of the clusters are arbitrary on the plane, one cannot expect a perfect global correlation between distances in the plane and distances in attribute space.

Lumer and Faieta [225] suggest that their algorithm is "halfway between a cluster analysis—insofar as elements belonging to different concentration areas in their n-dimensional space end up in different clusters—and a multidimensional scaling, in which an intracluster structure is constructed." Note that in the present example, the exact locations of the various clusters on the low-dimensional space are arbitrary, whereas they usually have a meaning in classical factorial analysis.

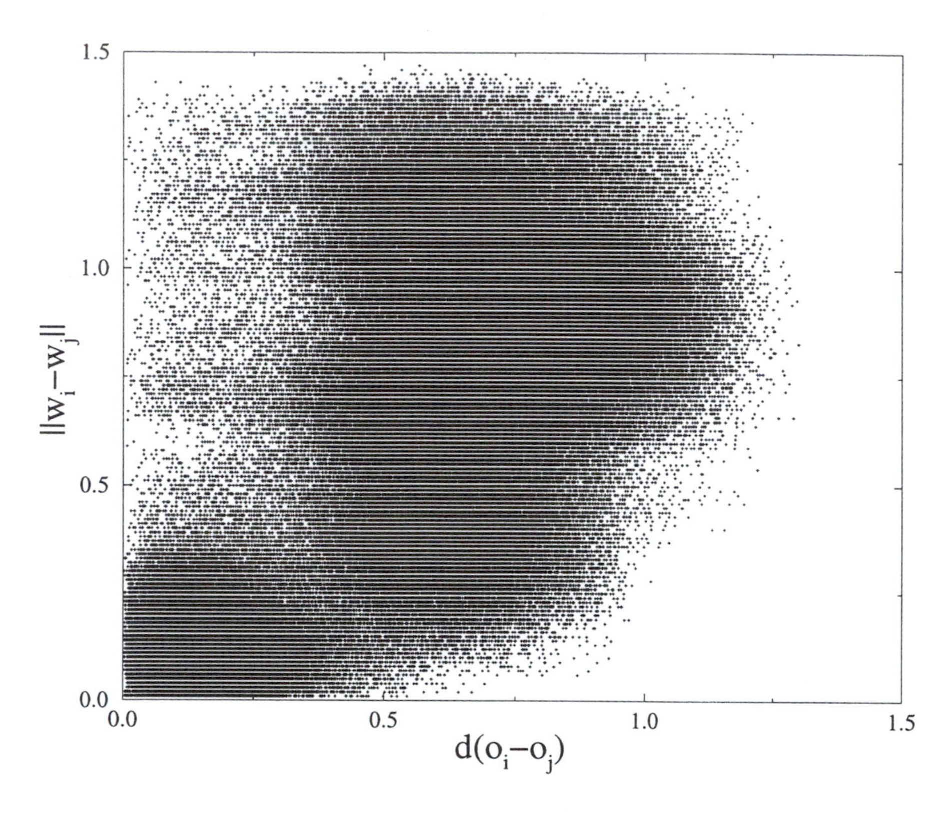

FIGURE 4.8 $||w_i - w_j||$ as a function of $d(o_i, o_j)$ for all possible pairs (o_i, o_j) for the spatial distribution of items obtained at $t = 1,000,000$ in the previous simulation (Figure 4.7).

In many cases, it is not useful to have any information about the locations of the clusters (especially in the context of textual databases), and, as noted by Lumer and Faieta [225], relaxing the global positioning constraints allows the clustering process to speed up significantly. The LF algorithm is therefore nonparametric, both with respect to the number of emerging clusters and to their positions on the representation plane. We shall see in the next chapter how to make the LF algorithm parametric with respect to these two aspects: although it is sometimes a good thing to have a nonparametric approach, it might also be useful to be able to construct the parametric counterpart of this algorithm.

The LF algorithm clusters the data and provides a visualization of the obtained clustering, as well as of the whole clustering process. When compared with clustering techniques, this algorithm is based on a mixture of two approaches: hierarchical and partitioning algorithms [226]. In hierarchical algorithms, all elements are considered clusters, and pairs of clusters that are most similar are iteratively merged into larger and larger clusters, providing solutions to the clustering problem at different levels of granularity. By contrast, in partitioning algorithms, an initial partition of the data is generated, and then iteratively improved by moving elements to their nearest cluster centroid. Such algorithms, although they are computationally more efficient than hierarchical algorithms, are usually more parametric: in particular, they require the number of clusters to be determined in advance.

Lumer and Faieta's [225, 226] algorithm is similar to a hierarchical algorithm in that an element is included in a cluster after comparison of this element's attributes with those of the elements in the cluster, whereas a partitioning algorithm would compare the element's attributes with the average attributes of the elements of the cluster. But the algorithm does not allow clustering at different levels. It is also similar to a partitioning algorithm in that elements are moved from cluster to cluster, thereby refining the partition through several passes on the data. The visualization part of the algorithm is similar in spirit to multidimensional scaling (MDS) techniques, which map a (usually) high-dimensional space onto a low dimensional space in which the data can be visualized. However, because cluster positions in the low-dimensional space are arbitrary, the algorithm can only allow qualitative conservation of dissimilarity relations: two objects which are close in the low-dimensional space have close attributes; the other way round however is not necessarily true.

Lumer and Faieta [226] went beyond the simple example of the gaussian distributions in the plane, and applied their algorithm to interactive "exploratory database analysis," where a human observer can probe the contents of each represented point and alter the charateristics of the clusters. They showed that their model provides a way of exploring complex information spaces, such as document or relational databases, because it allows information access based on exploration from various perspectives.

They applied the algorithm to a database containing the "profiles" of 1650 bank customers. Attributes of the profiles included marital status, gender, residential status, age, a list of banking services used by the customer, etc. Given the variety of attributes, some of them qualitative and others quantitative, they had to define several dissimilarity measures for the different classes of attributes, and to combine them into a global dissimilarity measure. For example, for attributes such as marital status or gender, they assigned a dissimilarity of 0 to pairs having the same value of the attribute, and a value of 1 to pairs with different values. For age, the dissimilarity was some decreasing function of the age difference in the pair.

Figure 4.9 shows a schematic representation of the clusters after applying the algorithm. Some arbitrarily selected clusters in Figure 4.9 are represented with numbers to illustrate how the algorithm works. Cluster 1 comprises primarily customers who are single, either male or female, age about 20, whose most popular banking product is interest checking and that mostly live with their parents. Most people in cluster 2 are married, some of them widowed, female owners of a house with no mortgage, some of them tenants, whose average age is 57. Cluster 3 comprises primarily married male tenants of average age 44. These clusters have a clear

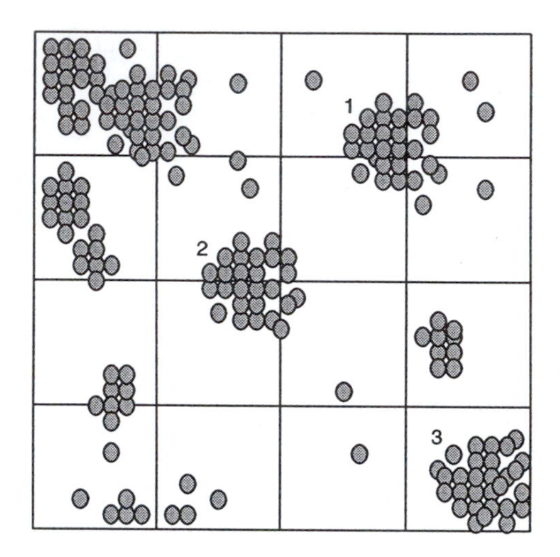

FIGURE 4.9 Schematic layout of the data points on the grid after application of the LF algorithm. Clusters numbered 1, 2, and 3 correspond to those described in the text.

significance, suggesting that the algorithm has been able to find interesting underlying common properties. In addition to what is shown here, they implemented their system in such a way that the user could, by the click of a mouse, access the information associated with any point in the representation plane and move it to another location.

Despite interesting results, it is not obvious that this algorithm has a future. It does not appear to be very efficient in terms of computation time. It does, however, provide the user with a simple way of implementing multidimensional scaling. A systematic comparison between the LF algorithm and other algorithms that perform the same type of task should be undertaken, with respect to both computation time and quality of solutions.

4.5 GRAPH PARTITIONING

Kuntz and Layzell [210] and Kuntz, Layzell, and Snyers [211] (hereafter KLS) have proposed an interesting application of Lumer and Faieta's [225] idea to the problem of graph partitioning as an alternative to other methods such as quadratic placement (QP) and multidimensional scaling (MDS). Whereas classical graph partitioning methods work directly on the graph, the KLS algorithm, like QP and MDS, considers the problem of graph partitioning through embedding into a euclidian space. Such approaches provide a geometric representation of the graph, so that visualization is enhanced if the representation is low dimensional. In addition, euclidian metrics can be used, and natural clusters in the graph can be detected.

All these features can be of particular importance in the context of computer-aided design (CAD), where space is obviously relevant at many levels, from transistors to logical gates to higher-level components. KLS's method is an original way of finding a reasonably good representation of a graph, where natural components or clusters in the graph emerge as a result of local actions taken by agents—the simple agents that we described in the previous sections. Indeed, the KLS algorithm can also be seen as a graph-drawing algorithm [307]. Complex networks of relationships, which arise in numerous disciplines, can often be represented by graphs. Such graphs can be weighted, possibly with negative weights, directed or undirected, etc. It is usually convenient to be able to visualize graphs in the form of diagrams, preferably embedded in a two-dimensional space, where it's easier for researchers to interpret. Automated procedures for drawing graphs are extremely useful, especially for large graphs. Many graph-drawing algorithms have been designed in the past three decades [307], but few of those algorithms deal with arbitrary graphs [192]. As we shall see, the KLS algorithm cannot be used to draw arbitrary graphs, but only graphs that contain natural clusters.

Let $G = (V, E)$ be a graph, where $V = \{v_i, i = 1, \ldots, N\}$ is the set of N vertices and E the set of edges. We assume that G is nondirected, that is, E is symmetric. Let the adjacency matrix be denoted by $A = [a_{ij}]; a_{ij} \neq 0$ if and only if $(v_i, v_j) \in E$. We shall only treat the cases where $a_{ij} = 0$ or 1, which correspond to $(v_i, v_j) \notin E$ and $(v_i, v_j) \in E$, respectively. The case of weighted edges can be readily derived. Let δ_i be the degree of vertex i, that is, the number of edges that connect to v_i. More generally, $\delta_i = \sum_j a_{ij}$. Let δ be the diagonal matrix, the elements of which are the δ_i's.

Each vertex of the graph is initially assigned a set of random coordinates in $\mathcal{R}^n$, and the task to be performed by the algorithm is to change the vertices' coordinates in $\mathcal{R}^n$ so that

1. clusters present in the graph are located in the same portion of space,
2. intercluster edges are minimized, and
3. different clusters are clearly separated.

The precise formulation of the usual problem of graph partitioning is to find k nonempty components such that the total intercomponent weight is minimized. For arbitrary k, the decision problem is NP-complete, and for fixed k, a solution can be found in $O(|V|^{k^2})$ steps, where $|V|$ denotes the number of elements in V. If the additional constraint of having the same number of vertices in every cluster is added, the problem is NP-complete for fixed k [142].

It may also be useful to see the KLS algorithm as an approach to graph drawing, or graph representation. In this case, criteria are more difficult to define because aesthetic factors play a role: a graph representation must not only allow for the easy extraction of the characteristic features of the graph, it must also be pleasing to the eye. For example, minimizing the number of edge crossings in a planar embedding of a graph can be an objective of the graph-drawing algorithm, but it does not always lead to representations that highlight the graph's properties [193]. This difficulty explains why general graph-drawing algorithms are rare.

Let $(x_{im})_{m=1,\ldots,n}$ be the coordinate of vertex i in $\mathcal{R}^n$, and $X = [x_{im}]_{(i=1,\ldots,N)(m=1,\ldots,n)}$ the $N \times n$ matrix describing the coordinates of all vertices. One interesting formulation of the graph-partitioning problem involves a quadratic cost function C (quadratic placement), where the problem is to minimize

$$C = \sum_{i=1}^{N} \sum_{j=1}^{N} \sum_{m=1}^{n} a_{ij}(x_{im} - x_{jm}) \tag{4.8}$$

under the constraint that $X^t X = I_n$, where I_n is the identity matrix in $\mathcal{R}^n$. This constraint is a way of avoiding the collapse of all points onto a single coordinate. It can be shown that this problem is equivalent to minimizing $tr(X^t Q X)$, where $Q = \Delta - A$ is called the laplacian matrix of A and $tr(A)$ is the trace of A, still under the constraint that $X^t X = I_n$. This method then relies heavily on matrix diagonalization, which can be costly computationally.

The KLS heuristic method is based on the idea of Lumer and Faieta [225]. Here, o_i is replaced by v_i, and the dissimilarity function $d(o_i, o_j)$ by a distance $d(v_i, v_j)$ to be defined. As was the case with LF, the embedding is into a grid, that is, a subspace of Z^2. The ant-based clustering model is essentially a dynamical system where vertices move in the plane and are attracted to or rejected by clusters: one has to choose a distance such that the system will converge toward a good representation of the graph or a reasonable solution of the partitioning problem, with k unknown. One such distance is

$$d(v_i, v_j) = \frac{|D(\rho(v_i), \rho(v_j))|}{|\rho(v_i)| + |\rho(v_j)|}, \tag{4.9}$$

where $\rho(v_i) = \{v_j \in V; a_{ij} \neq 0\} \cup \{v_i\}$ is the set of vertices adjacent to v_i, including v_i, D is the symmetric difference of two sets $D(A, B) = (A \cup B) - (A \cap B))$, and $|\ldots|$ denotes the number of elements of a set. When two vertices v_i and v_j have a lot of common adjacent nodes, $(\rho(v_i) \cup \rho(v_j)) - (\rho(v_i) \cap \rho(v_j))$ is a small set, and $d(v_i, v_j)$ is small: this is good, because v_i and v_j should be represented by close locations in the embedding space. When two vertices v_i and v_j have few or no adjacent edges in common, $(\rho(v_i) \cup \rho(v_j)) - (\rho(v_i) \cap \rho(v_j))$ is a large set, and $d(v_i, v_j)$ is large: the points representing v_i and v_j in the embedding space should be further apart. The denominator of the righthand term of Eq. (4.10) is a kind of "normalization" factor that weighs the contribution of the symmetric difference by the total number of vertices that are adjacent to the two vertices, so that d always lies between 0 and 1. For example, $d(v_i, v_j) = 1$ when the two vertices are not adjacent and have no adjacent edge in common: this is because $D(\rho(v_i), \rho(v_j)) = \rho(v_i) \cup \rho(v_j)$ and $|\rho(v_i)| + |\rho(v_j)| = |\rho(v_i) \cup \rho(v_j)|$. At the other extreme, $d_1(v_i, v_j) = 0$ if the two vertices are adjacent and have all their adjacent vertices in common: this is because $\rho(v_i) \cup \rho(v_j) = \rho(v_i) \cap \rho(v_j)$. In the case where $a_{ij} = 0$ or 1, $d(v_i, v_j)$ is simply a "normalized" version of the L_1 distance $d_{L_1}(v_i, v_j)$ between two nodes in the graph:

$$d(v_i, v_j) = \frac{\sum_{k=1}^{n} |a_{ik} - a_{jk}|}{\sum_{k=1}^{n} |a_{ik}| + \sum_{k=1}^{n} |a_{jk}|} = \frac{d_{L_1}(v_i, v_j)}{\sum_{k=1}^{n} |a_{ik}| + \sum_{k=1}^{n} |a_{jk}|} \tag{4.10}$$

Similarly to the LF algorithm, one has to define probabilities of picking up and depositing items, here vertices. Let us assume that an agent is located at site r at time t and finds a vertex v_i at that site. The local density with respect to vertex v_i at site r is given by

$$f(v_i) = \begin{cases} \frac{1}{s^2} \sum_{v_j \in \text{Neigh}_{(s\times s)(r)}} \left[1 - \frac{d(v_i, v_j)}{\alpha}\right] & \text{if } f > 0\,, \\ 0 & \text{otherwise}\,. \end{cases} \tag{4.11}$$

$f(v_i)$ is a measure of the average distance within the graph of vertex v_i to the other vertices v_j present in the Z^2 neighborhood of v_i, α defines the scale for dissimilarity and is set to 1 in our simulations. Let us briefly examine the behavior of f by looking at the extreme cases. If, for example, all s^2 sites around r are occupied by vertices that are connected to v_i and have all their adjacent vertices in common with $v_i(\forall v_j \in \text{Neigh}_{(s\times s)}(r), d(v_i, v_j) = 0)$, then $f(v_i)$ is equal to 1 and the vertex should not be picked up. If all s^2 sites around r are occupied by vertices which are not at all connected to v_i or to one of v_i's neighbors, then $f(v_i) = 0$, and the vertex should be picked up. If all sites around r are empty, then, obviously, $f(v_i) = 0$, and the object should be picked up with high probability. Picking up and dropping probabilities of vertex v_i are defined as follows:

$$p_p(v_i) = \left(\frac{k_1}{k_1 + f(v_i)}\right)^2 , \tag{4.12}$$

$$p_d(v_i) = \left(\frac{f(v_i)}{k_2 + f(v_i)}\right)^2 , \tag{4.13}$$

where k_1 and k_2 are two constants similar to k_1 and k_2 in the BM and the LF algorithm.

The KLS algorithm was tested on random graphs $\Gamma(n, c, p_i, p_e)$ [141], which have been used in the context of VLSI to test iterative improvement partitioning algorithms. Such graphs are composed of c clusters of n vertices, p_i is the probability that two vertices within a cluster are connected, and p_e is the probability that two vertices that belong to different clusters be connected. Figure 4.10 illustrates the fact that the KLS algorithm is able to find "natural" clusters in a graph $\Gamma(25, 4, 0.8, 0.01)$. Figure 4.10(a) shows the initial random distribution of nodes on the plane. Figure 4.10(b) shows that, after $T = 2 \cdot 10^6$ steps, vertices distribute themselves in space in such a way that exactly $c = 4$ clusters of vertices appear, which correspond to the four clusters of the graph. The positions of vertices are arbitrary on the plane.

The quality of the solution found by the KLS algorithm can be evaluated if one considers the algorithm a mapping from graph space onto the plane: we expect

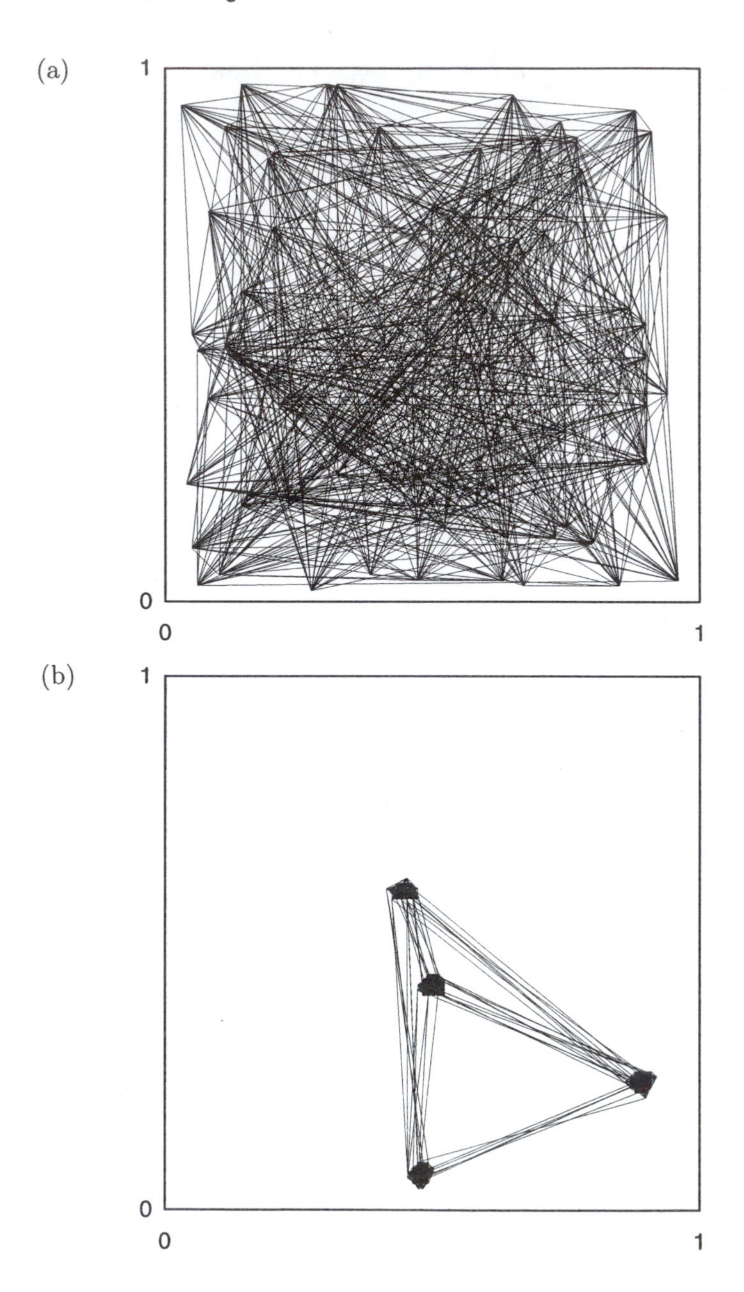

FIGURE 4.10 (a) Initial distribution of vertices on the portion of plane $[0,1] \times [0,1]$, for a random graph $\Gamma(25, 4, 0.8, 0.01)$. (b) The algorithm, with 10 agents, is able to find "natural" clusters in this graph. Vertices distribute themselves in space in such a way that exactly $c = 4$ clusters of vertices appear, which correspond to the 4 clusters of the graph. $k_1 = 0.03$, $k_2 = 0.1$, $\alpha = 1$, $s^2 = 9$, $T = 2 \cdot 10^6$ steps.

Algorithm 4.2 High-level description of the KLS algorithm

```
/* Initialization */
For every vertex v_i do
   place v_i randomly on grid
End For
For all agents do
   place agent at randomly selected site
End For
/* Main loop */
For t = 1 to t_max do
   For all agents do
      If ((agent unladen) and (site occupied by vertex v_i)) then
         Compute f(v_i) and p_p(v_i)
         Draw random real number R between 0 and 1
         If (R ≤ p_p(v_i)) then
            Pick up item v_i
         End If
      Else If ((agent carrying item v_i) and (site empty)) then
         Compute f(v_i) and p_d(v_i)
         Draw random real number R between 0 and 1
         If (R ≤ p_d(v_i)) then
            Drop item
         End If
      End If
      Move to randomly selected neighboring site not occupied by other agent
   End For
End For
Print location of vertex
Draw edges between vertices
/* Values of parameters used in experiments */
k_1 = 0.3, k_2 = 0.1, α = 1, s^2 = 9, T = 2.10^6 steps.
```

nodes which are far apart in the graph to lie far away from each other on the plane, and vice versa. In a random graph of the type studied here, we expect to have two characteristic distances in the graph: the distance between vertices that belong to the same cluster and the distance between vertices that do not belong to the same cluster. Let w_i be the coordinates in the grid of vertex v_i. Figure 4.11(a) shows $||w_i - w_j||$ as a function of $d(v_i, v_j)$ for all pairs of vertices for the same graph and the same distribution of vertex positions in the plane as in Figure 4.10(a): the two expected clouds of points are observed, and there is, of course, clearly no correlation between $||w_i - w_j||$ and $d(v_i, v_j)$. Figure 4.11(b) shows the same relationship after $T = 2 \cdot 10^6$ time steps (vertex positions are those of Figure 4.10(b)): the distance between the positions of vertices in the plane representation is an increasing function of their distance in the graph, with two clouds of points corresponding to the two characteristic distances in the graph. The algorithm is able to generate a good mapping of the graph onto the plane.

At the time this section was being written, no quantitative comparision had been made with other graph partitioning methods or graph-drawing algorithms, which may perform faster and/or converge toward higher-quality solutions. The KLS algorithm may turn out to be "no more" than a proof-of-concept that it is possible to solve problems with a swarm-based approach (as many examples in this book may turn out to be), but, even so, it is an extremely valuable example, and may even be able to find original solutions that other algorithms have been unable to find.

It would also be useful to study this algorithm in the context of either ratio cut partitioning and clustering [163] or edge crossing minimization [146]. The ratio cut partitioning problem [163] is an extension of the partitioning problem that allows for some level of imbalance between clusters, so that the best solution is the one that offers the best tradeoff between total intercluster cut and cluster balance. Let us consider, for example, the bipartitioning problem, where given a graph $G = (V, E)$, one has to find a partition of V into two disjoint clusters U and W such that $|U| = |W|$ and the total edge weight $e(U, W)$ between U and W is minimized. The associated ratio cut problem consists of finding a partition of V into two disjoint clusters U and W such that $e(U, W)/|U||W|$ is minimized. It is intuitively clear that the ratio cut cost function provides the freedom to find more "natural" partitions in graphs by relaxing the constraint of perfect balance. This problem, despite this additional freedom, is still NP-complete, as can be shown by reduction from the problem of bounded min-cut graph partitioning [145].

The KLS algorithm may be relevant to the ratio cut problem because its underlying self-organizing dynamics makes it more likely to converge toward a solution that is "natural" with respect to the graph being partitioned or clustered. Let us introduce the family of graphs $\Gamma(P_m^n, c, p_i, p_e)$, which are identical to graphs of the type $\Gamma(n, c, p_i, p_e)$ except for the number of vertices per cluster, m which is a random variable that follows a discrete probability distribution P_m^n the expectation of which is given by $\sum_{m=1}^{\infty} mP_m^n = \langle P_m^n \rangle = n$. Such graphs are characterized by clusters of m vertices, where m fluctuates around its average value n. Whe applied to such graphs, KLS's algorithm finds a two-dimensional representation of the graph where clusters are still clearly separated despite their different sizes. Figure 4.12 shows an example of a graph representation found by the algorithm for a graph comprising four clusters of 25 vertices on average, the probability distribution being gaussian. This indicates that the algorithm may certainly be more appropriate for ratio cut clustering.

Finally, the KLS algorithm is particularly useful when there actually are clusters in a graph and can be used as an extraction tool. However, it may not perform well, either in terms of representation or of partitioning, on other types of graphs where there are no obvious clusters. Geometric graphs illustrate this limitation of the algorithm well. In geometric graphs, which are used in the context of numerical modeling, vertices, characterized by euclidian coordinates, are located on a plane

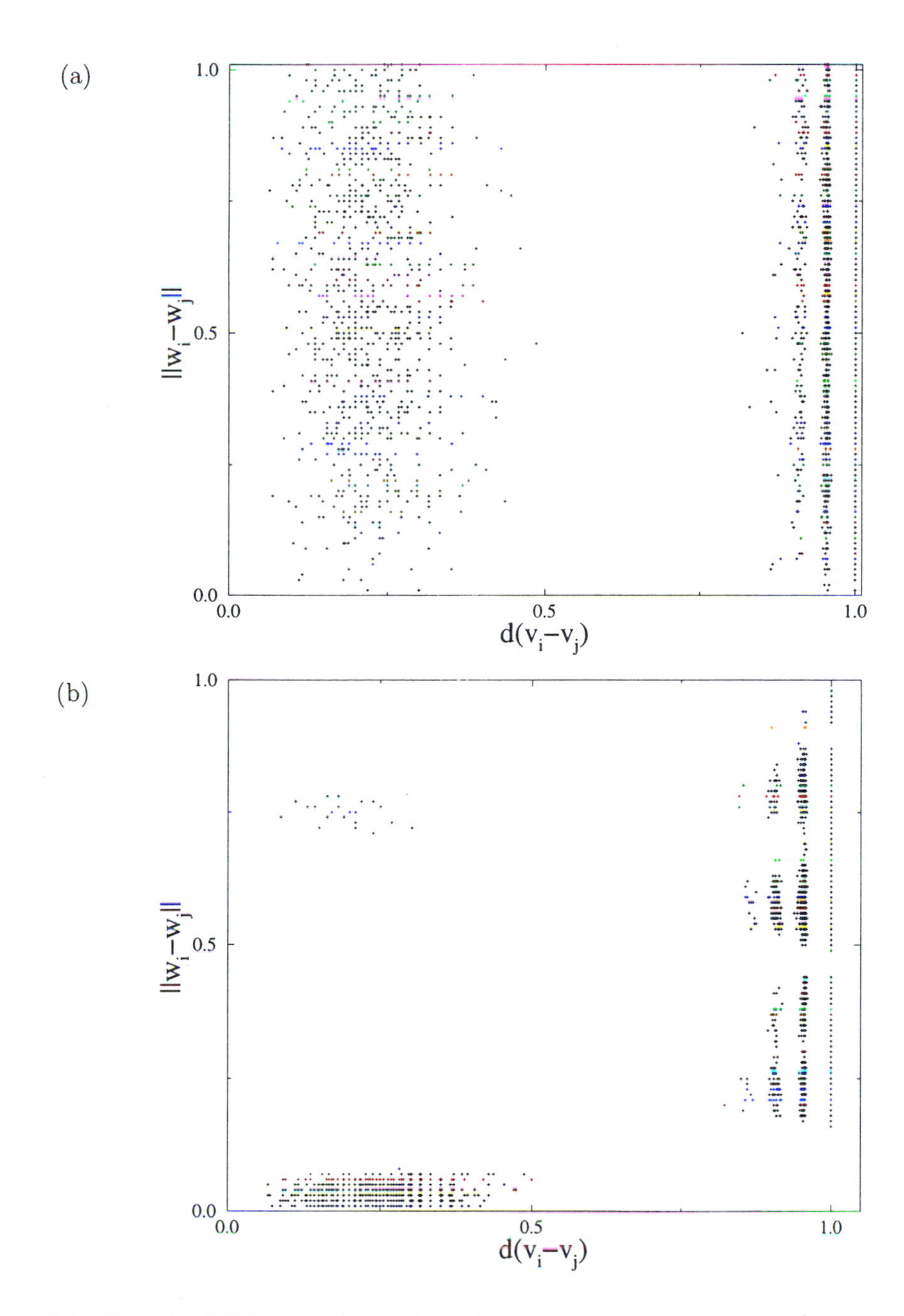

FIGURE 4.11 (a) $||w_i - w_j||$ vs. $d(v_i, v_j)$ for the graph representation of Figure 4.10(a). $||\ldots||$ is the L_1 norm in the plane, and $d(\ldots)$ is the (normalized) L_1 norm for graphs. (b) $||w_i - w_j||$ vs. $d(v_i, v_j)$ for the graph representation of Figure 4.10(b).

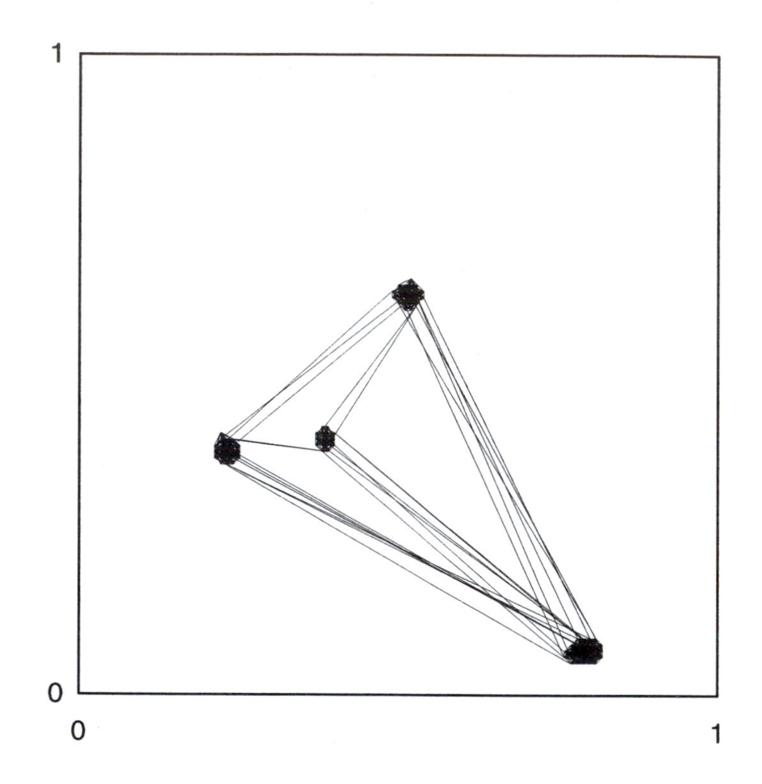

FIGURE 4.12 Representation of a graph $\Gamma(P_m^n \propto e^{-(m-n)^{2/49}}, c = 4, p_i = 0.1, p_e = 0.01)$ found by KLS's algorithm, $n = 25$.

and are locally connected to other vertices that lie within a radius of connectivity R. Figure 4.13 shows an example of a geometric graph with 200 vertices randomly distributed in a $[0, 1] \times [0, 1]$ square area. Vertices are connected to all vertices that lie within a disc of radius $R = 0.18$. Figure 4.14 shows the spatial distribution of vertices and their connections found by the KLS algorithm after $T = 5 \cdot 10^6$ time steps. It is clearly difficult to say that the algorithm has been able to extract the structure of the geometric graph.

4.6 ROBOTIC IMPLEMENTATIONS

4.6.1 INTRODUCTION

Although designing robots that implement the process of distributed clustering may not be particularly useful in itself in the context of engineering, several groups have done it [12, 147, 148, 235, 305], because it is *conceptually* useful. Showing that it

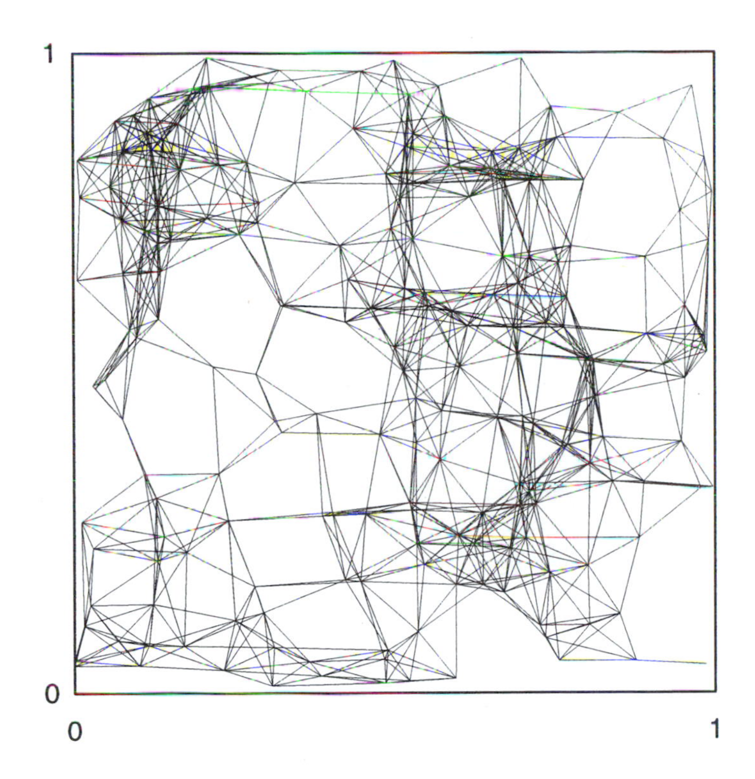

FIGURE 4.13 Randomly generated geometric graph of 200 vertices in a square area of linear size 1, where the radius of connectivity is $R = 0.18$ (L_1 norm in the plane).

is possible to design a collection of extremely simple robots that are capable of clustering items in a decentralized way on the basis of purely local perception is an extremely important step. Of course, ultimately, one wishes to see more complex applications of swarm intelligence in the field of robotics, and indeed, collective robotics is flourishing (see, for example, Cao et al., [61]).

4.6.2 EXAMPLE

Among the first to design a collection of robots to implement distributed clustering are Beckers and his colleagues (Beckers et al., [12]). The problem was to design simple robots that could displace items (pucks). Their robots are equipped with C-shaped grippers with which they can push items. A gripper is also used as a mechanical sensor to evaluate resistance to pushing. This mechanism allows the local density of pucks to be measured: high resistance means high local density, whereas

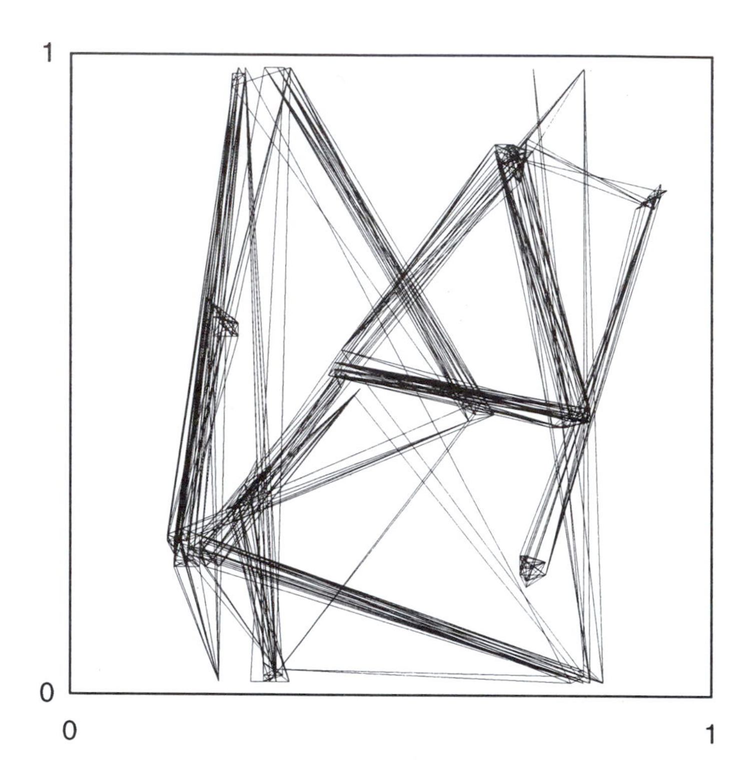

FIGURE 4.14 Representation of the geometric graph of Figure 4.13 found by KLS's algorithm. $k_1 = 0.03$, $k_2 = 0.1$, $\alpha = 1$, $s^2 = 9$, $T = 5 \cdot 10^6$ steps.

weak resistance means low local density. The gripper activates a microswitch when the perceived resistance to pushing exceeds a threshold.

Each robot is equipped with two infrared sensors for obstacle avoidance, and a microswitch which is activated by the gripper when a certain number of pucks to be pushed is exceeded (here, three pucks are sufficient to trigger the microswitch). Many robotic systems of this type rely on a common principle, introduced by Brooks [44, 45]: the subsumption architecture. The idea is to decompose any behavior into atomic, hierarchically organized subbehaviors which can be activated when necessary. Rather than modularize perception, world modeling, planning, and execution, this approach builds control systems in which many individual modules (that incorporate their own perceptual, modeling, and planning requirements) directly generate some part of the behavior of the robot. An arbitration scheme, built within the framework of the modules, controls which behavior-producing module has control of which part of the robot at any given time. All possible combinations of atomic subbehaviors can generate many robot behaviors. This recent approach

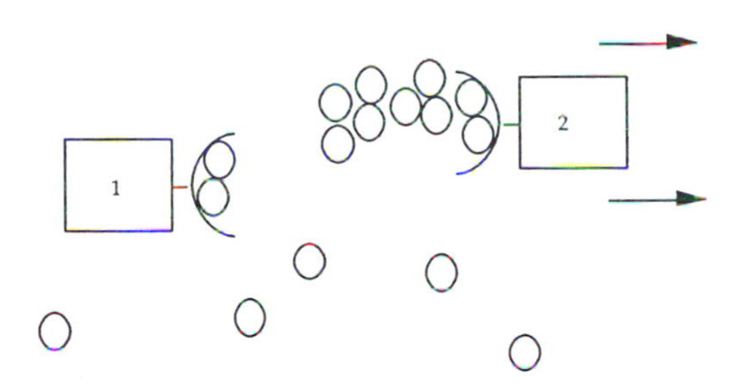

FIGURE 4.15 Robot 1, because it has only two pucks in its C-shaped gripper, continues to move in a straight line. Robot 2, with three pucks in its gripper, is encountering a cluster: its microswitch is activated by the gripper, so that the robot leaves the pucks next to the cluster.

in robotics can be opposed to a more classical one, based not on behavioral modules but on reasoning steps and functional modules. The subsumption approach relies on a strong assumption: that it is possible, with an appropriate combination of stimuli-responses, to solve most simple robotic problems. This idea is central to the reactive approach to robotics.

Here, there are three possible behaviors for the robots, and only one can be active at any time:

1. A robot moves in a straight line if no sensor is activated, until an obstacle is detected or the microswitch is activated. Notice that pucks are not detected as obstacles because they are too low. Moving in a straight line is the default behavior.
2. When the robot detects an obstacle, it turns away from the detected obstacle by a random angle and starts moving in a straight line in the new direction. Thanks to the C-shaped gripper, any puck that is carried by the robot when it is executing its obstacle avoidance behavior is retained in the gripper throughout the turn (Figure 4.15).
3. When the robot pushes three or more pucks (either when there are three or more pucks in the gripper, or when the robot bumps into a cluster of three or more pucks), the microswitch is activated by the gripper: this triggers a reversal of the robot's motors for one second, so that the robot steps back and then executes a turn by a random angle before moving in a straight line in the new direction. The pucks have been "dropped" at a location of perceived high puck density.

The obstacle-avoidance behavior has priority over the "puck-dropping" behavior, which itself has priority over the default behavior. Beckers et al. [12] have reproduced the spatial clustering process of Deneubourg et al. [88] with groups of robots comprised of one to five robots.

Beckers et al.'s [12] robotic experiments have been performed with 81 pucks regularly spaced in an experimental square arena (250×250 cm^2). Robots are initially placed in the center of the arena. Every ten minutes of run time, the robots are stopped manually, for the positions of the pucks and the sizes of clusters to be recorded (a cluster being defined as a set of pucks separated by no more than one puck diameter). Three successive phases are observed, irrespective of the number of robots:

1. In the early stage of the clustering process, robots move, collect pucks and drop them when three pucks have been collected, which leads to the formation of miniclusters comprised of three pucks. Such miniclusters rapidly become large enough not to be easily destroyed.
2. During the second phase, the robots remove a few pucks from existing clusters by striking them at an appropriate angle, and depositing the pucks next to other clusters. Some clusters rapidly grow during this phase, leading to the formation of few relatively large clusters.
3. The third phase is the one that has the longest duration. From time to time, a puck is removed from a large cluster and added to another large cluster, or, most often, to the same cluster. This competition among clusters eventually leads to the formation of a single cluster. The equilibrium state is a dynamic one, since the robots continue to remove and add pucks to this last cluster. It is however surprising that the system always converges toward a single cluster. Large clusters are more likely to lose pucks, but at the same time, they are also more likely to receive pucks. This latter process seems to be dominant, which intuitively explains the convergence property.

This convergence to a single large cluster is surprising, and relies on a sophisticated implementation of stigmergy: in effect, a puck can be withdrawn from a cluster by a robot only when the robot is approaching the cluster tangentially, whereas a frontal approach rather tends to trigger a deposition behavior. This indicates that cluster geometry is extremely important in the clustering process, and obviously additions and removals of pucks influence cluster geometry. The stigmergic process is, therefore, more complex than one might think.

The clustering process has also been tested with an initial seed, that is, an initial cluster. In fact, in order to prevent the seed cluster from being broken up during the early stage of clustering, Beckers et al. [12] used a large saucer. They found that three robots were able to form a single cluster around the saucer, thereby showing that robots can be induced to form a cluster at a predefined location. The time taken to converge toward this cluster induced by the seed is, however, longer (126 min) than the time it takes to reach a single cluster without a seed (105 min).

Finally, Beckers et al. [12] studied the performance of the group of robots as a function of group size (Figure 4.16). In their setup, they found that the optimal number of robots, with respect to both average convergence time and efficiency (expressed in robot $\cdot$ minutes required to converge), is three. A degradation of

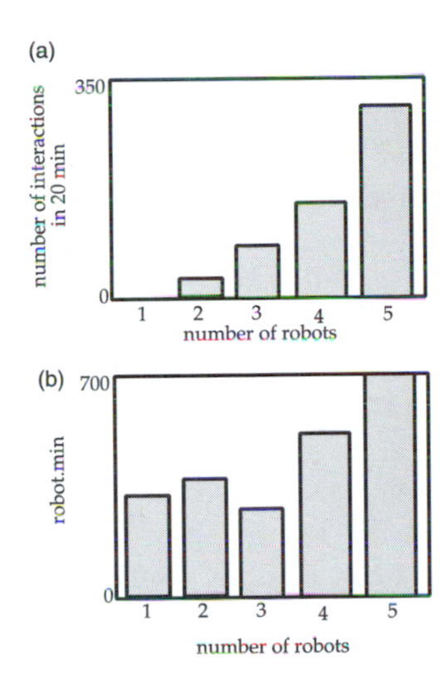

FIGURE 4.16 (a) Number of interactions (avoidance) among robots during a 20 minute experiment as a function of the number of robots. (b) Efficiency, expressed in robot · minutes, as a function of the number of robots: a group of three robots seems to be optimal for this particular setup [12].

the performance is observed for more than three robots, because of the increasing number of useless interactions among robots. For up to three robots, the robots work in parallel, largely unaffected by each other's behavior, except for indirect interactions through modifications of clusters, whereas for more than three robots, they start to interact directly. Not only do robots lose time when they encounter and have to avoid each other, they can also lose their pucks in the avoidance process (although this should be rare, thanks the C-shaped gripper), and may destroy existing clusters in the process of turning away from each other, for the avoidance behavior has priority over the puck dropping behavior.

4.6.3 OTHER SELECTED WORKS

As mentioned previously, several other groups have worked on the same problem. For example, Gaussier and Zrehen [147, 148] have used a slightly more sophisticated "gripper" and a more complicated puck-dropping behavior, but the principles of their robotic implementation, based on the Khepera robot (a miniature mobile robot designed to perform "desktop" experiments [235, 236, 241]), are basically identical to those of Beckers et al. [12]. A hook mounted on the robot's back allowed pucks to be grasped and shifted, with the number of pucks carried by the hook triggering picking up and dropping behaviors in a way similar to Beckers et al. [12]. A preprogrammed Winner-Take-All-type neural net is used to implement the perception of items and obstacles, and picking up and dropping behaviors are also preprogrammed so as to favor cluster building rather than cluster destruction. Although this robotic implementation of collective clustering was the first or one of the first with that of Beckers et al. [12], Gaussier and Zrehen [147, 148] did not attempt to evaluate the performance of their system, which is why we have presented the work of Beckers et al. [12] in more detail.

Martinoli et al. [235, 236] have performed more experiments of the same type, refining in particular the communication architecture among teammates (with a possible "supervisor" unit) to avoid the problem of communication bottleneck in larger teams of robots, and introducing the idea of programmable "active" seeds, that is, pucks that actively emit signals or actively respond to signals emitted by robots. Active pucks extend the sensorial capabilities of the robots without adding to the robots themselves. The grippers of individual robots are also more sophisticated (Figure 4.17). When a robot's sensors are activated by an item, the robot either avoids it if it detects a large obstacle, such as another robot, a wall or a cluster of seeds, or, if currently not carrying a seed, picks up the item with its gripper. Then, in either case, it turns by an angle of about 180 degrees. In addition, robots are equipped with IR reflecting bands, so that their size as perceived by other robots' proximity sensors is increased, thereby facilitating early recognition by the other robots. Experiments were performed with one to five robots equipped with grippers, and 20 pucks initially randomly scattered in a 80×80 cm^2. Figure 4.18 shows the mean cluster size as a function of time for group sizes from 1 to 5. The implementation of a special supply floor, whereby robots can transfer energy from an external supply by electrical contact regardless of their positions, allows study of the clustering process over time scales of several hours. In three two-hour experiments with two robots, it is found that cluster size does not vary a lot after 80 minutes. Figures 4.19(a) and 4.19(b) show typical initial and final (after two hours with two robots) arrangements of pucks in the experimental arena.

4.6.4 FROM CLUSTERING TO SORTING

Melhuish et al. [237] recently went one step further than clustering. They show that it is possible to perform a rough type of sorting, or, more precisely, segregation, with a simple extension of the robotic implementations decribed in sections 4.6.2 and 4.6.3. They use two types of items—red and yellow frisbees. Picking up and dropping behaviors are essentially identical for both types of frisbees, and are similar to the behaviors described in section 4.6.2, except for one aspect: when dropping a yellow frisbee, a robot pulls the frisbee backwards for some distance before releasing it. In other words, yellow frisbees, and only yellow frisbees, are not deposited hard up against other frisbees, but some little way off. This mechanism may be a good model of differential space requirements of the different brood types in ants. This simple addition seems to be sufficient to generate a segregation of yellow and red frisbees, with red frisbees in the center of the cluster and yellow frisbees at the periphery (Figure 4.20). This distribution is similar to the one found in *Leptothorax* ants

FIGURE 4.17 Three Khepera robots equipped with different combinations of modules (from left to right): a robot with a gripper and IrDA modules, a robot with IrDA and radio modules, and a robot with a gripper and radio modules. Three active seeds are also represented. The robots lie on a supply floor board developed by Martinoli et al. [236], whereby robots can take advantage of the external supply source regardless of their positions. The IrDA and radio modules correspond to different communications tools. See Martinoli et al. [236] for a more detailed description. Reprinted by permission.

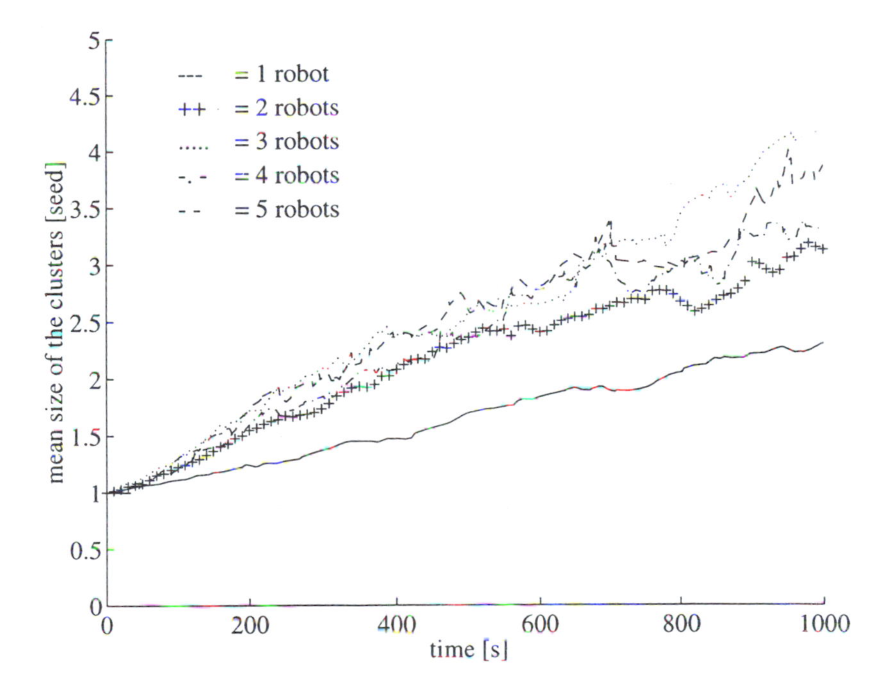

FIGURE 4.18 Average cluster size (over five experiments) as a function of time for robot group sizes from 1 to 5. After Martinoli et al. [236]. Reprinted by permission.

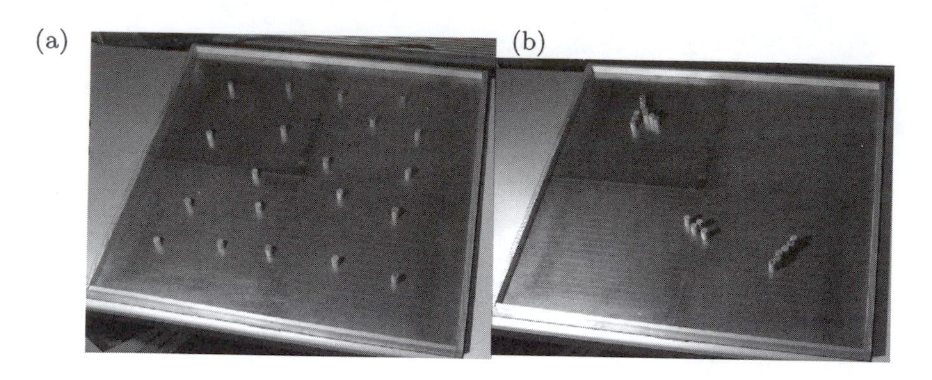

FIGURE 4.19 (a) Initial configuration of 20 pucks in the experimental arena. After Martinoli et al. [236] (b) Configuration of pucks after two hours with two robots. After Martinoli et al. [236]. Reprinted by permission.

(Figure 4.2), where brood items that have larger space requirements are located at the periphery of the brood pile.

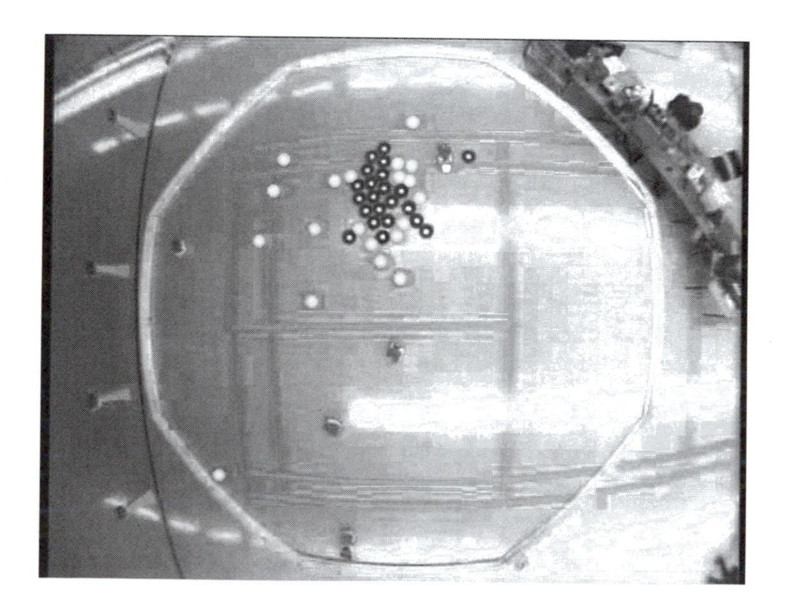

FIGURE 4.20 Configuration of red and yellow frisbees obtained by Melhuish et al. [237] after 8 hours and 35 minutes, using their pullback algorithm.

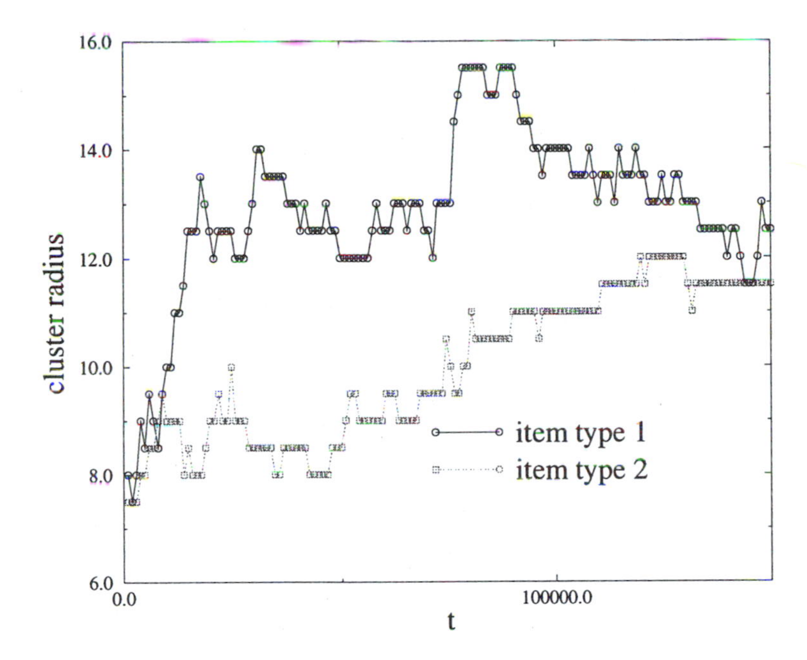

FIGURE 4.21 Average cluster sizes as a function of time when two types of items are sorted. For type 1 items: $k_1 = 0.1$, $k_2 = 0.15$; for type 1 items: $k_1 = 0.03$, $k_2 = 0.15$.

4.6.5 A PROPOSAL

Given the current stage of collective robotics, it is certainly possible to envision a distributed robotic implementation, in the context of clustering, of the models of task succession presented in section 3.3.6, especially model (1). We propose that such an implementation could be the next step in the design of swarm-based robotic systems because it would constitute a significant leap forward in the functional complexity of the system. The robotic implementation can be sketched as follows (we leave the details of the implementation to the imagination of roboticists): assume that several types of pucks are present in the arena, each type being characterized by some property that is recognizable by the robots. Such a property can be weight, color, or, in the case of active seeds, radio frequency and so forth. The robots may be programmed with different, but nonzero, probabilities of picking up the various items. If the probabilities are significantly different, it is likely that we will observe the emergence of clusters of the most easily picked up items, followed by the formation of clusters of items in the second category, etc. In other words, sequential cluster formation should be observed, or, if robots are programmed to take items back to their base, sequential processing of the various types of items. This corresponds to model 1 in section 3.3.6.

Model 2 could also be implemented. Figure 4.21 shows the dynamics of average cluster sizes when two types of items are involved: a type 1 item is more likely to be picked up when isolated than a type 2 item; probabilities of being dropped are identical for both types of items. It can be seen that the average cluster size of type 1 items increases more rapidly than that of type 2 items, indicating that cluster formation proceeds sequentially: when clusters of type 1 items have been formed, type 2 items are available for pick up.

4.7 POINTS TO REMEMBER

- Corpse aggregation and brood sorting are commonly observed behaviors in ants, but they have been studied in few species. A simple agent-based model can account for the formation of clusters which contain only one type of item: an item is dropped by a laden agent when the agent is surrounded by items of the same type; an item is picked up by an unladen agent when the agent perceives no (or few) item(s) of the same type in its neighborhood. In this model, two items are either similar or different: in other words, the dissimilarity measure between two items can take only two values, 0 or 1.
- The agent-based model can be extended to deal with items which have variable attributes in a possibly continuous, possibly high-dimensional, space. The dissimilarity measure between two items can take any positive value. An item is dropped by a laden agent when the agent is surrounded by items which are similar (with respect to the dissimilarity measure) to the item it is carrying; an item is picked up by an unladen agent when the agent perceives items in its neighborhood which are dissimilar from the item to be picked up. This extended model can be applied to map high-dimensional data onto a two-dimensional representation.
- Nodes in a graph are particular examples of objects with high-dimensional attributes. When the extended model is applied to the nodes of a graph, the agents sort the nodes in such a way that clusters of nodes in the graph translate into clusters of points in the representation space. This application can be used to partition or draw graphs.
- The clustering model has been used to test the feasibility of swarm-based robotic systems. All robotic implementations are based on local perception and action. Interference among robots may become a problem if there are too many robots in a small area. Although one recent example of sorting by a swarm of robots has been described, swarm-based robotics remains confined to apparently trivial tasks.

CHAPTER 5

Self-Organization and Templates: Application to Data Analysis and Graph Partitioning

5.1 OVERVIEW

The biological phenomena described in the previous chapter were corpse aggregation and brood sorting by ants. The clusters of items obtained with the models introduced in sections 4.3.1 and 4.3.2 emerged at *arbitrary* locations. The underlying self-organizing process, whereby large clusters grow even larger because they are more attractive than smaller clusters, does not ensure the formation of clusters at specific locations.

In the two biological examples described in this chapter, the self-organizing dynamics of aggregation is constrained by *templates*. A template is a pattern that is used to construct another pattern. The body of a termite queen or a brood pile in ants are two examples of structures—the second one resulting from the activities of the colony—that serve as templates to build walls. Walls built around the termite queen form the royal chamber; walls built around the brood pile form the ant nest.

When a mechanism combines self-organization and templates, it exhibits the characteristic properties of self-organization, such as snowball effect or multistability, and at the same time produces a perfectly predictable pattern that follows the template.

The two nonparametric algorithms presented in chapter 4, one for multidimensional scaling and the other for graph partitioning, can be made parametric through the use of templates. The number of clusters of data points or vertices can be predefined by forcing items to be deposited in a prespecified number of regions in the space of representation, so that the number of clusters and their locations are known in advance.

5.2 THE INTERPLAY OF SELF-ORGANIZATION AND TEMPLATES

In the previous chapter, we saw how the attractivity of corpses or the differential attractivity of items of different types could lead to the formation of clusters of specific items. Self-organization lies in this attractivity, which induces a snowball effect: the larger a cluster, the more likely it is to attract even more items. But self-organization can also be combined with a template mechanism in the process of clustering. A template is a kind of prepattern in the environment, used by insects—or by other animals—to organize their activities. For example, in the context of building, the shape to be built is predefined by the prepattern, and insects build along the prepattern. This prepattern can result from natural gradients, fields, or heterogeneities that are exploited by the colony. Many ant species (for example, *Acantholepsis custodiens* (Figure 5.1) [43], *Formica polyctena* [63], and *Myrmica rubra* [64]) make use of temperature and humidity gradients to build their nests and spatially distribute their brood.

The prepattern can also be the body shape of an animal, as illustrated by the construction of the royal chamber in termites (Figure 5.2). The physogastric queen of *Macrotermes subhyalinus* emits a pheromone that diffuses and creates a pheromonal template in the form of a decreasing gradient around her (Figure 5.3) [48]. It has experimentally been shown that a concentration window exists, or a threshold, that controls the workers' building activities: a worker deposits a soil pellet if the concentration of pheromone is within this window or exceeds the threshold. Otherwise, workers do not deposit any pellet, or even destroy existing walls. If one places a freshly killed physogastric queen in various positions, walls are built at a more or less constant distance from the queen's body, following its contours, while a wax dummy of the queen does not stimulate construction. What defines the pattern that will be built is the queen's building pheromone which creates a chemical template. Building following this template ensures that the size of the royal chamber is adjusted to the size of the queen.

If the template formed by the queen's body determines the overall shape of the chamber, other factors play a role: tactile stimuli and pheromones other than the queen's building pheromone, such as cement and trail pheromones, facilitate the recruitment, coordination, and orientation of individual workers, and determine the detailed shape of the reconstructed chamber. Moreover, the pheromonal template alone cannot explain the two following observations:

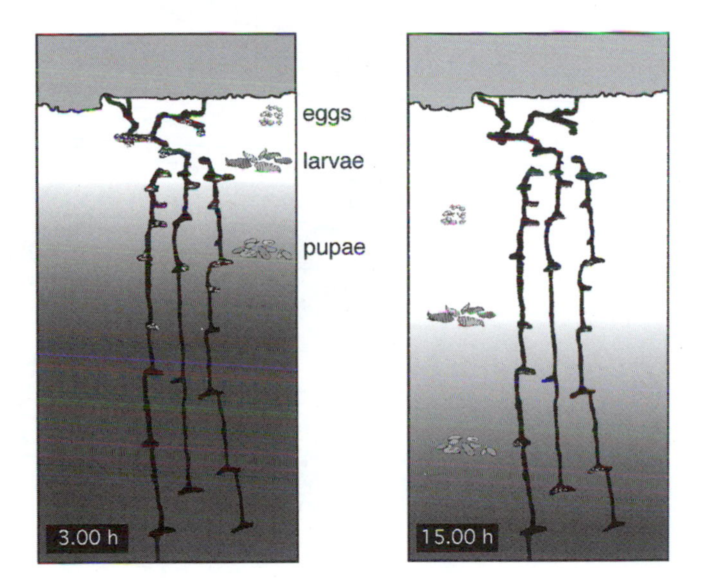

FIGURE 5.1 The spatial distribution of eggs, larvae, and pupae in the ant *Acantholepsis custodiens* depends on the temperature gradient along the depth axis. The gradient changes between 3:00 a.m. (left) and 3:00 p.m. (right).

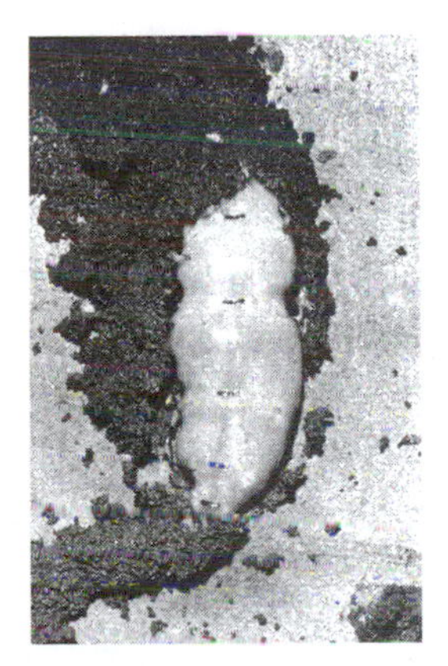

FIGURE 5.2 Chamber being built around the queen (after Grassé [158], p. 536, reprinted by permission © *Masson.*), 1h18 after the colony has been placed in a Petri dish and provided with soil pellets. The queen was 9 cm long, the Petri dish 25 cm in diameter and 4 cm high. 1200 individuals were present in the Petri dish.

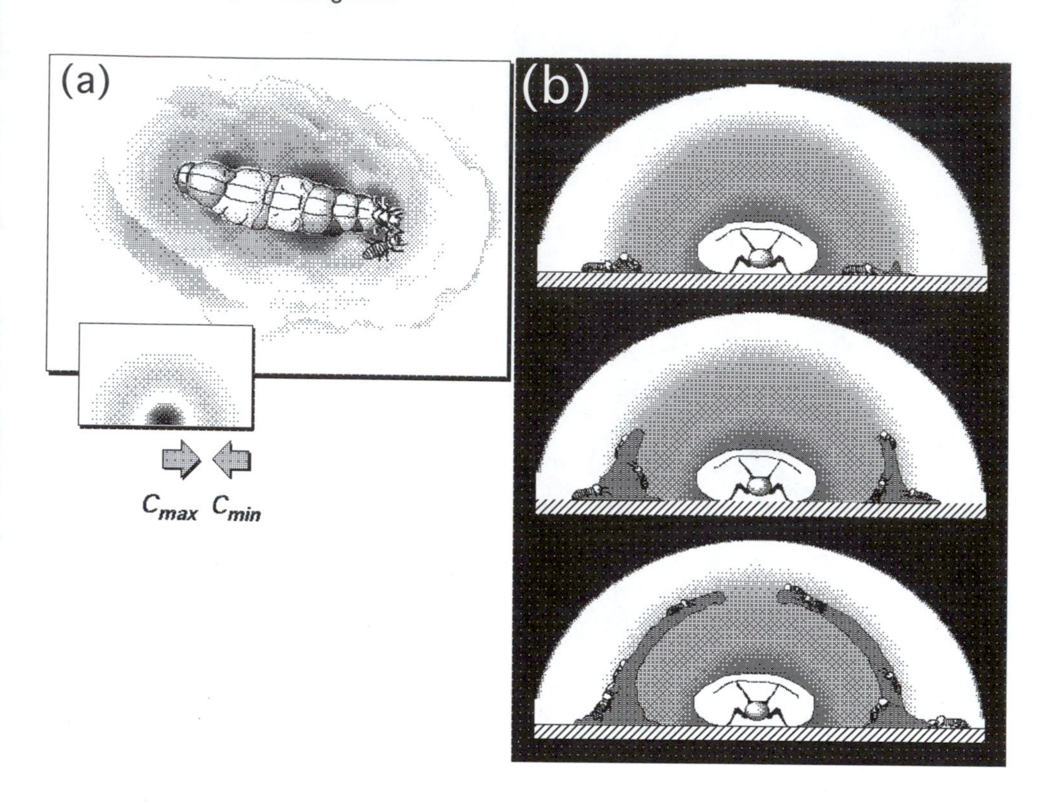

FIGURE 5.3 (a) Sketch of pheromonal template, representing the physogastric queen and the king. Different pheromone concentrations are represented by different grey levels. $[C_{\min}, C_{\max}]$ is the concentration window within which depositions are most likely to occur. (b) The template is three dimensional. Workers use the template as well as the queen's body, over which they walk to build a roof protecting the queen.

1. Walls are not built uniformly along the chemical template from the beginning. Schematically, pillars are first formed, and then the space between pillars is filled to produce walls. This process was briefly described in section 1.2.3.
2. At the group level, one of the most noticeable dynamic effects during the construction process is that the rate of building increases very rapidly, especially in large groups. The *per capita* rate of building increases disproportionately with the number of builders. Sixty minutes after the beginning of the construction of the royal chamber, there are about 0.2 depositions/worker on average in a group of 20 workers, whereas after a similar interval a group of 80 workers makes about 2.5 depositions/worker on average (Figure 5.4(a)). The saturation in this increase observed beyond 80 workers certainly means that the intrinsic maximum rate of building per worker has been reached. Figure 5.4(b) shows the mean number of

depositions (during 10 min) per worker added 50 min after the beginning of the experiment, in which worker groups of different sizes are put in the experimental arena with the queen to construct a new royal chamber. The number of depositions per added worker increases significantly with the number N of workers already present from $t = 0$ to $t = 50$ min ($N = 10, 20, 40, 80, 160, 320$).

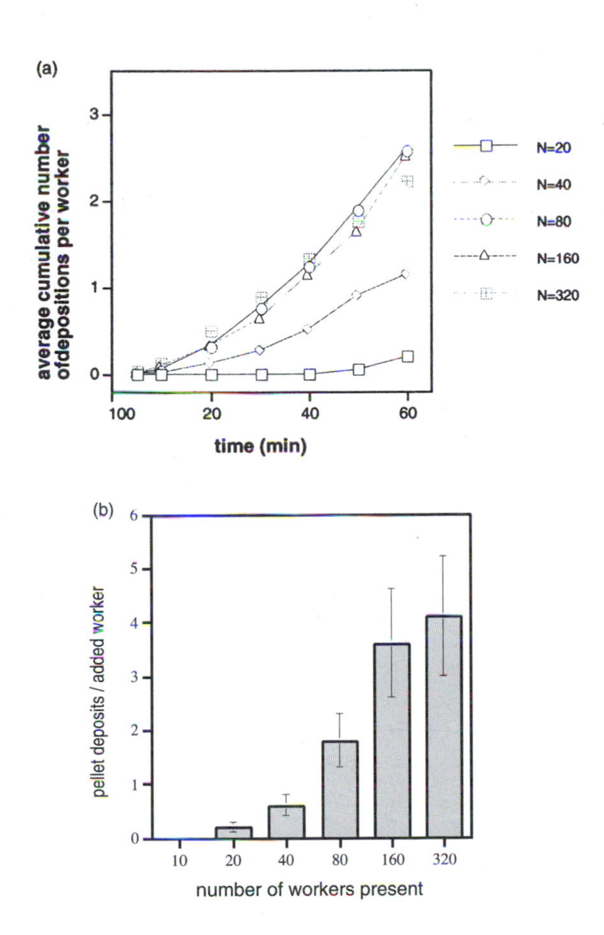

FIGURE 5.4 (a) Mean number (average over 6 to 10 experiments; error bars not shown) of depositions per worker during 60 min for groups of different sizes (10, 20, 40, 80, 320) (after Bruinsma [48], pp. 14 and 36). (b) Mean number of depositions per worker introduced 50 min after the beginning of Bruinsma's [48] experiment, where worker groups of different sizes (10, 20, 40, 80, 160, 320) were put in the experimental arena with the queen. The measures were taken in the 10 min following introduction. Error bars result from 30 replicates of the experiment (after Bruinsma [48], p. 37, reprinted by permission).

Positive feedback mechanisms, involving a cement pheromone in the pellets that triggers a chemotactic[1] behavior from workers and spatial heterogeneities, such as small obstacles, which attract workers within a small radius and stimulate them to deposit pellets, can explain these two observations [33]. The more pellets at one location, the more termites are attracted toward this location and deposit pellets. A model of this positive feedback effect is given in the next section. It is very likely that the construction of the royal chamber relies on a combination of template and self-organizing mechanisms.

5.3 A REACTION-DIFFUSION MODEL OF THE CONSTRUCTION OF THE ROYAL CHAMBER

The first step toward modeling the construction of the royal chamber is a simple model of the emergence of pillars, briefly introduced in section 1.2.3 [83]. This model shows how the different parameters characterizing the random walk of the termites, the attractivity of the cement pheromone, the diffusion of the pheromone, etc., determine the characteristic properties of the spatial distribution of pellets and the possible emergence of pillars. Let $H(r,t)$ be the concentration, at location r and time t, of the cement pheromone that is emitted by the deposited material and diffuses freely in air. Equation (5.1) describes the dynamics of H:

$$\partial_t H = k_2 P - k_4 H + D_H \nabla^2 H \,, \tag{5.1}$$

where k_2 is the amount of pheromone emitted per unit of deposited material per unit time, the total production being the product $k_2 P$, with P the amount of deposited material that is still active. Pheromone decay is represented by $-k_4 H$, and $D_H \nabla^2 H$ accounts for pheromone diffusion, with D_H being the diffusion coefficient, assumed to be constant.

To determine the attractiveness of the cement pheromone, the dynamics of the loaded termites, the density of which is denoted by C, has to be described. We assume that the path of a termite results from the addition of two processes: random walk and response to the pheromone gradient, or chemotactic behavior. The simplest model of chemotactic behavior assumes that the response is proportional to the gradient: the greater the gradient, the more attracted termites are. Greater gradients attract termites toward peaks of concentration, which correspond to zones where a lot of material has been deposited. The equation then reads:

$$\partial_t C = \Phi - k_1 C + D_C \nabla^2 C - \gamma \nabla (C \nabla H) \,. \tag{5.2}$$

In Eq. (5.2), it is assumed that $\gamma \nabla (C \nabla H)$ describes the attractiveness of the pheromone gradient, γ denotes the intrinsic strength of this attractiveness (γ is

[1]Chemotaxis is a gradient-following behavior: individuals move along a concentration gradient, toward regions of higher concentration.

assumed to be positive: this corresponds to a case in which regions of higher H concentration are indeed attractive): for the same value of the gradient, the greater γ, the greater the attractiveness. There is also a random component in individual motion, described by $D_C \nabla^2 C$, where D_C is the "diffusion" constant of termites. It is further assumed that there is a spatially and temporally constant flow Φ of loaded termites into the system (that is, the flux of insects bringing new building material into the modeled zone), and that the rate of unloading per termite per time unit is a constant k_1. Finally, Eq. (5.3) describes the dynamics of the active material P: the amount of material P deposited per unit of time is equal to $k_1 C$, and the rate of disappearance of P is $k_2 P$, the total production of pheromone emitted in space per time unit (it is assumed the cement pheromone contained in pellets dissociates from the pellets and can then diffuse freely in air):

$$\partial_t P = k_1 C - k_2 P . \tag{5.3}$$

When material is dropped, cement pheromone is emitted and diffuses, thereby attracting more termites toward its area of origin, where more material is therefore dropped. This accumulation of material induces a stronger and stronger emission of pheromone which attracts more of the termites. This positive feedback at different sites also gives rise to competition between close pillars, leading to a regular spatial distribution of pillars.

This mechanism of attractiveness does not always lead, however, to the emergence of regularly spaced pillars, and quantitative conditions are required to produce such a pattern: when the density of termites or the rate of deposition is too low, the development of pillars may be impossible, and only a spatially homogeneous stationary state is observed, that is, a uniform distribution of pellets in space. In this homogeneous stationary state (C_0, H_0, P_0), both spatial and temporal derivatives are equal to 0, so that:

$$C_0 = \frac{\Phi}{k_1}, \qquad H_0 = \frac{\Phi}{k_4}, \qquad P_0 = \frac{\Phi}{k_2} . \tag{5.4}$$

Deneubourg [83] studied the condition under which a diffusive instability could occur. He performed a classical linear stability analysis [247], studying how a small perturbation of the form $e^{\omega t} e^{ikx}$ on top of the homogeneous stationary state could evolve into a spatially heterogeneous stationary state. The condition for the growth of perturbations is $Re(\omega) \geq 0$, which is usually satisfied for some values of the spatial wavenumber k within a certain range of the parameters governing the nonlinear terms of the equations. The dispersion relation $\omega(k)$ depends on the values of the parameters. For values of k such that $Re(\omega(k)) \geq 0$, a perturbation of that particular wavelength can grow in principle, but generally, the diffusive instability will have the wavenumber k for which $Re(\omega(k))$, is maximal (since the instability grows faster at this wavenumber). The relevant parameter here is the chemotactic

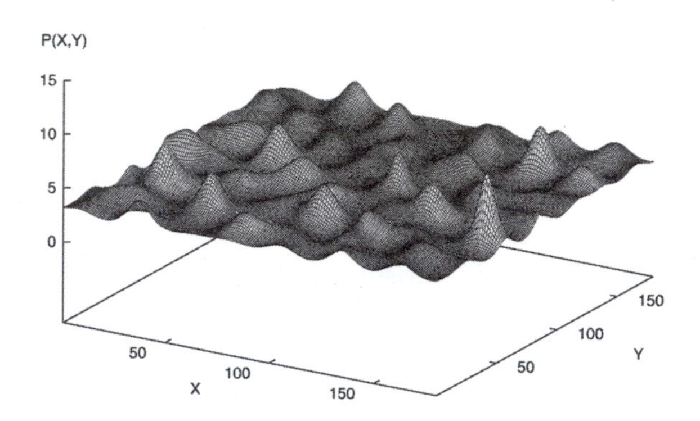

FIGURE 5.5 Spatial distribution of P for a 2D system (180 × 180) and $k_1 = k_2 = k_4 = 0.8888$, $D_C = 0.01$, $D_H = 0.000625$, $\Phi = 3$, $\gamma = 0.004629$, $t = 100$. The distribution of pillars is only statistically regular, because of the random initial distribution of P. The initially fastest-growing modes do not necessarily continue to be dominant after some time, as structures can emerge locally and modify the physics of the system.

parameter γ. Deneubourg [83] has shown that a value

$$\gamma_c = \frac{\left((k_4 D_C)^{1/2} + (k_1 D_H)^{1/2}\right)^2}{\Phi} \tag{5.5}$$

of γ exists, such that no perturbation can grow for $\gamma < \gamma_c$. For $\gamma = \gamma_c$, there exists a wavenumber $k_c = (k_1 k_4 / D_C D_H)^{1/4}$ for which $\omega(k) = 0$, by means of numerical integration, whereas $\omega(k) < 0$ for any other k: a perturbation with $k = k_c$ at $\gamma = \gamma_c$ is marginal. Deneubourg [83] has studied the dynamics of a one-dimensional system with a value of γ close to the marginal state, and found that the system converges to a spatially periodic state. The same observation is true for a two-dimensional system (Figure 5.5). Finally, let us briefly remark that the condition $\gamma > \gamma_c$ for the existence of coherent structures can be expressed as $\Phi > \left((k_4 D_C)^{1/2} + (k_1 D_H)^{1/2}\right)^2 / \gamma$, which is a condition of minimal flux given γ: this corresponds to the empirical observation that a minimal number of individuals is required to produce a pattern of pillars [158].

Let us now assume that there is a self-maintained pheromonal template because of the queen. Let $T(x, y)$ be the amount of queen pheromone at location (x, y) in

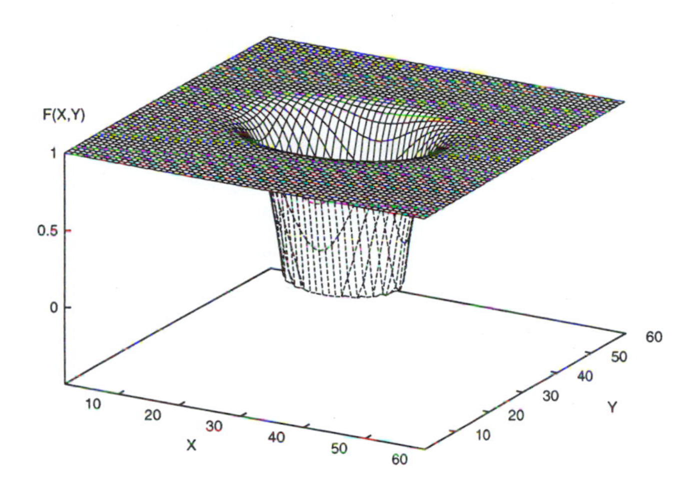

FIGURE 5.6 Simulated pheromonal template created by the queen. $T(x,y) = \exp(-[((x-x_0)/\lambda_x)^2+((y-y_0)/\lambda_y)^2])$, $\lambda_x = 7$, $\lambda_y = 5$, $x_0 = y_0 = 30$, $F(x,y) = 1-T(x,y)$ where $T(x,y) = \exp(-[((x-x_0)/\lambda_x)^2 + ((y-y_o)/\lambda_y)^2])$.

the two-dimensional arena. Let us assume that the "geometric center" of the queen is located at (x_0, y_0), and that the shape of the queen is roughly elliptic with the principal axis along the x-direction, so that the pheromone template she creates by emitting pheromone is given by

$$T(x,y) = e^{-[((x-x_0)/\lambda_x)^2+((y-y_0)/\lambda_y)^2]}, \tag{5.6}$$

where λ_x and λ_y are characteristic distances for the decay of the pheromonal pattern, that can be assumed to be proportional to the size of the queen in the x and y directions. The same chemotactic term as the one used to describe the attractive effect of cement pheromone can be used here, so that the full equation reads:

$$\partial_t C = \Phi - k_1 C + D_C \nabla^2 C - \gamma \nabla(C\nabla H) - v\nabla(C\nabla T), \tag{5.7}$$

where v is the force of attraction of the queen's pheromonal template. Bruinsma [48] has shown that a mechanism inhibits the deposit of pellets when the concentration of pheromone exceeds some threshold.

An inhibition function $F(x,y) = 1-T(x,y)$ prevents depositions from occurring when pheromone intensity is too large: this is precisely the definition of a template, which organizes the building activity at a constant distance from the queen. The template mechanism can be included by modifying Eqs. (5.2) and (5.3) as follows:

$$\partial_t C = \Phi - Fk_1 C + D_C \nabla^2 C - \gamma \nabla(C\nabla H) - v\nabla(C\nabla T). \tag{5.8}$$

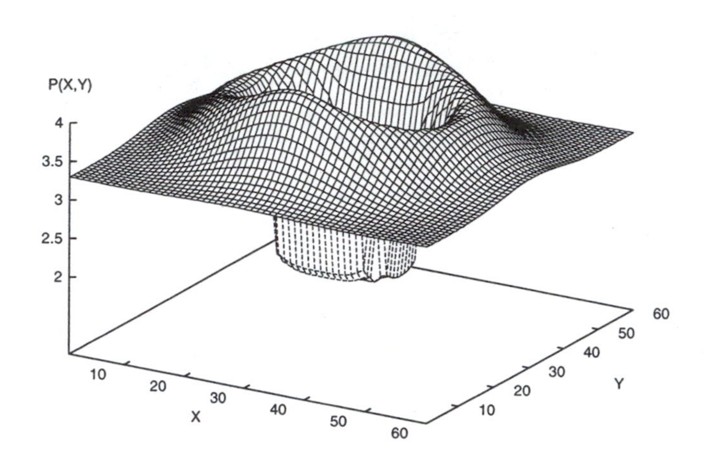

FIGURE 5.7 Spatial distribution of P at $t = 100$ for a 2D system and $k_1 = k_2 = k_4 = 0.8888$, $D_C = 0.01$, $D_H = 0.000625$, $\Phi = 3$, $\gamma = 0.004629$, added chemotactic motion toward the pheromonal template represented in Figure 5.6, with $v = 0.02$. A chamber forms around the simulated template.

$$\partial_t P = Fk_1 C - k_2 P\,. \tag{5.9}$$

Since $0 < T(x,y) \leq 1$, $F(x,y) = 1 - T(x,y)$ also satisfies $0 \leq F(x,y) = 1 - T(x,y)$ where $T(x,y) = \exp(-[((x-x_0)/\lambda_x)^2 + ((y-y_o)/\lambda_y)^2])$. Bruinsma's [48] description of this phenomenon is slightly more complicated than was assumed here: workers are indeed first attracted toward the queen, and then, after some time, move away from the queen until they deposit a pellet. In the model, it was assumed that individuals do not exhibit this "double motion," first toward the queen and then away from the queen. But, as a first approximation, the model describes a similar dynamics, since individuals move toward the queen and deposit their pellet *as soon as* they find a site to deposit a pellet. Figure 5.7 shows that a chamber can form around the queen following the pheromonal template. Let us also mention that the force of attraction v of the queen pheromone plays the role of a bifurcation parameter: chamber "walls" can appear along the trail only if $v > v_c$.

In addition to generating walls around the queen as illustrated by Figure 5.7, this model can also explain the two observations that could not be accounted for by the template alone: the formation of pillars and the snowball effect in deposition behavior [48]. The formation of pillars is a property of the model when certain conditions are met (such as $\gamma \geq \gamma_c$ and $v \geq v_c$). The snowball effect results from the fact that pellet deposits stimulate even more pellet deposits because of the cement pheromone. To make the model more realistic, tactile stimuli and trail pheromone should also be taken into account.

From this example, we see that the combination of self-organization and a template exhibits the properties of self-organization, such as the snowball effect, and at the same time produces a perfectly predictable pattern, the royal chamber which is formed by walls built following the queen's pheromonal template.

5.4 WALL BUILDING IN *LEPTOTHORAX albipennis*

Another example of a template, possibly in combination with self-organization, is the construction of nest walls by the ant *Leptothorax albipennis* (formerly *tuberointerruptus*) [131, 132]. These ants construct simple perimeter walls in a two-dimensional nest (Figure 5.8). The walls, consisting of aggregated grains of sand (or fragments of stone, or particles of earth), are constructed at a given distance from the tight cluster of ants and brood, which serves as a chemical or physical template. The exact nature of this template remains unknown. Here again, the template mechanism allows the size of the nest to be regulated as a function of colony size, including both workers and brood. Each worker always has about 5 mm^2 of floor area in the nest [131]. But, as was the case with the construction of the termites' royal chamber, there may be an additional stigmergic self-organizing mechanism on top of the template: grains attract grains (in the same way that corpses attract corpses in the formation of cemeteries), so that deposition behavior is influenced by two factors, the local density of grains and the distance from the cluster of ants and brood. The probability of depositing a brick is highest when both the distance from the cluster is appropriate and the local density of bricks is large; it is lowest when the cluster is either too close or too far and when the local density of bricks is small. When the distance from the cluster does not lie within the appropriate range, deposition can nevertheless be observed if bricks are present; conversely, if the distance from the cluster is appropriate, deposition can take place even if the number of bricks is small.

This process can easily be simulated on a grid with the same ingredients as in the simulation of cemetery formation (section 4.3.1), that is, probabilities of picking up and depositing grains, combined with a template that represents the worker-queen-brood cluster's template. The probability of depositing a grain is given by $p_d p_t$, where p_d depends only on the perceived number of grains and p_t accounts for the effect of the template represented in Figure 5.9. The template function p_t exhibits a maximum at some distance from the center of the cluster, which is assumed to be circular. Grain deposits are most likely to happen when both p_d and p_t are close to 1. Similarly, the probability of picking up a grain is defined by $p_d(1 - p_t)$, where p_p depends only on the perceived number of grains. Grains are more likely to be removed from outside the template. p_p and p_d are given by Eqs. (4.1) and (4.2), respectively, where f is the perceived fraction of grains in the agent's neighborhood. Figures 5.10(a), 5.10(b), and 5.10(c) show the result of

FIGURE 5.8 *Leptothorax albipennis* (formerly *tuberointerruptus*) nest in the laboratory. The ants have been given a 0.8 mm deep cavity between 80 × 80 mm glass plates. They have built a dense wall from sieved sand. The worker ants, each approximately 2.5 mm in length, are densely clustered around the central queen, the largest individual present, and the white brood. © Nigel R. Franks, printed with permission.

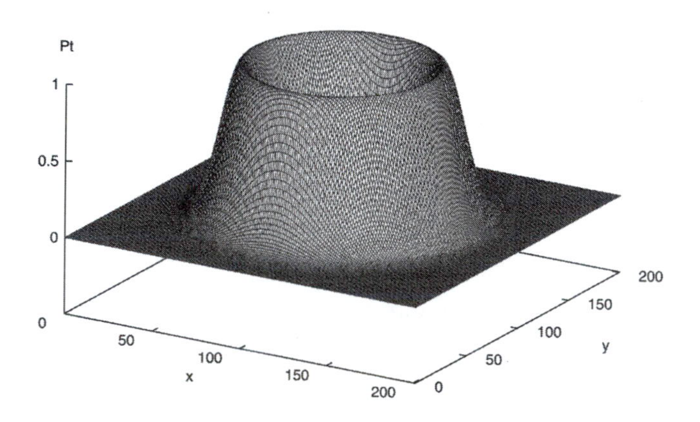

FIGURE 5.9 Template used for the simulation. The template represents the effects of the cluster in which the queen, the brood, and most of the workers are aggregated.

such a simulation at three different stages: $t = 50$, $t = 1{,}000$, and $t = 10{,}000$ time steps; the four heaps at the four corners represent four reservoirs of grains. We see that walls are progressively built along the template.

This double mechanism (template + self-organization) may be sufficient to explain the hysteretic[2] phenomenon observed by Franks and Deneubourg [132]. In experiments in which approximately half of the adult ants in each nest were removed and later reunited after the building was completed by the first half of the colony, there was no increase in nest size following reunification. When three-quarters of the adult ants were removed and returned after the building was completed by the remaining quarter, nest size increased in three of the four experiments, and actually decreased slightly in one experiment.

Franks and Deneubourg [132] propose a mechanism that combines self-organization and template to explain these observations. Their model is based on grain picking up and dropping behavior, both of which are assumed to be under the control of both amplifying effects and the influence of the template provided by the cluster of brood and workers in the new nest site. An unladen ant (the density of which is denoted by U) meets a grain (the density of which is denoted by S) and can pick it up and become a laden ant (the density of which is denoted by L). The rate of transformation of U into L when meeting S at location r is given by $P(r)US$, where $P(r)$ represents the influence of the template. $P(r)$ is close to 1 in regions from which grains should be removed and close to 0 where grains should be deposited. Laden ants drop their grains at a rate $D(r)L(1 - S/K)$, where K is the "carrying" capacity per unit area, that is, the maximum number of grains that can be found per unit area. Here again, $D(r)$ represents the direct influence of the template: it should be close to 1 at an appropriate distance from the cluster of brood and workers to stimulate grain deposits, and should be close to 0 otherwise. The number of grains present in a laden ant's perception area influences its probability of depositing or picking up grains. The dropping rate is assumed to be transformed into $D(r)G(S)L(1-S/K)$, where $G(S) = (g_1 + g_2 S)$ is a linearly increasing function of the number of grains. g_1 and g_2 are, respectively, parameters that characterize spontaneous grain dropping and dropping next to another grain. The picking-up rate is assumed to be transformed into $P(r)F(S)SU$, where $F(S) = (g_1 + g_2 S)^{-1}$ is a decreasing function of the number of grains. g_1 and g_2 are assumed to be the same for both G and F, so as to reduce the number of parameters. The resulting dynamics of the population of grains in a given small area, neglecting loading and unloading times, is given by:

$$\partial_t S = D(r)G(S)L\left(1 - \frac{S}{K}\right) - P(r)F(S)SU\,. \tag{5.10}$$

[2]Hysteresis occurs when the properties of a system depend not only on the system's state, but also on its history. Another name for hysteresis is path-dependence. Here, the size of the nest cannot be inferred from the sole knowledge of colony size, but requires knowledge of how the colony has been formed.

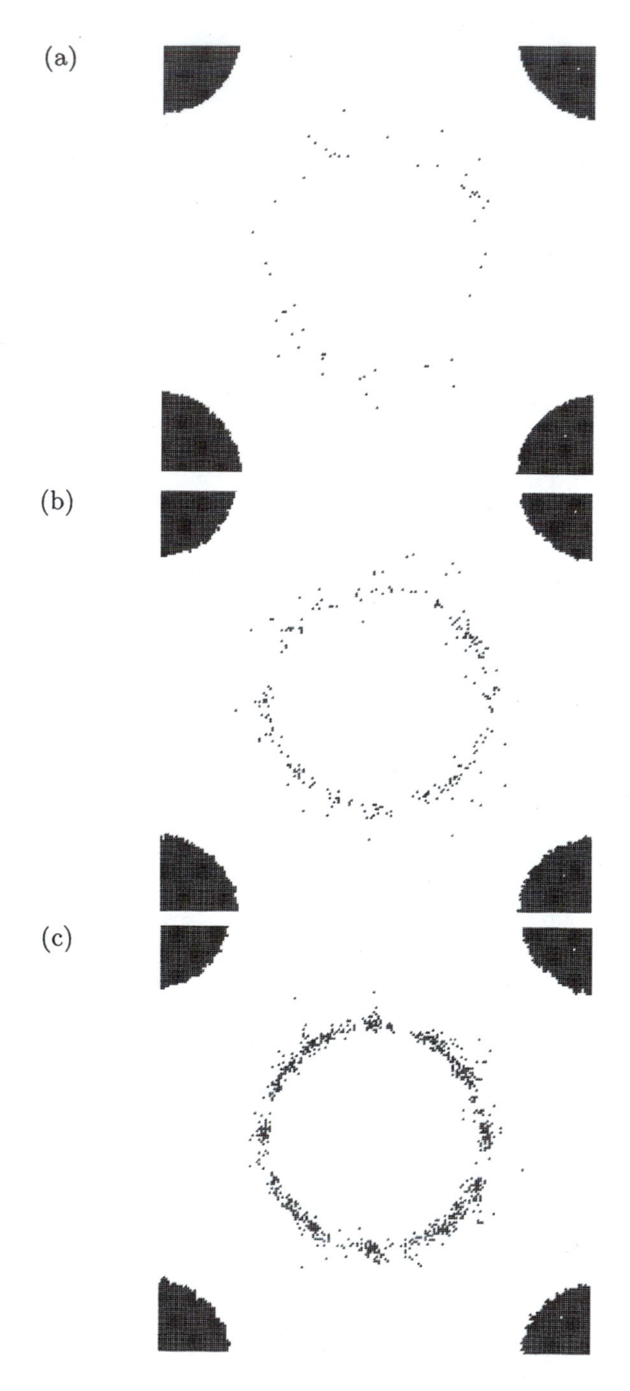

FIGURE 5.10 (a) Simulation of wall building with 10 agents at $t = 50$, $T = 50$, $k_1 = 0.1$, $k_2 = 0.3$, 10 agents. (b) Same as (a) at $t = 1,000$. (c) Same as (a) at $t = 10,000$.

In the stationary regime, characterized by $\partial_t S = 0$, it is found that

$$\eta = \frac{S}{(g_1 + g_2 S)^2}\left(1 - \frac{S}{K}\right)^{-1}, \tag{5.11}$$

where

$$\eta \equiv \frac{D(r)L}{P(r)U}. \tag{5.12}$$

This formulation allows the system to be analyzed. η is a measure of the tendency of the ants to build in a particular area. If D and L are large, η is large, and if P and U are large, η is small. The influence of the template, reflected in D and P, is easy to analyze. But what determines the respective numbers of laden and unladen ants?

As experiments show, unladen ants are essentially ants that move outward from inside the nest. If the wall perimeter is small and the number of workers inside is large, U is large, leading to a small value of η. If, for the same population, the wall is built further away, U is smaller and η is larger. In short, the higher the density of individuals inside the nest, the further away the wall is built because grain deposits are prevented when population density is large. Franks and Deneubourg [132] provide estimates for most of the parameters: $L/U < 0.1$ in the zone of active building, with $U \approx 2\ \text{cm}^{-2}$, $g_1 \approx 25g_2$, $g_2 \approx 10^{-2}s^{-1}$, and $P \approx 10^{-4}s$ (in the building zone). They find that the model exhibits multistationarity as a function of g_1/g_2, when the stationary value of S is expressed as a function of η (because population dynamics occurs on a faster time scale than grain dynamics, it can be assumed that S reaches its stationary value within a time where η remains constant).

1. When $g_1/g_2 > 0.125K$, the solution of S is unique, and S grows with η (Figures 5.11(a) and 5.11(c)).
2. When $g_1/g_2 < 0.125K$, several solutions can be reached for $\eta > \eta_c(g_1/g_2)$, depending on the history of the system (Figure 5.11(b)). This phenomenon is called hysteresis.

η is directly related to local building activity and colony characteristics: for a given zone initially located far from the center of the nest, η is small, but grows as population increases and may eventually decrease when the population becomes so large that walls should be built beyond this zone.

Let us consider η as a function of population size only: $\eta = \eta(n)$, where n is the size of the population. If one assumes that $g_1/g_2 < 0.125K$, which, according to Franks and Deneubourg [132], is indeed the case, S may lie on a given branch of solutions when a fraction of the workers is removed and may or may not jump to another branch of solutions when the missing individuals are returned. The perturbation associated with multiplying the population by 2 (when 50% of the colony is removed and returned) may not be sufficient to induce nest increase, whereas the perturbation associated with multiplying the population by 4 (when

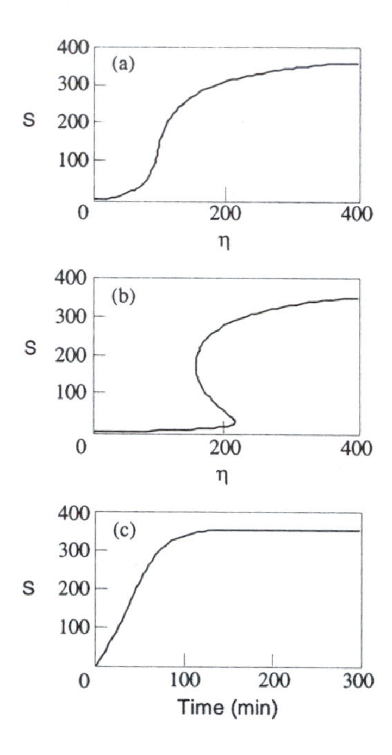

FIGURE 5.11 (a) Behavior of Eqs. (5.11) and (5.12): S is shown as a function of η. S is the number of grains and η is a measure of the tendency of the ants to build in a particular area. Here, $g_1 = 0.75$, $g_2 = 0.01$, $K = 400$. The relationship between S and η is simply sigmoidal: the system does not have multiple stationary states. After Franks and Deneubourg [132]. (b) The same as in Figure 5.11(a), with $g_1 = 0.1$, $g_2 = 0.01$, $K = 400$. The relationship between S and η is a classic hysteresis curve: the system has multiple stationary states. After Franks and Deneubourg [132]. (c) The temporal dynamics of grain dropping in the zone for the set of parameters corresponding to Figure 5.11(a). $g_1 = 0.75$, $g_2 = 0.01$, $K = 400$, $D = 0.4$, $P = 0.0002$, $L = 0.2$, $U = 2$. Initial condition $S = 0$. After Franks and Deneubourg [132]. Reprinted by permission © *Academic Press.*

75% of the population is removed and returned) may be sufficient to induce an increase of the nest, but not necessarily, depending on the history of the system. In the experiment where 75% of the adult population was removed and then returned without an increase in nest size, Franks and Deneubourg [132] observed that the nest built by the remaining 25% of the population was unusually large, which may explain why the nest was not enlarged after reunification. This model may therefore explain the empirical observations.

In this example, again, the combination of self-organization and template exhibits the properties of self-organization, such as hysteresis, and at the same time produces a perfectly predictable pattern—the nest walls constructed at a given distance from the cluster of ants.

5.5 APPLICATIONS

The idea that template mechanisms can be combined with self-organizing mechanisms, or that a system can *self-organize along a template*, has applications in the context of the data analysis and graph partitioning models of the previous chapter. These models were intrinsically nonparametric (with respect to the number of emerging classes of objects in the case of data analysis, and with respect to the number of emerging clusters in the case of graph partitioning). Being nonparametric can be a great strength on some occasions, but can be problematic in others. For example, in the customer profile application sketched in section 4.4 (Figure 4.9), the bank might want to divide its customers into a predefined number of categories. In electronic circuit design, using a graph partitioning method requires the ability to control the number of clusters that will be formed. Each cluster corresponds to a particular region on a chip, and the goal is to distribute elements among these regions to minimize connections between regions and balance the number of elements per region. The number of regions is known in advance.

Combining the underlying self-organizing mechanisms of the algorithms described in section 4.3 with templates is an elegant, and biologically inspired, way of restoring parametricity: for example, it is possible to impose that all items be deposited in some particular regions of space, and objects that are deposited within a given region are considered to belong to the same class or cluster. In this way, the desired number of classes or clusters can be predefined. In order to define a template, we have to introduce a "template" probability $P_t(x, y)$, where (x, y) are the planar coordinates of the agent. If we assume for simplicity that projection space or clustering space is restricted to $[0,1] \times [0,1]$, and we wish to have four separated dropping zones, $P_t(x, y)$ can simply be defined by, for example:

$$P_t(x,y) = a\left[e^{-\frac{x^2+y^2}{\sigma^2}} + e^{-\frac{(x-1)^2+y^2}{\sigma^2}} + e^{-\frac{(x-1)^2+(y-1)^2}{\sigma^2}} + e^{-\frac{x^2+(y-1)^2}{\sigma^2}} - b\right], \quad (5.13)$$

where σ^2 determines the steepness of the template, and a and b are parameters that have to be adjusted to σ^2 so that $0 \leq P_t(x, y) \leq 1$. Figure 5.12 shows the template given by Eq. (5.13). The four corners of the square are the zones where dropping behavior is favored.

The template probability $P_t(x, y)$ can now be used in combination with the two algorithmic applications of the previous chapter, namely data analysis and graph representation and partitioning.

5.5.1 DATA ANALYSIS

Modifying the LF algorithm to include a template is straightforward. Let $(r_i = (x_i, y_i)$ be the location of item o_i. The two Eqs. (4.5) and (4.6) of section 4.4 are transformed into:

$$p_p(o_i) = 0.7\left(\frac{k_1}{k_1 + f(0_i)}\right)^2 + 0.3(1 - P_t(r_i)), \quad (5.14)$$

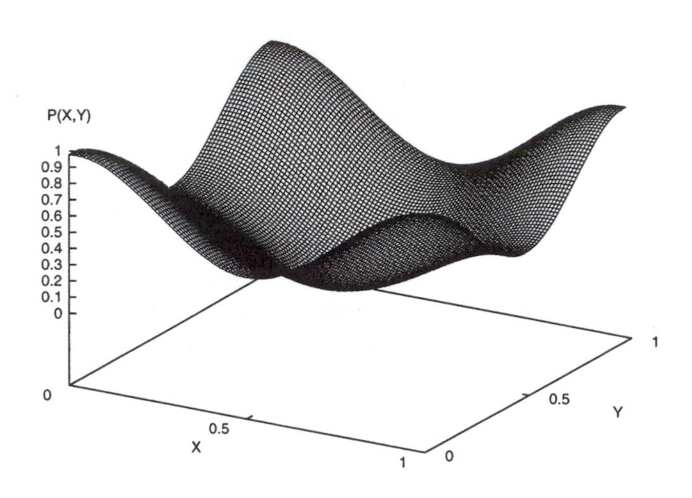

FIGURE 5.12 Template function $P_T(x, y)(a = 1.9, b = 0.5, \sigma^2 = 0.1)$.

FIGURE 5.13 Distribution of items at $t = 500,000$ obtained with the LF algorithm supplemented with the template probability represented in Figure 5.12. The simulation parameters are identical to those of Figure 4.6.

$$p_d(o_i) = \begin{cases} 2P_t(r_i)f(o_i) & \text{if } f(o_i) < k_2 \\ P_t(r_i) & \text{if } f(o_i) \geq k_2 \end{cases} \tag{5.15}$$

so that picking-up behavior is favored where $P_t(r_i)$ is close to 1, and dropping behavior is favored where $P_t(r_i)$ is close to 0. This procedure can be tested with the same simple example as in section 4.4. Four clusters of 100 points each are generated in attribute space, with x and y distributed according to normal distributions $N(\mu, \sigma)$: $[x \propto N(0,2),\ y \propto N(0,2)]$, $[x \propto N(0,2),\ y \propto N(8,2)]$, $[x \propto N(8,2),\ y \propto N(0,2)]$, $[x \propto N(8,2),\ y \propto N(8,2)]$ (Figure 4.5). The data points were then assigned random locations on a 100×100 lattice, with 1 lattice unit = 0.1, and the clustering algorithm was run with 10 ants. Figure 5.13 shows the system at $t = 500,000$. The items distribute themselves in the grid according to both the dissimilarity measure and the template.

5.5.2 GRAPH PARTITIONING

Several graph partitioning problems exist. One problem of particular interest is when one has to partition the graph into c clusters of (approximately) equal size while minimizing the number of intercluster connections. This problem is NP-complete. Adding a template mechanism to the KLS algorithm solves this particular problem. Inclusion of the template is straightforward. Equations (4.12) and (4.13) in section 4.5 are transformed into:

$$p_p(v_i) = 0.7\left(\frac{k_1}{k_1 + f(v_i)}\right)^2 + 0.3\,(1 - P_t(w_i))\ , \tag{5.16}$$

$$p_d(v_i) = P_t(w_i)\left(\frac{f(v_i)}{k_2 + f(v_i)}\right)^2\ , \tag{5.17}$$

where w_i is the location on the plane of vertex v_i. The template function given by Eq. (5.13) allows us to solve the problem of partitioning the graph into four clusters. Figure 5.14(a) shows the resulting distribution of vertices on the plane for a random graph of the type $\Gamma(25, 4, 0.8, 0.01)$ (see section 4.5 for detail). We observe exactly four well-segregated clusters of vertices that correspond to the four clusters of the graph. This spatial distribution reflects on the plot of $||w_i - w_j||$ vs. $d(v_i, v_j)$ (Figure 5.14(b)). Of course, this instance is particularly adapted to the template, since the graph contains four clusters (the same number as there are regions in the template) with the same number of vertices. Frustration may be observed when there are more clusters in the graph than regions in the template or vice versa. As regards the number of vertices in each region, it can be controlled with the help of an additional regulating mechanism that inhibits picking up (respectively, dropping) in regions which do not contain enough (respectively, contain too many) vertices. This "load-balancing" mechanism allows an equal repartition of the vertices into the regions. For example, in the absence of load balancing, if the template contains four clusters and the graph contains only three clusters, one might observe the result shown in Figure 5.15, where one corner of the template is not filled.

(a)

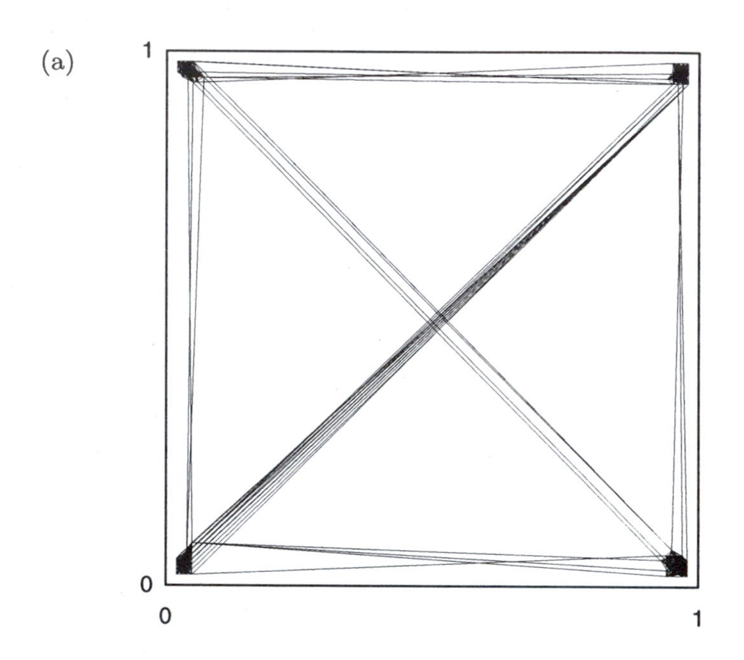

(b)

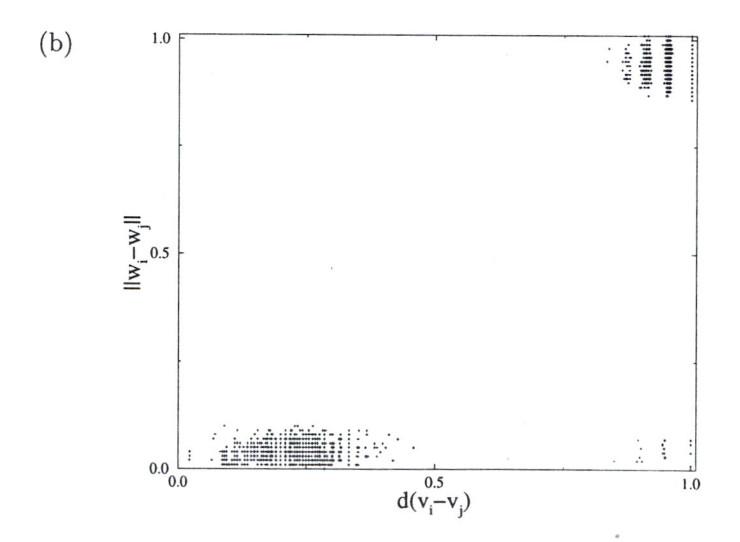

FIGURE 5.14 (a) The algorithm has isolated the four clearly separated clusters of the graph. Vertices distribute themselves in space, following the template function. $k_1 = 0.03$, $k_2 = 0.1$, $s^2 = 9$, $T = 2.10^6$ 6 steps. (b) $||w_i - w_j||$ vs. $d(v_i, v_j)$ for the graph representation of Figure 5.14(a).

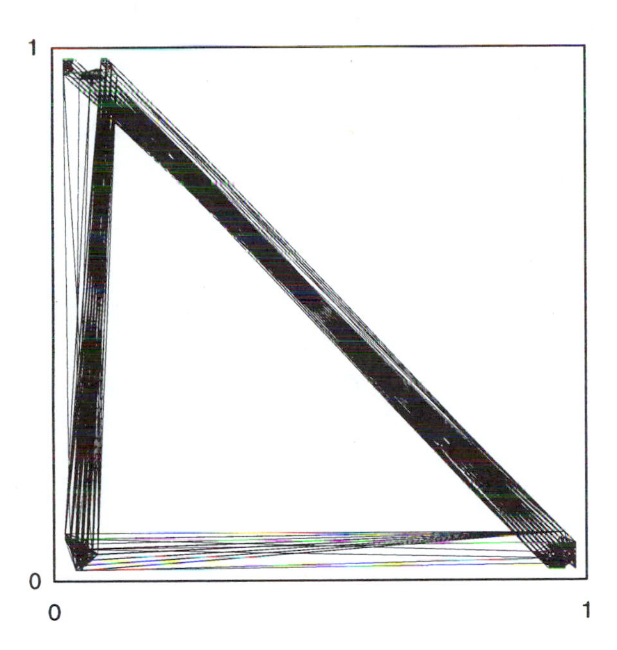

FIGURE 5.15 Result obtained with a template that has four corners (represented in Figure 5.14) for a graph $\Gamma(33, 3, 0.8, 0.01)$ that has only three clusters. Vertices distribute themselves in space, only partially following the template function: three clusters can be clearly observed. $k_1 = 0.03$, $k_2 = 0.1$, $s^2 = 9$, $T = 2.10^6$ steps.

5.6 POINTS TO REMEMBER

- A template is a pattern used by insects to organize their activities. Building following a template is typical. The construction of the royal chamber in *Macrotermes* termites consists of depositing soil pellets at a certain distance from the queen, following a chemical template emitted by the queen. Wall building in *Leptothorax* ants consists of depositing grains of sand at a certain distance from the central cluster that contains the workers, the queen, and the brood, following the contour of the cluster.
- In these two examples, the deposition of items following the template is self-organized: pellets attract pellets, and grains attract grains, leading to snowball effect and multistability.
- The combination of self-organization and template is useful to transform the non-parametric data analysis and graph partitioning algorithms presented in chapter 4 into parametric algorithms. The number of clusters, as well as the locations of these clusters, can be prespecified by constructing a template function characterized by as many regions as desired clusters.

CHAPTER 6

Nest Building and Self-Assembling

6.1 OVERVIEW

Social insect nest architectures can be complex, intricate structures. Stigmergy (see section 1.2.3), that is, the coordination of activities through the environment, is an important mechanism underlying nest construction in social insects. Two types of stigmergy are distinguished: quantitative, or continuous stigmergy, in which the different stimuli that trigger behavior are quantitatively different; and qualitative, or discrete stigmergy, in which stimuli can be classified into different classes that differ qualitatively. If quantitative stigmergy can explain the emergence of pillars in termites, the building behavior of the paper wasps *Polistes dominulus* seems to be better described by qualitative stigmergy.

In this chapter, a simple agent-based model inspired by discrete stigmergy is introduced. In the model, agents move in a three-dimensional grid and drop elementary building blocks depending on the configuration of blocks in their neighborhood. From the viewpoint of bricks, this model is a model of self-assembly. The model generates a large proportion of random or space-filling forms, but some patterns appear to be structured. Some of the patterns even look like wasp nests.

The properties of the structured shapes obtained with the model, and of the algorithms that generate them, are reviewed. Based on these properties, a fitness function is constructed so that structured architectures have a large fitness and unstructured patterns a small fitness. A genetic algorithm based on the fitness function is used to explore the space of architectures.

Several examples of self-assembling systems in robotics, engineering, and architecture are described. Self-assembling or self-reconfigurable robotic systems, although they are not directly inspired by nest construction in social insects, could benefit from the discrete-stigmergy model of nest building. The method of evolutionary design, that is, the creation of new designs by computers using evolutionary algorithms, is a promising way of exploring the patterns that self-assembling models can produce.

6.2 NEST BUILDING IN SOCIAL INSECTS

Many animals can produce very complex architectures that fulfill numerous functional and adaptive requirements (protection from predators, substrate of social life and reproductive activities, thermal regulation, etc.). Among them, social insects are capable of generating amazingly complex functional patterns in space and time, although they have limited individual abilities and their behavior exhibits some degree of randomness (see examples in Wilson [328] and Hansell [165]). Among all activities performed by social insects, nest building is undoubtedly the most spectacular, as it demonstrates the greatest difference between individual and collective levels. Figures 1.5, 1.6, and 1.7 in chapter 1 show some examples of nest architectures built by several wasp and termite species. One question that arises is: how can insects in a colony coordinate their behavior in order to build these highly complex architectures?

There is no evidence that the behavior of an individual in a social species is more sophisticated than that of an individual in a solitary species. Yet, the difference between nests produced by solitary and social species can be enormous. The first hypotheses put forward to explain complex nest building in social insects were anthropomorphic: individual insects were assumed to possess a representation of the global structure to be produced and to make decisions on the basis of that representation. Nest complexity would then result from the complexity of the insects' behavior.

Insect societies, however, are organized in a way that departs radically from the anthropomorphic model in which there is a direct causal relationship between nest complexity and behavioral complexity. Recent work suggests that a social insect colony is a decentralized system composed of cooperative, autonomous units that are distributed in the environment, exhibit simple probabilistic stimulus-response behavior, and have access to local information [85, 31]. Insects are equipped with a sensory-motor system (including chemoreceptors, mechanoreceptors, thermoreceptors, hygroreceptors, etc.) that enables them to respond to stimuli which are either emitted by their nestmates or originate from the environment. Although such signals are not equivalent to signs which could have symbolic value—these signals are simply attractive or repulsive, activating or inhibiting—they affect behavior in a way that depends on their intensity and on the context in which they are released. Nest complexity may therefore result from the variety of stimuli that surround insects in a social context, as stimuli include not only environmental cues but also interactions among nestmates. The set of stimuli need not be stationary: for example, as a colony grows, starting from a single unhelped foundress, more and more stimuli are likely to appear because of the emergence of new individuals, thereby forming a richer and richer stimulatory environment for the insects. When applied to nest construction, this idea is very attractive: the stimuli that initially trigger building behavior may be quite simple and limited in number, but as construction proceeds, these stimuli become more complex and more numerous, thereby inducing new types of behavior. Construction can best be seen as a morphogenetic process during which past construction sets the stage for new building actions. This principle can be coupled with the above-mentioned demographic effects: as the nest gets bigger, the greater the variety of signals and cues it is likely to encompass (the probability of finding heterogeneities also increases when the colony expands its territory). This may explain, for example, why the most populous termite societies have the most complex nests [158].

We have seen in the last chapter that templates are widely used by social insects to organize and coordinate their building activities: a "blueprint" of the nest "already exists" in the environment in the form of physical or chemical heterogeneities. More complex types of templates may also exist: those heterogeneities that result from the colony's activities and that, in turn, influence the colony's future activities. Indeed, a single action by an insect results in a small modification of the environment that influences the actions of other insects: this form of indirect communication through the environment is an important aspect of collective coordination and has been coined "stigmergy" by Grassé [157]. Briefly described in chapter 1 (section 1.2.3), stigmergy has been present, explicitly or implicitly, in every chapter.

Stigmergy offers an elegant and stimulating framework for understanding the coordination and regulation of building activities. The main issue is to determine how stimuli are organized in space and time to result in a robust and coherent construction—for example, colonies of a given species build qualitatively similar architectural patterns. Stigmergy is basically just a mechanism that mediates or implements worker-worker interactions; therefore, it has to be supplemented with a mechanism that makes use of these interactions to coordinate and regulate collective building. Recent investigations indicate that at least two such mechanisms play a role in the building activities of social insects: self-organization (SO) [85, 31] and discrete stigmergy [310, 311].

6.2.1 STIGMERGY AND SELF-ORGANIZATION

Deneubourg [83] has shown that SO, combined with stigmergy, can explain the construction of pillars in the termites studied by Grassé [157]. Deneubourg's [83] chemotaxis-based model was described in the previous chapter (sections 5.2 and

5.3). The pillar construction example was also used to illustrate bifurcation in a self-organizing system in section 1.2.2 and stigmergy in section 1.2.3. In section 1.2.2, we learned that one of the main ingredients of SO is multiple interactions. Such interactions can be direct or indirect: stigmergy covers all indirect interactions through the environment. In the example of the emergence of pillars, stigmergy is merely an ingredient of SO. SO is made possible by the fact that the intensity of the stigmergic interactions among termites can take on a continuous spectrum of values: the more intense the interactions (that is, the higher the pheromone concentration), the more likely the self-organizing effect (or the snowball effect) is to engage. Because coordination is facilitated by quantitative variations in the intensity of the interactions, the combination of stigmergy and self-organization is also called *quantitative stigmergy*.

6.2.2 DISCRETE STIGMERGY

6.2.2.1 Introduction

Qualitative or *discrete stigmergy* differs from quantitative stigmergy in that individuals interact through, and respond to, *qualitative* or *discrete* stimuli: when termites build pillars they respond to *quantitative* stimuli—pheromone fields and gradients. Discrete stigmergy in insects is based on a discrete set of stimulus types: for example, an insect responds to a type-1 stimulus with action A and responds to a type-2 stimulus with action B. In other words, qualitatively different stimuli result in different responses. It is easy to see how such a mechanism is still compatible with stigmergic interactions: for example, a type-1 stimulus triggers action A by individual I_1; action A transforms the type-1 stimulus into a type-2 stimulus that triggers action B by individual I_2. It is more difficult, however, to see how coordination and regulation can be achieved with discrete stigmergy than with SO: because stimuli are qualitatively different, no positive feedback effect can amplify a stimulus to transform it into a more intense version of the same stimulus. A stimulus is transformed into another, qualitatively different, stimulus under the action of an insect. Also, in principle, there is no such thing as the intensity of a stimulus: all stimuli are equally likely to elicit a behavioral response. Building in social wasps seems to provide an example of qualitative stigmergy (Fig. 6.1). In reality both types of stigmergy can, and are likely to, coexist if, for example, stimuli have a variable intensity within a given stimulus type.

Imagine that each individual is carrying a building element of a certain type, such as a soil pellet or a load of wood pulp, and deposits this element only when specific "qualitative" conditions are met, for example, when that individual is surrounded by three walls: the building element will be added to the individual's current location under these conditions. Therefore, from the viewpoint of the building element, a form of self-assembly, mediated by the individual, is taking place: building elements self-assemble whenever appropriate. Theraulaz and Bonabeau [310, 311] and Bonabeau et al. [26] have introduced a model of self-assembly, inspired by the building behavior of wasps (see also Karsai and Pen-

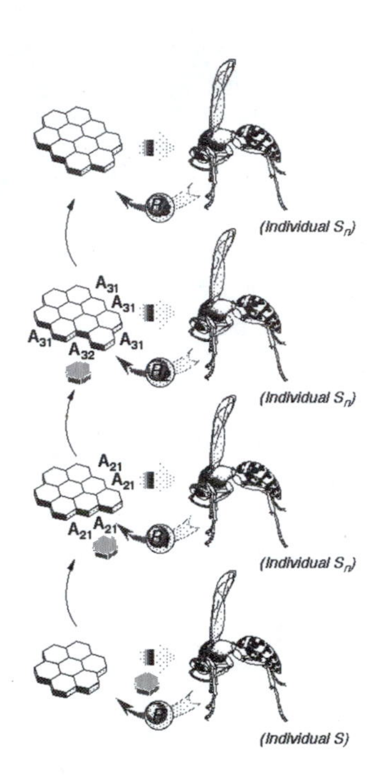

FIGURE 6.1 The equivalent of Figure 1.13 for wasps: local configurations of cells around the nest are modified each time a new cell is added by a wasp. Only a small number of these configurations trigger the addition of a new cell.

zes [194]), that illuminates how coordination may emerge. This model is described in the next subsection.

6.2.2.2 Construction in social wasps Social wasps have the ability to build nests that range from very simple to highly organized. Nests are made of plant fibers that are chewed and cemented together with oral secretion. The resulting carton is then shaped by the wasps to build the various parts of the nest (pedicel, combs of cells, or external envelope). Wenzel [324] has classified wasp nest architectures and found more than 60 different types, with many intermediates between extreme forms. A mature nest can have from a few cells up to a million cells packed in stacked combs, the latter generally being built by highly social species. Modularity is another widespread feature of nest architectures: a basic structure is repeated.

Previous studies showed that individual building algorithms consist of a series of if-then decision loops [89, 114, 115]. The first act usually consists, with only a few exceptions, of attaching the future nest to a substrate with a stalk-like pedicel of

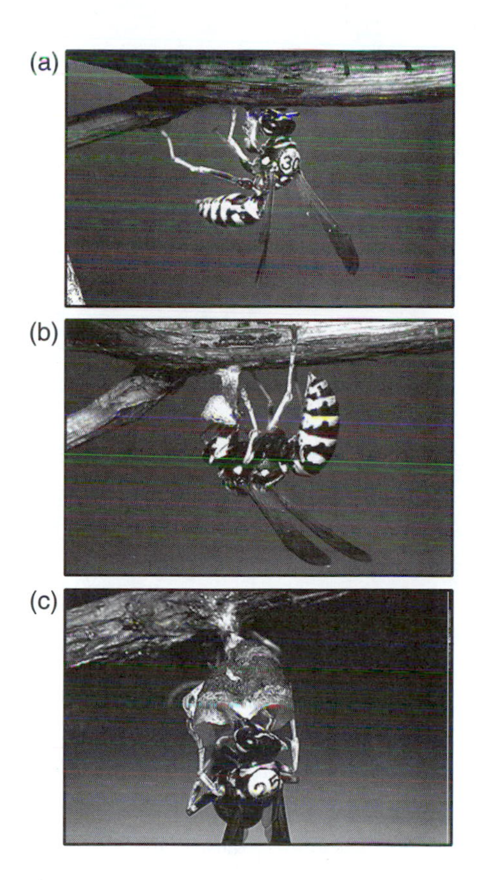

FIGURE 6.2 The first steps of nest building in a *Polistes dominulus* (Christ) wasp society: the three pictures show the initiation of the nest, with the construction of a pedicel (6.2(a)), of the first cell (6.2(b)), and of a few subsequent cells (6.2(c)). Papers of different colors are offered to the wasps at regular intervals to track the building dynamics. After Theraulaz and Bonabeau [311]. Reprinted by permission © *Academic Press.*

wood pulp [324] (Figure 6.2). Then, wasps initiate the nest by building two cells on either side of a flat extension of the pedicel. Subsequent cells are added to the outer circumference of the combs between two previously constructed cells. As more cells are added to the evolving structure, they eventually form closely packed parallel rows of cells and the nest generally has radial or bilateral symmetry around these initial cells. A wasp tends to finish a row of cells before initiating a new row, and rows are initiated by the construction of a centrally located first cell.

The number of potential sites where a new cell can be added increases significantly as construction proceeds: several building actions can, in principle, be

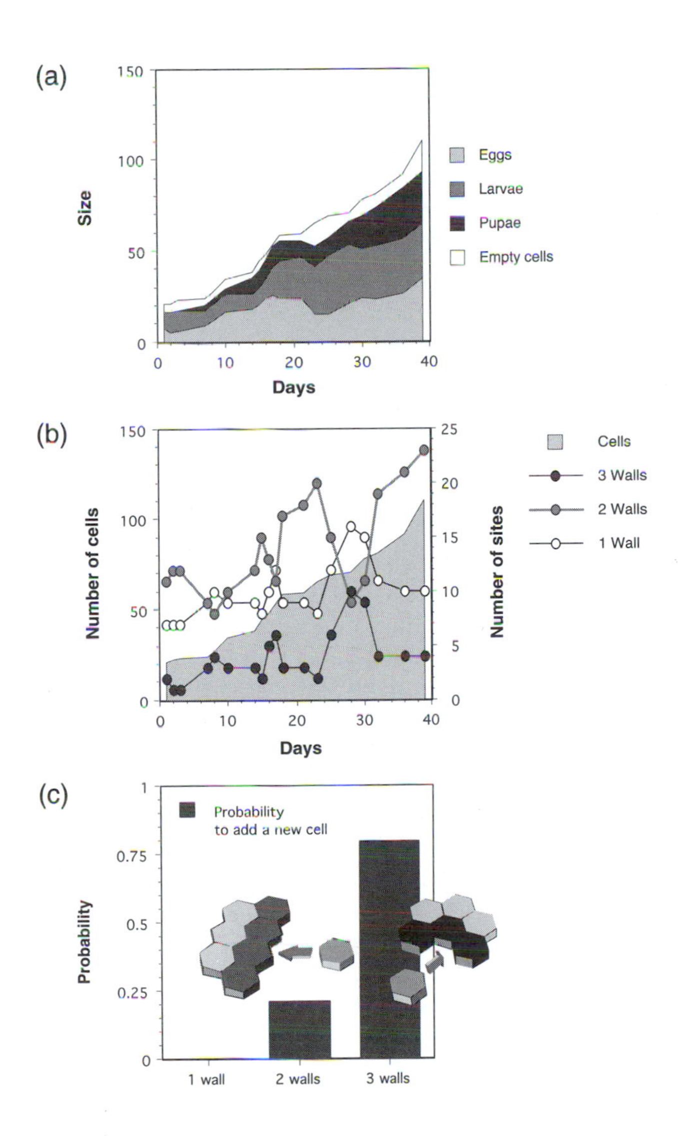

FIGURE 6.3 (a) Number of cells of different types as a function of time in a paper wasp nest (*Polistes dominulus*). (b) Development of nest structure and dynamics of the local building configurations. (c) Differential probability to add a new cell to a corner area where three adjacent walls are present and on the side of an existing row where two adjacent walls are present.

made in parallel—and this is certainly an important step in the emergence of complex architectures in the course of evolution. Parallelism, however, could deorganize the building process by introducing the possibility of conflicting actions being performed simultaneously. But the architecture seems to provide enough constraints to canalize the building activity. It can be seen, with a careful study of the dynamics of local configurations of cells during nest development (Figure 6.3(a)) in the primitively eusocial wasp *Polistes dominulus*, that there are not equal numbers of sites with one, two, or three adjacent walls. The great majority of them is composed of sites with two adjacent walls (Figure 6.3(b)). Cells are not added randomly to the existing structure: wasps have a greater probability to add new cells to a corner area where three adjacent walls are present, than to initiate a new row by adding a cell on the side of an existing row (Figure 6.3(c)). Therefore, obviously, wasps are influenced by previous construction, and building decisions seem to be made locally on the basis of perceived configurations in a way that possibly constrains the building dynamics.

6.3 MODEL OF SELF-ASSEMBLY

6.3.1 MODEL

In order to explore the potential of discrete stigmergy, Theraulaz and Bonabeau [310, 311] introduced an ensemble of algorithms that could hardly be made simpler: asynchronous automata that move in a three-dimensional discrete space and behave locally in space and time on a pure stimulus-response basis. The deposit of an elementary building block (hereafter called a brick) by an agent depends solely on the local configuration of bricks in the cells surrounding the cell occupied by the agent. When a stimulating configuration is encountered, the agent deposits a brick with probability one (brick deposits are deterministic). Figure 6.4 shows a schematic representation of the surrounding cells that an agent can perceive in a cubic lattice. Two types of bricks can be deposited. No brick can be removed once it has been deposited. All simulations start with a single brick: in the absence of this initial brick in the environment, one rule would have to allow the spontaneous deposit of bricks when no other brick is present, but such a rule would result in random brick deposits that would fill space. We call microrule the association of a stimulating configuration with a brick to be deposited, and we call algorithm any collection of compatible microrules. Two microrules are not compatible if they correspond to the same stimulating configuration but result in the deposition of different bricks. An algorithm can therefore be characterized by its microrule table, a lookup table comprising all its microrules, that is, all stimulating configurations and associated actions. See Algorithm 6.1.

A single agent in this model is capable of completing an architecture. In that respect, building is a matter of purely individual behavior. But the individual building behavior, determined by the local configurations that trigger building actions, has to be organized in such a way that a group of agents can produce the same

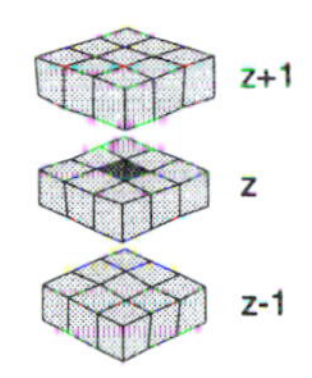

FIGURE 6.4 Schematic representation of the perception range of an agent in a cubic lattice. After Theraulaz and Bonabeau [311]. Reprinted by permission © *Academic Press*.

Algorithm 6.3 High-level description of the construction algorithm

```
/* Initialization */
Construct lookup table /* identical for all agents */
Put one initial brick at predefined site /* top of grid */
For k = 1 to m do
  assign agent k a random unoccupied site /* distribute the m agents */
End For
/* Main loop */
For t = 1 to t_max do
  For k = 1 to m do
    Sense local configuration
    If (local configuration is in lookup table) then
      Deposit brick specified by lookup table
      Draw new brick
    Else
      Do not deposit brick
    End If
    Move to randomly selected, unoccupied, neighboring site
  End For
End For
/* Values of parameters used in simulations */
m = 10
```

architecture as well. Some natural wasp species face the same problem, since nest construction is generally first achieved by one female, the founder, and is then taken over by a group of newly-born workers. The group of agents has to be able to build the architecture without the combined actions of the different agents interfering and possibly destroying the whole activity of the swarm. In other words, individual activities have to be coordinated to ensure coherent construction.

The model can also be considered a generalization of diffusion-limited aggregation (DLA) and other growth models. In DLA, a randomly moving brick sticks to a cluster as soon as the cluster is encountered. DLA is a specific case of the model, where more complicated microrules can be implemented. DLA, however, is not efficiently coded in the present model: $63 \times 2^{20} (\approx 66.1 \times 10^6)$ microrules are required to code DLA in three dimensions, and 240 microrules in two dimensions. This shows that very simple physical laws may be very hard to implement in the model, unless some kind of alternative coding is adopted, such as don't-care-cells, that is, cells whose state does not alter the decision to deposit a brick, that is, a brick is deposited or not deposited, irrespective of the state of don't-care-cells: if such a procedure is used, the coding of DLA requires only 6 microrules in three dimensions and 4 microrules in two dimensions. Conversely, it also indicates that some architectures which appear to require numerous microrules may be simply and more compactly expressed in terms of meaningful physical laws.

6.3.2 SIMULATIONS

A random exploration of the space of the algorithms defined by the model of section 6.3.1 yields no interesting result but only random or space-filling shapes. However, it is possible to produce complex shapes, some of them strikingly similar to those observed in nature, with algorithms belonging to this simple family [26], by working backward from some shapes to be generated to their corresponding algorithms. Theraulaz and Bonabeau [310, 311] found *a posteriori* that structured shapes can be built only with special algorithms, coordinated algorithms, characterized by emergent coordination: stimulating configurations corresponding to different building stages must not overlap, thereby avoiding the deorganization of the building activity. This feature creates implicit handshakes and interlocks at every stage, that is, constraints upon which structures can develop consistently. This coordination is essential, since parallelism, that has a clear adaptive advantage by allowing a real colony to build at several locations at the same time, introduces the possibility of conflicts. This approach (which is described at length in Bonabeau et al. [26] and Theraulaz and Bonabeau [310, 311]), therefore shows how the nest itself can provide the constraints that canalize a stigmergic building behavior into producing specific shapes.

In Figure 6.5, all architectures, except 6.5(g) and 6.5(h), have been obtained with coordinated algorithms. The difference between (6.5(g), 6.5(h)) and all other architectures is striking, though hard to formalize. One given coordinated algorithm always converges toward architectures that possess similar features. On the other hand, algorithms that produce unstructured patterns may sometimes diverge: the same algorithm leads to different global architectures in different simulations. As an illustration of this fact, architectures 6.5(d) and 6.5(e) result from two different simulations using the same coordinated algorithm, and architectures 6.5(g) and 6.5(h) result from two different simulations of the same noncoordinated algorithm. We see that there is a high degree of structural similarity between 6.5(e) and 6.5(d), in sharp contrast with 6.5(g) and 6.5(h). This tendency of noncoordinated algorithms to diverge results from the fact that stimulating configurations are not organized in time and space and many of them overlap, so that the architecture grows in space without any coherence. It is not true, however, that every unstructured architec-

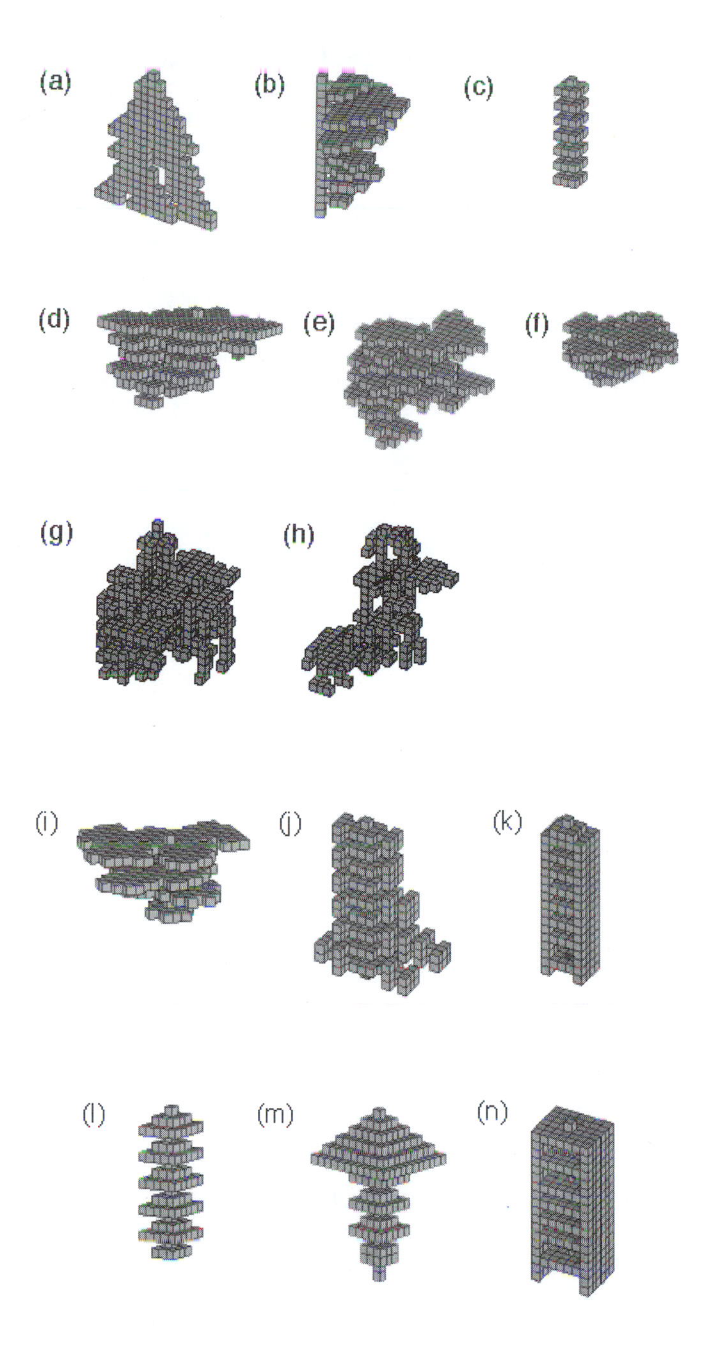

FIGURE 6.5 Caption on next page.

FIGURE 6.5 Simulations of collective building on a 3D cubic lattice. Simulations were made on a 20 × 20 × 20 lattice with 10 wasps. For those architectures which are reminiscent of natural wasp nests, we give the name of the genera exhibiting a similar design. After Theraulaz and Bonabeau [311]. Reprinted by permission © *Academic Press.* (a) Nest-like architecture (*Parapolybia*) obtained after 20,000 steps. (b) Nest-like architecture (*Parachartergus*) obtained after 20,000 steps. (c) Achitecture obtained after 20,000 steps. (d) Nest-like architecture (*Stelopolybia*) obtained after 20,000 steps. (e) Nest-like architecture (*Stelopolybia*) obtained after 20,000 steps. (f) Nest-like architecture (*Stelopolybia*) obtained after 20,000 steps. (g) Architecture obtained after 15,000 steps using a noncoordinated algorithm. (h) Architecture obtained after 15,000 steps using the same noncoordinated algorithm as in Figure 6.5.g. (i) Nest-like architecture (*Vespa*) obtained after 20,000 steps. (j) Architecture obtained after 80,000 steps. (k) Nest-like architecture (*Chatergus*) obtained after 185,000 steps. (l) Architecture obtained after 125,000 steps. (m) Architecture obtained after 85,000 steps. (n) Nest-like architecture (*Chatergus*) obtained after 100,000 steps.

ture is generated with a nonconvergent algorithm: some of them can be produced consistently in all simulations. Moreover, even in shapes built with coordinated algorithms, there may be some degree of variation, which is higher in cases where the number of different choices within one given building state is large. For example, we see that 6.5(d) and 6.5(e) are similar but not perfectly identical: this is because there are several possible options at many different steps in the building process, or, in other words, the process is not fully constrained. Architectures 6.5(k), 6.5(l), 6.5(m), and 6.5(n) provide examples of fully constrained architectures: two successive runs produce exactly the same shapes.

Some of the structured architectures of Figure 6.5 are reminiscent of natural wasp nests. Among these architectures the presence of plateaus is observed in 6.5(b), 6.5(d), 6.5(e), 6.5(f), 6.5(i), and the different levels of the nest are linked together with either a straight axis (6.5(b), 6.5(i)) or with a set of artificial one-brick "pedicels" (6.5(d), 6.5(e), 6.5(f)). Some others possess an external envelope: architectures 6.5(k) and 6.5(n) are shown with a portion of the front envelope cut away to allow for the visualization of the nest's interior, and correspond to nests found in the *Chartergus* genera. Figures 6.5(c), 6.5(j), 6.5(l) and 6.5(m), are examples of structured patterns that do not look like real nests.

One limitation, among others, of the model (if considered as a model of wasps' building behavior) is that the simulated swarms live in an abstract space using cubic bricks to build architectures, while natural wasp nests are built with hexagonal cells as constituent units: symmetry properties are not equivalent. To overcome this limitation, simulations have been performed with bricks that are hexagonal in the xy-planes (horizontal planes are triangular lattices). Figure 6.6 shows the equivalent of Figure 6.4 for a hexagonal lattice. Figure 6.7 presents a few of the architectures obtained with this particular topology. We can see that these shapes look more "biological" than shapes generated with cubic bricks.

6.3.3 ANALYSIS

Obtaining "structured" architectures with the very simple model introduced in sections 6.3.1 and 6.3.2 is an interesting result. But still more interesting is to go one step further and to try to understand the properties of those algorithms that generate structured architectures. In that respect, several useful observations have been made by Theraulaz and Bonabeau [310, 311]. They noticed:

1. The smoothness, or more precisely the consistency, of the mapping from algorithms to architectures in the structured subspace: two close-structured architectures appear to be generated by two close-coordinated algorithms. This result extends to unstructured patterns generated by coordinated algorithms. That two close algorithms generate two close architectures could not be fully tested but is certainly not generally true: for example, removing a microrule from a coordinated algorithm's microrule table may result in a deorganization of the building process. Moreover, outside the subspace of coordinated algorithms, one given algorithm may produce quite different patterns in different runs, so that comparing noncoordinated algorithms among themselves is irrelevant. Therefore, we expect the algorithmic landscape to be generally rugged but somewhat smoother in the coordinated subspace.
2. The compactness of the subspace of structured shapes: a factorial correspondence analysis [19, 215] showed that not only are such shapes rare, but they

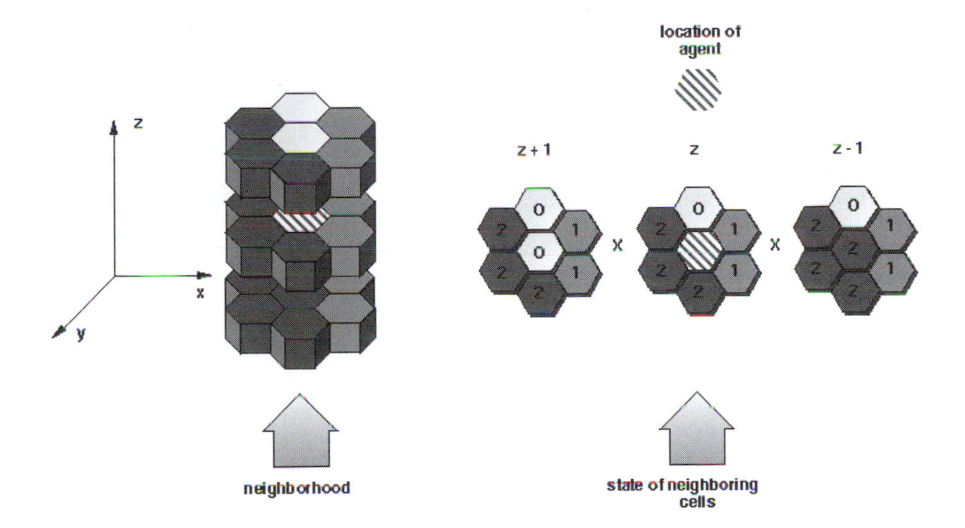

FIGURE 6.6 Schematic representation of the perception range of an agent in a hexagonal lattice. Symbol 0 represents an empty site. Symbols 1 and 2 represent sites occupied by bricks of type-1 and type-2, respectively.

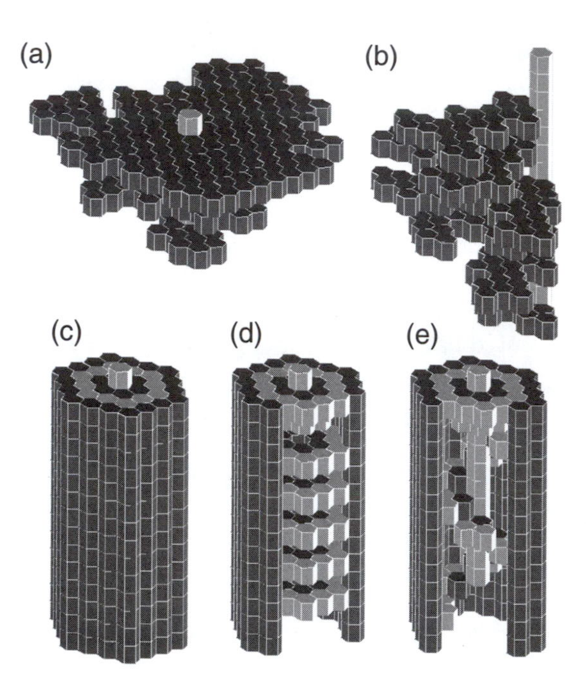

FIGURE 6.7 Simulations of collective building on a 3D hexagonal lattice. Simulations were made on a $20 \times 20 \times 20$ lattice with 10 wasps. Some architectures are reminiscent of natural wasp nests and exhibit a similar design. (a) Nest-like architecture (*Vespa*) obtained after 20,000 steps. (b) Nest-like architecture (*Parachartergus*) obtained after 20,000 steps. (c,d) Nest-like architecture (*Chatergus*) obtained after 100,000 steps. A portion of the front envelope has been cut away in 6.7(d). (e) Lattice architecture including an external envelope and a long-range internal helix. A portion of the front envelope has been cut away.

also all lie in the same, small region (Figure 6.8). This property suggests that the type of shapes that can be produced with this family of algorithms is highly constrained.

To better understand what a coordinated algorithm is, let's review an example—depicted in Figure 6.9. This figure represents the successive steps of the construction of an architecture which looks like nests built by *Epipona* wasps. The transition between two successive building steps is shown to depend on a given number of local configurations that stimulate the deposit of a brick. Once all the bricks in the current step have been deposited, the building process goes on to the next step. Steps 1

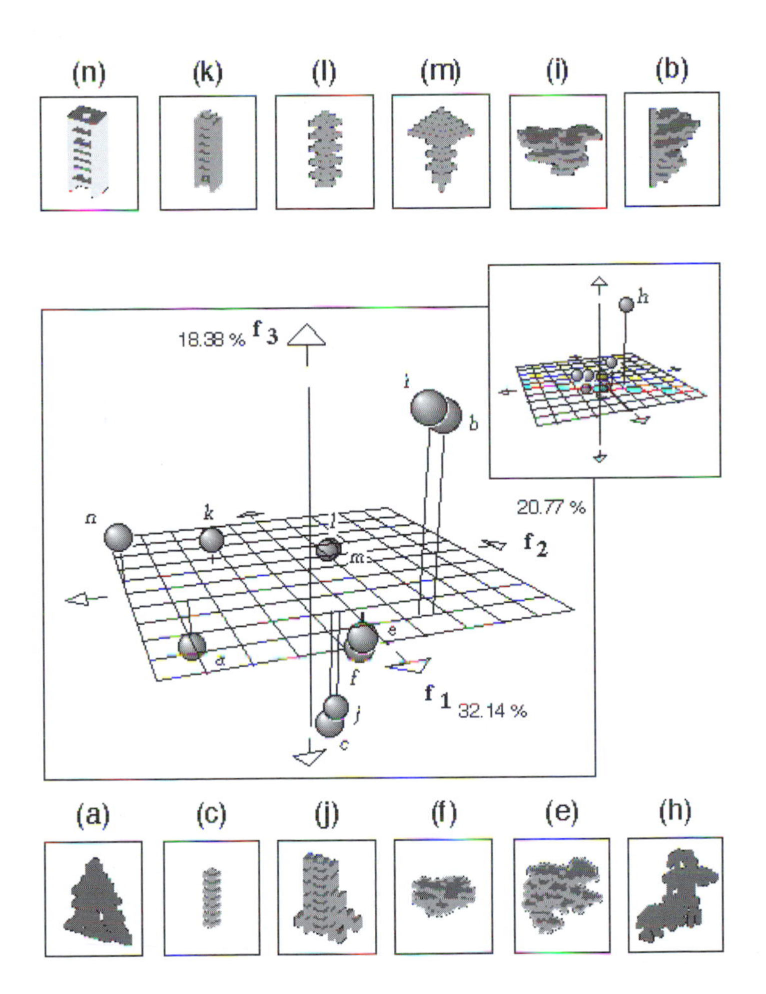

FIGURE 6.8 Factorial Correspondence Analysis (FCA). The three axes are the leading three eigenvectors of a matrix M the columns of which represent 12 architectures of Figure 6.5, and the rows of which represent the 211 associated stimulating configurations: for instance $M_{ij} = 0$ if architecture j has not been obtained with i as a stimulating configuration, and $M_{ij} = 1$ if i is a stimulating configuration necessary for the generation of architecture j. The subspace occupied by the architectures grown with coordinated algorithms is magnified. After Theraulaz and Bonabeau [311]. Reprinted by permission © *Academic Press.*

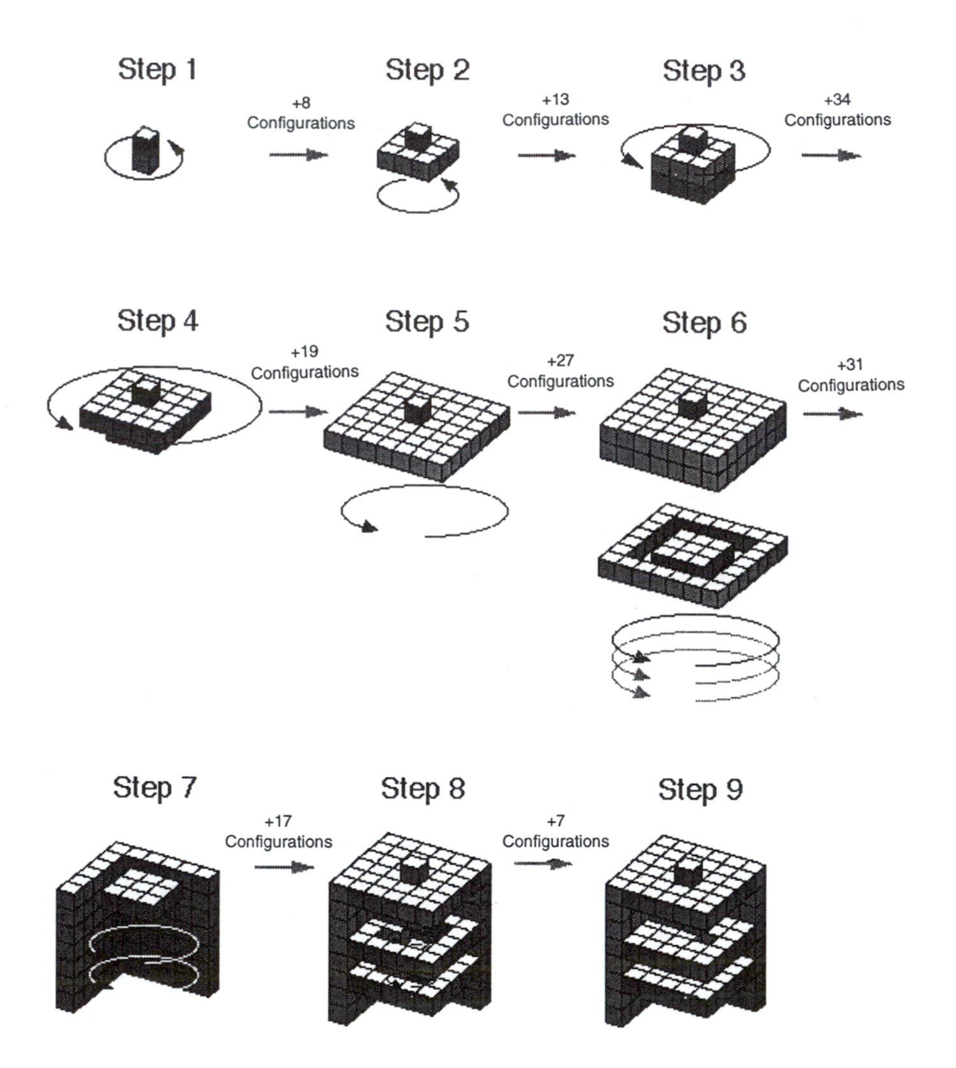

FIGURE 6.9 Successive building steps in the construction of an *Epipona* nest with a lattice swarm. The completion of each step gives rise to stimulating configurations belonging to the next step. All stimulating configurations are organized to ensure a regular building process. In particular, within a given step, the stimulating configurations do not have to be spatially connected and can occur simultaneously at different locations. In steps 7 to 9, the front and right portions of the external envelope have been cut away. After Theraulaz and Bonabeau [311]. Reprinted by permission © *Academic Press.*

to 5 correspond to the enlargement of the top of the nest, including the first sessile comb of cells (step 3). Step 6 represents the construction and lengthening of the external envelope, from which parallel combs will be built (steps 7 and 8). These steps determine the distinction between this nest, where the entrance and access holes at the different levels lie in the periphery of the comb, from the *Chartergus* nest represented in Figures 6.5(k) and 6.5(m), where the hole lies in the middle of the nest.

Coordination appears to be a necessary but not sufficient condition to produce structured architectures: there is coordination in many cases, but the architecture produced exhibits no obvious structure. Therefore algorithms must have additional properties that produce structured architectures to explain their ability to generate such architectures.

Figure 6.10 illustrates the concept of modularity. Recurrent states may appear, inducing a cyclicity in the group's behavior. From the architectural viewpoint, this corresponds to a modular pattern, where each module is built during a cycle. All modules are qualitatively similar. The two figures on the right of Figure 6.10 show how modularity is implemented in a nest with successive plateaus: each cycle corresponds to the addition of a plateau. Figure 6.10(a) shows the final architecture (*Stelopolybia*), with the various stages, while Figure 6.10(b) explicitly shows the successive steps. Figure 6.10(c) shows a complete architecture of *Chartergus* and Figure 6.10(d) how modularity is implemented.

6.3.4 EXPLORING THE SPACE OF ARCHITECTURES

6.3.4.1 Defining a fitness function In order to explore the space of architectures grown by the simple agents described in section 6.3.1, Bonabeau et al. [36] use a genetic algorithm (GA). The goal of a genetic algorithm (GA) [126, 152] is to explore a space of solutions with a fitness function (a solution being simply a point in that space: the goal is to find a point that has a high fitness), by a mechanism inspired by natural selection. At each generation, a pool of individuals, characterized by their genes or genotype, is selected according to the fitness function: only the fittest individuals, that is, those individuals that have a high fitness given the problem to be solved, are allowed to reproduce and be present in the next generation. During reproduction, two mechanisms may be applied: (1) mutations, which are usually implemented as random alterations of genotypes, and (2) crossover, which consists of exchanges of portions of genotypes between two selected individuals. These two mechanisms allow exploration of the space of solutions from existing good solutions: in effect, mutations are small alterations which, when applied to good genotypes, may produce nearby good solutions, and crossover relies on the idea that combinations of good solutions may result in even better solutions.

Owing to the difficulty of defining a fitness function for architectural phenotypes (it is difficult to define formally the adaptive value or the biological plausibility of the shape of a nest), Bonabeau et al.'s [36] fitness function is based on a set of subjective criteria: the properties of structured architectures, described in

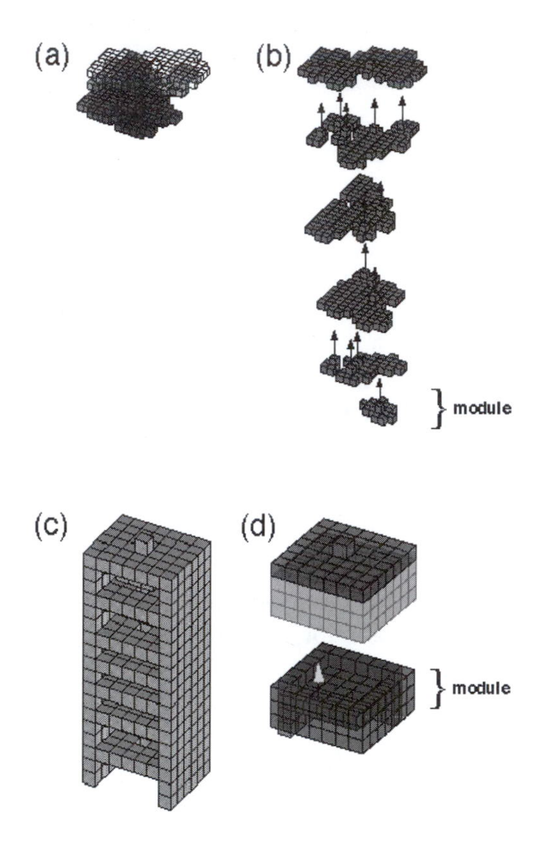

FIGURE 6.10 Formation of modular structures as a by-product of recurrent stimulating configurations in the architecture. A *Stelopolybia*-like nest (a, b) and a *Chartergus*-like nest (c, d) are depicted. After Theraulaz and Bonabeau [311]. Reprinted by permission © *Academic Press.*

sections 6.3.2 and 6.3.3, are transformed into quantitative traits measured during the construction of the architecture. The fitness function, which can be measured automatically without any human intervention, reflects the ratings of 17 human observers who were asked to evaluate the amount of "structure" in a set of 29 patterns [36]. The fitness function is based on the following observations:

1. In algorithms that generate coherent architectures many microrules are used, whereas in algorithms that generate structureless shapes only one microrule or a few microrules are actually used in the course of the simulation. A given al-

gorithm is defined by its microrule table, which assigns an action to be taken to each configuration of bricks. Some of the microrules may never be used, because the configurations of bricks they represent never occur. Only algorithms in which many microrules are actually applied in the course of the simulation can produce coherent architectures, because such architectures require a subtle and complex sequence of correlated deposits. Obviously, this condition is necessary but not sufficient, but a population of algorithms containing a large proportion of algorithms that satisfy this criterion is certainly a much better point to start from.

In relation to this observation, it is also interesting to mention that it usually takes more time to generate a complex architecture because of the constraints that are generated in the course of construction. Each constraint acts as a bottleneck, preventing any progress from being made until the constraint is satisfied. This result relates to the notion of logical depth, which is defined as the time it takes for a program to generate a given pattern [17]. Interesting architectures are logically deep.

2. Most structured architectures are compact in the following sense: bricks have adjacent faces with many of their neighbors. Building compactly requires collections of complementary, correlated microrules.
3. Structured architectures are most often characterized by modular patterns, that is, patterns that repeat themselves (see Figures 6.5 and 6.7). This modularity requires coordination. More complex architectures are characterized by more complex and larger atomic modules. Therefore one has to look for (1) large patterns that (2) repeat themselves.

6.3.4.2 Genetic algorithm A genetic algorithm was designed to make use of the observations made in section 6.3.4.1 [36]. Each generation is composed of 80 individuals. A straightforward fitness-proportionate selection GA is used: the probability for an individual to be selected into the reproductive pool is proportional to its fitness (roulette-wheel sampling [152]). New individuals are generated until a new generation of 80 individuals is formed. Algorithms are encoded with a variable-length genotype. A gene is the equivalent of a microrule. An algorithm that has n microrules in its rule table is encoded into n genes. Genes cannot be cut by crossover (Figure 6.11). The number of microrules of a newly generated algorithm in the initial population is a random variable. Its probability distribution is gaussian with mean $\mu = 50$ and standard deviation $\sigma = 10$.

The selection of microrules in the initial population, and when new random microrules are injected into subsequent generations, must be biased to allow construction to start:

1. Stimulating configurations containing a large number of bricks must be avoided because they may never be encountered in the course of one simulation. This

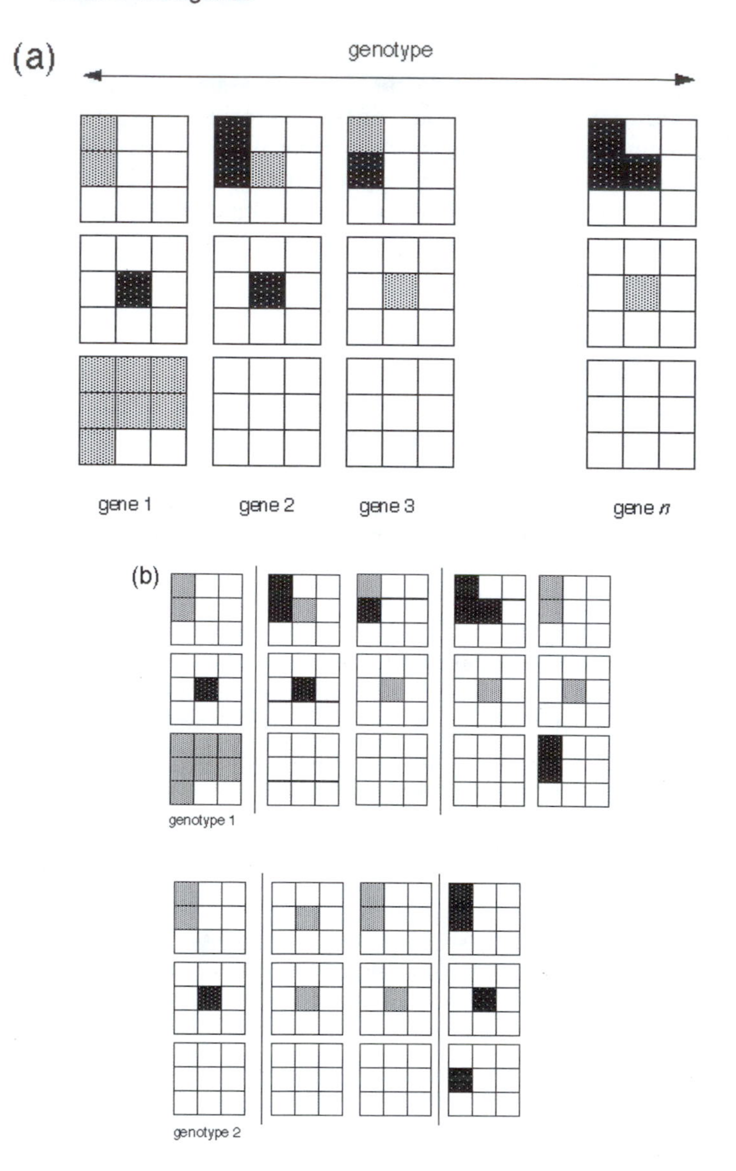

FIGURE 6.11 (a) Genotype of an n-gene individual that implements an algorithm with n microrules with two types of bricks, grey and black. Each gene corresponds to a microrule: the central site of the central 9×9 square is occupied by the brick to be deposited, where the agent is located; the remaining 26 sites are either empty or occupied by a neighboring brick. (b) Genotypes of a 5-gene individual and a 4-gene individual selected for crossover. Vertical bars indicate locations of crossover application. *continued.*

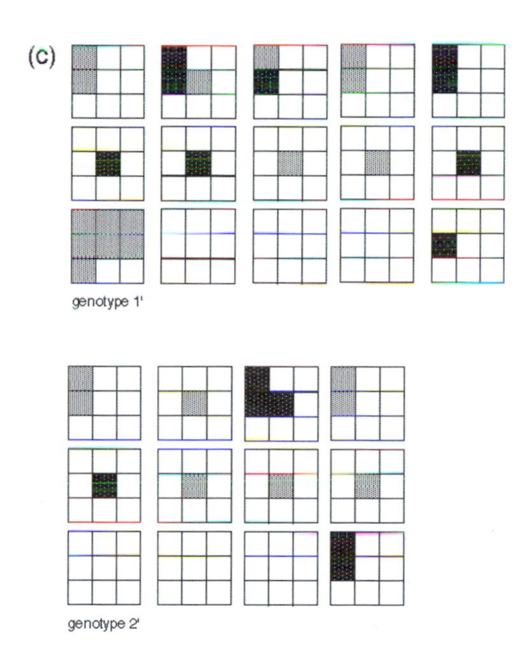

FIGURE 6.11 *continued.* (c) Genotypes obtained after application of the 2 point crossover operator.

bias is implemented by using probabilistic templates that, in addition to probabilistically ensuring that all directions are evenly covered, implement a rapidly decaying probability distribution for the number of bricks in stimulating configurations. A template simply defines the probability of putting a brick in each of the 26 cells that make a stimulating configuration. About 10 templates are used: for each stimulating configuration to be generated, one template is randomly selected.

2. Simulations start with one and only one brick in space, so that there must be one microrule in which the stimulating configuration contains only one brick.
3. Diagonal brick deposits are not allowed, because they often lead to quick random space filling and are never necessary to obtain interesting architectures. A diagonal brick deposit is one in which the deposited brick has only adjacent edges (as opposed to adjacent faces) with already present bricks. Only stimulating configurations which allow the deposit of bricks that have at least one face in common with at least one other brick are used.

In addition to randomly generated algorithms, the initial population also contains a small number of algorithms that are known to produce structured patterns.

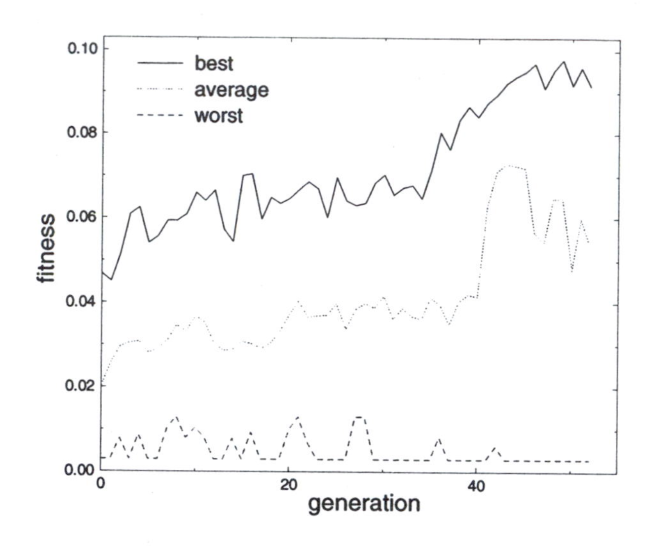

FIGURE 6.12 Best, average, and worst fitnesses as a function of generation number.

Such algorithms focus evolutionary exploration in the vicinity of coordinated algorithms.

Mutations are applied to every gene of every algorithm that results from the combination of two algorithms in the reproductive pool. Mutations are applied as follows: if a microrule has not been used during a simulation, its corresponding gene is replaced with a randomly generated microrule with probability p_1; if the microrule has been used, the same procedure is applied, but with probability p_2, with $p_2 \ll p_1$, in order to favor applicable microrules. In the simulations, $p_1 = 0.9$ and $p_2 = 0.01$.

Crossover is applied to randomly selected pairs of genotypes with probability $p_c = 0.2$. In the simplest version of the GA, a two-point crossover is performed at two randomly selected points (Figures 6.11(b) and 6.11(c)).

A given simulation stops either when a predefined number of bricks have been deposited in a $16 \times 16 \times 16$ grid, or after a predefined number of steps. In all simulations, the maximum number of bricks is 500 and the maximum number of iterations is 30000, where one iteration corresponds to all agents taking one action (10 agents are used in the runs). Most algorithms generate random space filling quite rapidly: limiting the number of bricks permits advancement to the next simulation without spending too much time waiting for the end of a run that would not produce anything interesting. On the other hand, simulations have to be long enough because it takes time to produce coherent architectures, owing to the numerous constraints they induce.

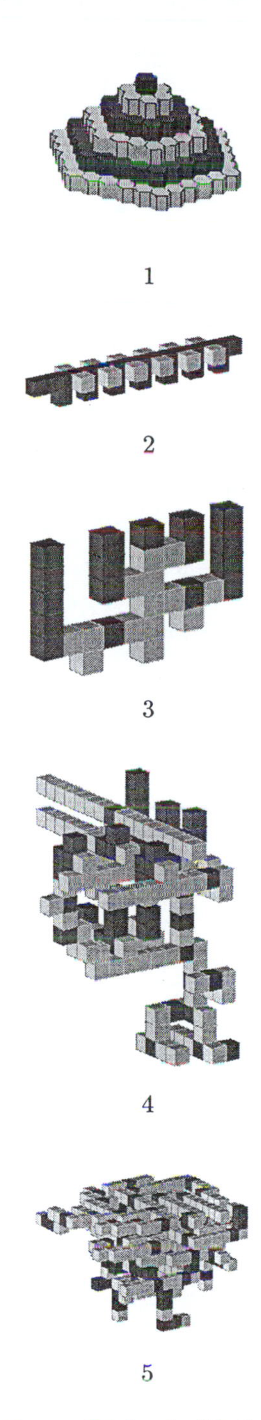

FIGURE 6.13 Example of patterns obtained with the genetic algorithm.

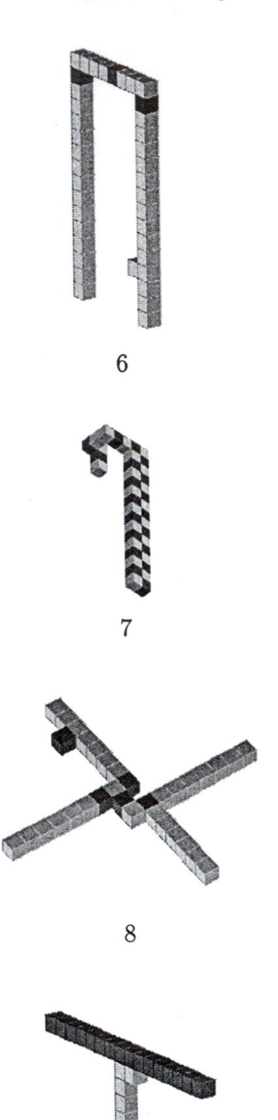

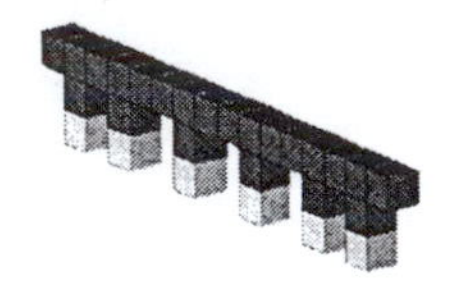

FIGURE 6.13 *continued.*

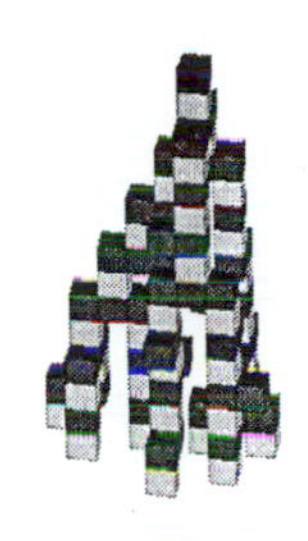

11

12

13

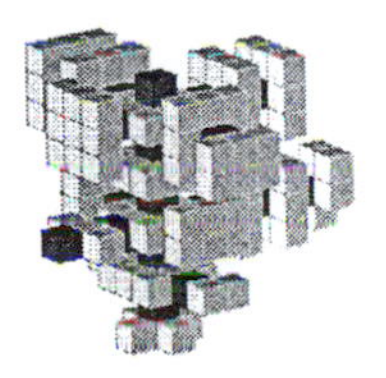

14

FIGURE 6.13 *continued.*

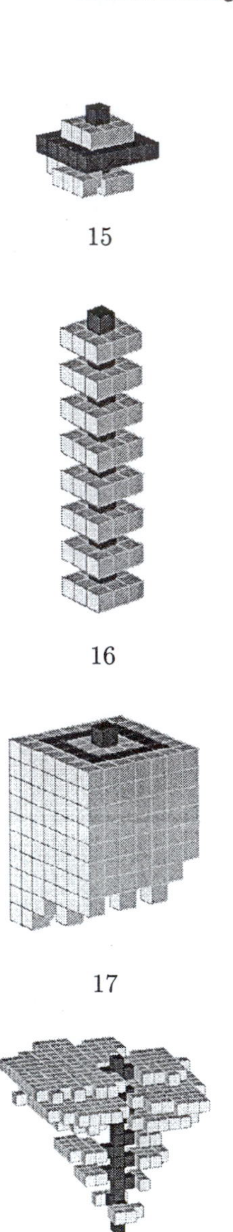

FIGURE 6.13 *continued.*

20

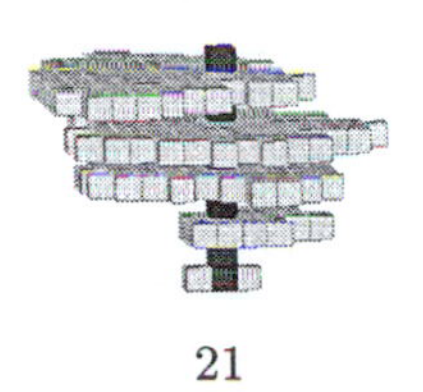

21

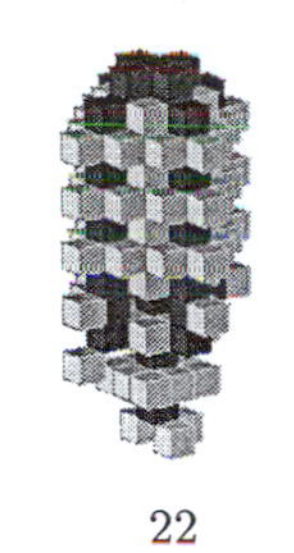

22

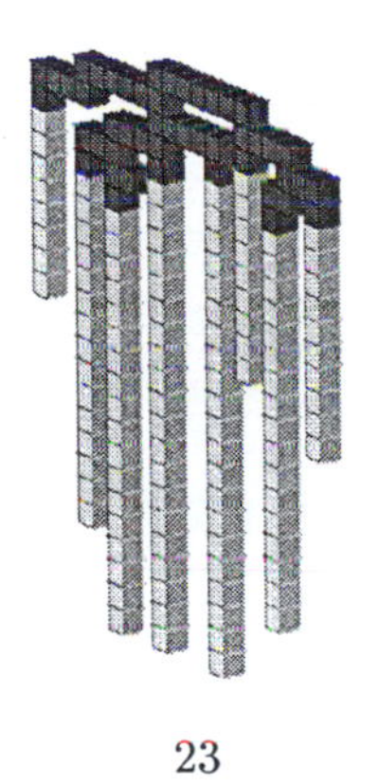

23

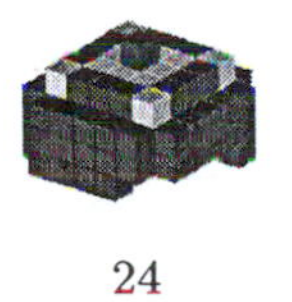

24

FIGURE 6.13 *continued.*

25

26

27

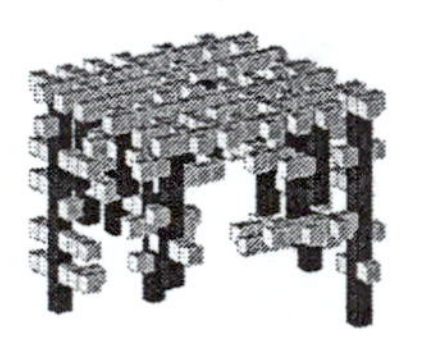

28

29

FIGURE 6.13 *continued.*

Figure 6.12 shows a typical example of how the fitness function evolves with generations. The average and best fitnesses increase, while the lowest fitness always remains close to 0, most often because of algorithms for which no microrule gets activated. The increase of the fitness function is slow, because of the epistatic interactions among genes. In effect, the application of some microrules requires the prior application of other microrules: the architectural fitness landscape is more rugged than the factorial correspondence analysis of section 6.3.3 suggests. The patterns generated by the model are inherently polygenic [249], that is, in the context of the model, they require the "cooperation" of several microrules. Other models that lead to similar conclusions (that whether a particular gene is perceived to be a major gene, a modifier gene or a neutral gene depends entirely on the genetic background in which it occurs, and that this apparent attribute of a gene can change rapidly in the course of evolution [249]) include Nijhout and Paulsen's [249] development model that uses a diffusion gradient and threshold mechanism, the parameters of which are affected by genes and in turn affect the phenotype; or Eggenberger's [118] model of multicellular organisms using cell division and differential gene expression in identical cells. In both cases, the fitness function is extremely simple. For example, Eggenberger's [118] fitness function is simply a function of the number of cells in the organism and of the symmetry of the organism with respect to a predefined axis. Despite the simplicity of the fitness function, understanding the evolutionary dynamics of these models is difficult because of interactions among genes. Another interesting conclusion is that epistatic processes reduce the need for a long genome [118]: in the model, interactions among microrules generate constraints and lead to the precise unfolding of the architecture without any need for the architecture to be encoded in its entirety.

Some structured patterns can, however, be found with the genetic algorithm. Figure 6.13 shows a few examples of patterns obtained with the genetic algorithm together with their associated fitness value F. Structured patterns include Figures 6.13(1), 6.13(5), 6.13(6), 6.13(7), 6.13(8), 6.13(9). It is particularly interesting that some of these patterns make use of submodules (Figures 6.13(5), 6.13(7), 6.13(9)).

6.3.4.3 Structure and function In the previous subsection, the genetic algorithm made use of a fitness function related to the structure of the generated pattern. One may argue that *function*, rather than *structure*, is the relevant property of social insect nest architectures. However, the level of functionality or adaptiveness of a model architecture is particularly difficult to test. Krink and Vollrath [204], in a remarkable work, have been able to do just that with a model of web construction behavior in orb web-building spiders: they measured the fitness associated with a given artificial web by computing costs and benefits. Costs were determined by construction time and the amount of silk used in the web, subdivided into sticky and nonsticky silk. Benefits were determined by measuring the response of the web to artificial prey items (total amount of prey caught, weighted by the impact position, size and value of each single item). Their model lends itself well to this type of fitness evaluation. The model presented in sections 6.3.1–6.3.3 is too abstract to allow functional features to be examined directly. In addition, it is difficult to define, for example, what the antipredator function exactly is, or how thermoregulation is achieved, all the more as ethologists themselves often do not know with certainty how such functions are actually implemented in real insect colonies. But ultimately, one hopes to be able to evolve architectures according to their functionality.

If function is difficult to formalize, an alternative choice for exploring the space of architectures consists of using an interactive GA, whereby a human observer evaluates the fitness of a pattern according to her taste or goal, and algorithms which generate the most "interesting" architectures would be selected [289]. The term interesting may refer to biological resemblance or to any other criterion, subjective or objective: biologists would prefer biologically plausible architectures and engineers may be looking for useful or simply appealing patterns. But, up to now, attempts to involve a human agent to interact with the evolutionary algorithm have been unsuccessful, perhaps because of the ruggedness of the landscape and/or the rarity of structured shapes.

6.4 BEYOND BIOLOGY

6.4.1 INTRODUCTION

The problem of coherent distributed building or self-assembly has counterparts in computer science and engineering. The model described in the previous sections can provide engineers with valuable insight into the design of programs or machines that exhibit collective problem-solving abilities, such as self-assembling machines or robots (e.g.: Bowden et al. [38], Hosokawa et al. [176, 177, 178, 179, 180], Fukuda et al. [136], Chirikjian [69], Murata et al. [245], Kokaji et al. [200], Tomita et al. [315], Yoshida et al. [336, 337]). Areas of current potential applications for self-assembly include microelectronics, optics, micro-electromechanical systems and displays.

Hosokawa et al. [176] studied the self-assembly of macroscopic triangular objects with magnets on two sides of the triangle. Hosokawa et al. [177, 178] designed a self-assembling system of mesoscopic objects (polyimide and polysilicon thin films, 400 μm wide) using surface tension as the bonding force. Bowden et al. [38] designed two-dimensional arrays made out of mesoscale objects (from 1 mm to 1 cm) by self-assembly. The objects are floating at the interface between perfluorodecalin and water and interact by lateral capillary forces. The shapes of the assembling objects and the wettability of their surfaces determines the structure of the arrays. What all these examples have in common is that they deal with objects that are either macroscopic or at least of the order of a half millimeter: in these examples, self-assembly relies on very different principles than in molecular systems [198, 326]. One step further, one promising area of application for macroscopic self-assembly is the design of self-assembling, reconfigurable robotic systems composed of numerous interacting macroscopic submodules. Self-assembling swarm-based robotic systems in which each robot is very small and endowed with limited perception and computation are perhaps the most promising application of swarm-based approaches in robotics. In section 6.4.2, we describe three examples of successful implementations of self-assembling robots.

In the design of self-assembling robotic systems, engineers face the same "inverse problem" as biologists: given a desired shape, find an algorithm (if possible a simple one) that can generate it under some material constraints. The self-assembling robotic systems described in the next section have not been inspired by the model of collective building developed in this chapter, but they do constitute an instance of self-assembly in the robotic domain. The model, by providing a better understanding of the dynamics of self-assembling systems in which each building block is a computational unit that can be programmed to respond to tunable specific local configurations, is obviously of relevance to these robotic systems. One cannot deny the technical *tour de force* that was necessary to implement these systems, but the development of "software" for such systems is still in its infancy. The model in this chapter provides guidelines for the development of such software.

Another area of potential application for the model described in this chapter is architecture [46, 75] or art [289]. A relatively recent trend in art, design, and architecture consists of using computer-based generative mechanisms. The artist, the designer, or the architect then has to pick from among all patterns generated by the computer those that are "appealing." This approach is described briefly in section 6.4.3.

Finally, the use of evolutionary algorithms to "evolve" useful self-assembling robotic systems or appealing forms is very promising. This is described in section 6.4.4.

6.4.2 SELF-ASSEMBLING ROBOTS

A pioneer of the idea of designing self-assembling machines, Penrose [260] was interested in self-reproduction. But the concept of self-assembling robots was seriously introduced into the robotics community only by Fukuda et al. [136]. Since then, several attempts have been made to design collective robotic systems that self-assemble. In this section we discuss four recent remarkable advances for generating self-assembling robots. These works are remarkable in that they combine a lot of technical skills to produce collective reconfigurable robotic systems, where each robot is a (relatively) small object with a set of simple behaviors. Although these works are not directly inspired by construction in social insects, they follow the same spirit as the model introduced in section 6.3 (simple agents, local information) and could benefit from the model.[1]

[1]The interesting work of Yim [334, 335], who designed modular robots composed of a few elementary building blocks that can be composed into complex robots for various types of locomotion, is not described in this book because it deviates from the spirit of self-assembly.

6.4.2.1 Metamorphic robots Chirikjian [69] defines a metamorphic robotic system as a collection of independently controlled mechatronic modules [69, 70, 257, 258], each of which has the ability to connect, disconnect, and climb over adjacent modules. Each module can be considered a robot. The four self-assembling robotic systems described in section 6.4.2 are of that type: motion is restricted to relative motion of the modules. Every module is always in contact with at least one other module, but has enough degrees of freedom to walk over or around other modules when constraints (connection with other modules) permit and reconfigure without any external help. The symmetry properties of the modules are such that lattices of modules can be formed which fill spatial regions with minimal gaps. Because of the permanent connectedness of the ensemble of modules, the system as a whole acts as a single physical object. All modules have the same physical structure (homogeneity), and each module is autonomous from the viewpoint of computation and communication.

Pamecha et al. [258] identify the following potential applications of metamorphic robotic systems:

1. Obstacle avoidance in highly constrained and unstructured environments.
2. Formation of bridges, buttresses, and other civil structures in times of emergency.
3. Envelopment of objects, such as recovering satellites from space.
4. Performing inspections in constrained environments such as nuclear reactors.

More generally, this type of system can be used for the self-organizing unfolding of structures without any human intervention, from spatial stations or satellites to scaffolds.

Pamecha et al. [257] introduce two types of modules, hexagonal and square, and explore how they can move with respect to one another while always keeping in contact. A planar hexagonal module is a closed six-bar linkage with three degrees of freedom controlled by placing actuators at alternate joints of the module (Figure 6.14).

The modules have electromechanical connectors or a coupling mechanism actuated by dc motors. Implementation issues due to the finite thickness of the bars (the axes of rotation of the joints of two mating links are coincident) were solved by designing ingenious error-tolerant connectors that connect modules and align mating links together. Each module carries male and female connectors of different polarities on alternate links. A male connector of a given module can only be connected with a female connector of an adjacent module. Figure 6.15 shows a schematic representation of "polarity matching" during reconfiguration, when one module is moving around another, fixed, module. Figure 6.16 shows a real robot (module) moving one step around a fixed robot. Refer to Pamecha et al. [257, 258] for a more detailed technical description of the connectors.

Unlike the hexagonal modules, which had the required kinematic degrees of freedom to move around neighboring modules by changing their joint angles, the square modules are rigid modules supplemented with a sliding mechanism that

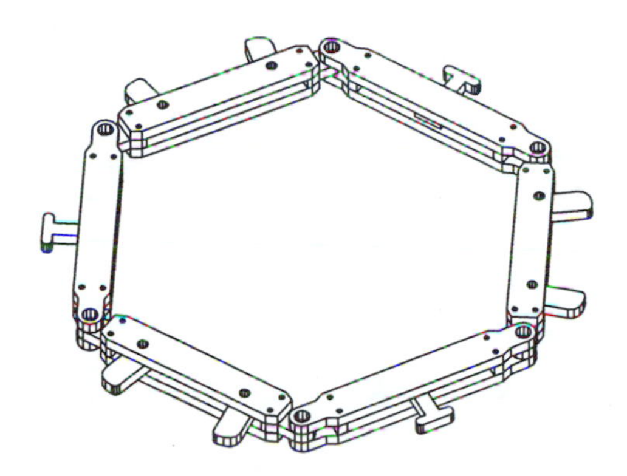

FIGURE 6.14 A planar hexagonal module. After Pamecha et al. [257]. Reprinted by permission © *ASME.*

allows relative motion of modules. Figure 6.17 shows how a square module moves diagonally along another square module. Figure 6.18 shows two real mating square modules.

These metamorphic robots set the stage for complex reconfigurations. But the modules have to be programmed to exhibit the desired global behavior. How does one find the appropriate "software"? One solution is to plan the complete sequence

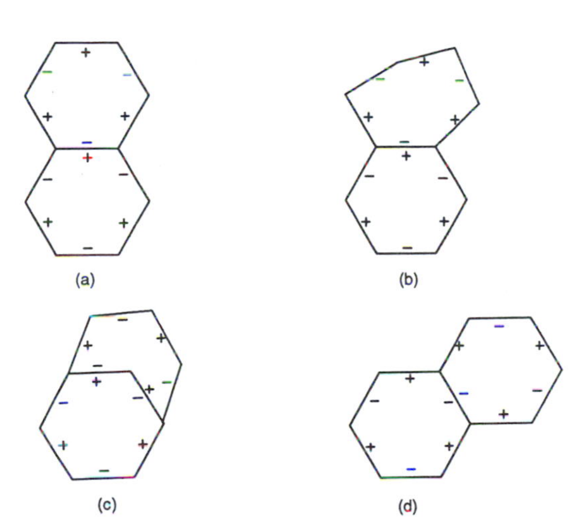

FIGURE 6.15 Polarity matching in the process of reconfiguration. After Pamecha et al. [257]. Reprinted by permission © *ASME.*

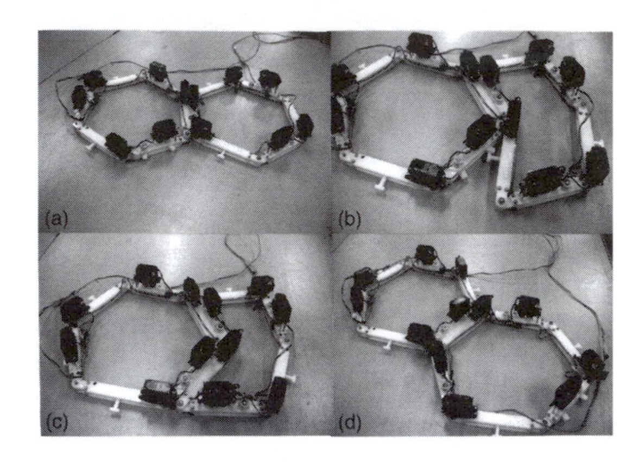

FIGURE 6.16 The locomotion of one module around another. One of the links of the mobile module always remains connected to the fixed module. After Pamecha et al. [258]. Reprinted by permission © *IEEE Press.*

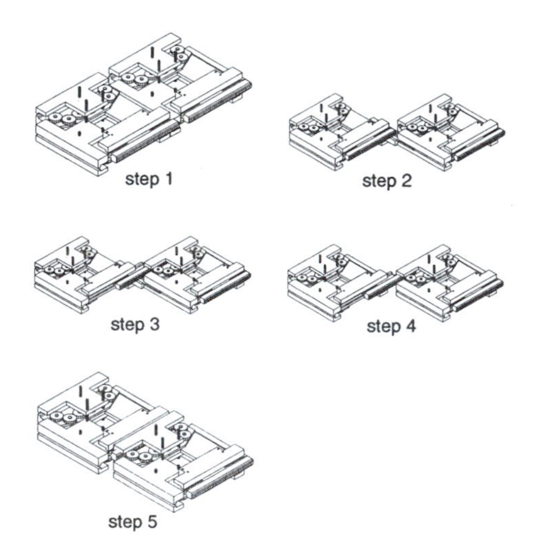

FIGURE 6.17 Illustration of the diagonal motion of a square module. After Pamecha et al. [257]. Reprinted by permission © *ASME.*

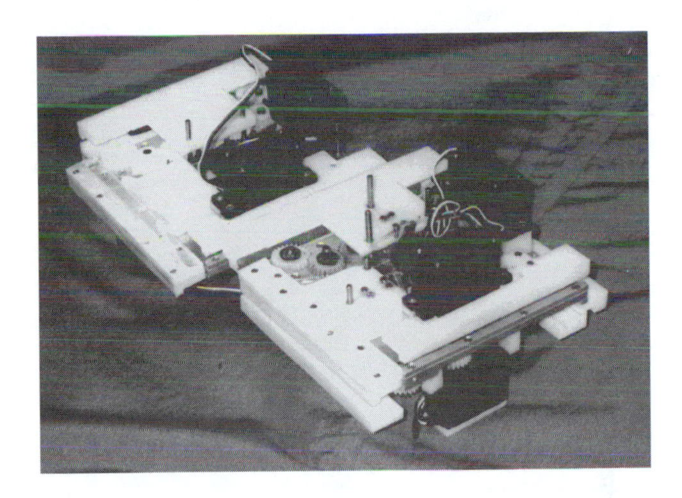

FIGURE 6.18 Two implemented mating square modules. After Pamecha et al. [257]. Reprinted by permission © *ASME.*

of actions in advance, but this is problematic when the environment is changing. Another solution is to endow the modules with the same local behavior and let the system self-assemble, which requires a knowledge of what can be done with a self-assembling system.

Chirikjian et al. [70] and Pamecha et al. [258] have developed a framework for studying motion planning and reconfiguration. The problem is to find a sequence of module motions from any given initial configuration to any given final configuration in a reasonable (close to minimal) number of moves, under the following constraints:

- A module can only move into a space which is accessible and not already occupied.
- Every module must remain connected to at least one other module. In the experimental setup used by Chirikjian et al. [70] and Pamecha et al. [258], one module, the base, remains fixed. At least one of the other modules must remain connected to the fixed base.
- At each time step only one module may move, by one lattice site only.

The problem of finding a minimal set of moves for reconfiguring metamorphic robots is computationally complex. The number of connected configurations of n modules increases exponentially with n. Chirikjian et al. [70] provide upper and lower bounds for the minimal number of moves required to change any initial configuration into any final configuration. These bounds are useful in evaluating the efficiency of reconfiguration heuristics. Pamecha et al. [258] define several distances between configurations that can be used to construct such heuristics. They suggest simulated annealing as a possible heuristic: the next configuration is selected

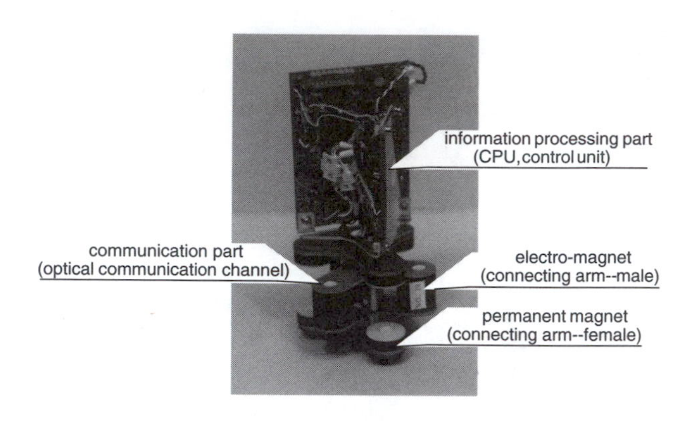

FIGURE 6.19 Murata et al.'s [245] fractum. After Yoshida et al. [337]. Reprinted by permission © *Springer-Verlag.*

among neighboring configurations on the basis of how far, according to the selected distance, it is from the final configuration. Neighboring configurations are those that are accessible in one move. After testing this approach with different metrics and combinations of metrics on a variety of problems (initial and/or final configurations that are simply connected or contain loops, obstacle envelopment), they conclude that "even though simulated annealing is a powerful technique, it has the uncertainties associated with a randomized approach" (Pamecha et al., p. 543 [258]) and suggest that it is best suited for off-line simulations. The best solution found by the algorithm can be used to reconfigure the real robotic system.

6.4.2.2 Fracta Murata et al. [245] have developed a remarkable metamorphic distributed robotic system composed of many identical units called *fracta*. Technical details can be found in Murata et al. [245], Kokaji et al. [200], Tomita et al. [315], and Yoshida et al. [336, 337]. Recent "self-repair" experiments by Yoshida et al. [337] involve 20 units. Each unit is hexagonal but in contrast with Chirikjian et al.'s [70] hexagonal units, they do not change their shape. A unit is a three-layered structure (Figure 6.19) with three pairs of permanent magnets (with a 120° angle between them) in the top and bottom layers and three electromagnets (rotated by 60° with respect to the magnets in the top and bottom layers).

When the polarity of an electromagnet is changed, the electromagnet is attracted toward, or repulsed from, the gap between the top and bottom layers of another unit. Two units change their connection by a specific sequence of electromagnet polarity changes. A unit can connect to at most six units. Each unit is equipped with serial optical channels for local bilateral communication.

A unit can be characterized by its connection pattern. There are 12 such connections patterns (other patterns can be readily derived from those 12 types by

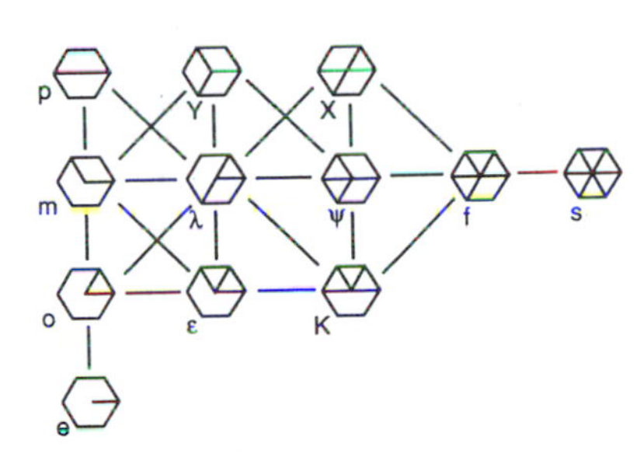

FIGURE 6.20 Connection patterns and how they are related. After Yoshida et al. [337]. Reprinted by permission © *IEEE Press.*

60°-rotations around the center of the hexagon), which are all represented in Figure 6.20. Links within units in Figure 6.20 represent connections with neighboring units. A movable unit is one that can move without carrying other units and keeps the whole system connected after the move. Connection patterns "e," "o," and "ε" correspond to movable units. Relationships between connection patterns are also depicted in Figure 6.20: a link between two connection patterns A and B indicates that it is possible to go from pattern A to pattern B with one move. The distance between two connection types A and B is the number of links necessary to go from A to B.

As was the case in section 6.4.2.1, the crucial issue is programming the units. Murata et al. [245] and Yoshida et al. [337] suggest using a "difference-based" algorithm, which makes use of purely local information. With this algorithm, each unit computes a "distance" or "difference" between the current state and the goal, computes the average distance to the goal of other neighboring units through inter-unit communication, and a unit that has a large distance to the goal moves into a space that reduces the distance.

First, the goal must be described. Yoshida et al. [337] use a 10-unit triangle as an example goal configuration. The triangle can be represented by "goal types," which are the connection patterns that the units should have in the final state, together with the final or desired connection pattern of their neighbors. Figure 6.21 shows the 10-unit triangle with the connection types of all units. For the 10-unit triangle, the goal types are:

- To have connection type "o" and be connected to two units of respective connections types "K" and "K," or
- to have connection type "K" and be connected to four units of respective connections types "o," "K," "K," and "s," or
- to have connection type "s" and be connected to six units that all have connection type "K."

FIGURE 6.21 Representation of the 10-unit triangle. After Yoshida et al. [337]. Reprinted by permission © *IEEE Press.*

The goal of a unit is to reach one of the goal types. At each time step, each unit computes the difference between the various goals and its current state, including the connection types of its neighbors. The difference $d_i(t)$ that is eventually taken into account by unit i at time t is the one that is minimal. Then unit i computes an average $x_i(t)$ of the distances computed by its neighbors, with a diffusion coefficient D and a "leak" constant L:

$$x_i(t) = x_i(t-1) - L + D \cdot \sum_{j \text{ connected to } i} \left[x_j(t-1) - x_i(t-1)\right]. \quad (6.1)$$

If $d_i(t)/x_i(t) > G$, where G is an activation threshold, unit i moves into a space that is selected probabilistically in such a way that a space that reduces the distance of a unit to the goals is more likely to be selected.

Because deadlocks may occur, frustrated nonmovable units are given a chance to force moves in their neighborhood [336, 337]. In addition, Yoshida et al. [336, 337] introduce a "self-repair" function by using extended goal descriptions that include spare units. The set of goals that include a spare unit has a lower priority than the set of goals that do not include a spare unit. Only when a unit fails is the lower-priority set of goals used. Figure 6.22 shows the self-assembly of an ensemble of 10 fracta units into a triangle, starting from an arbitrary initial condition. Figure 6.23 shows the self-repair function: a defective unit is removed from the system and self-repair takes place with the spare unit.fracta!programming—)

6.4.2.3 Vertically self-assembling robots Hosokawa et al. [179, 180] designed a self-assembling robotic system which is also two-dimensional but in the vertical plane. Although the technical issues related to gravity make this implementation very dif-

FIGURE 6.22 Four steps of self-assembly. After Yoshida et al. [337]. Reprinted by permission © *IEEE Press.*

ferent from the previous ones, the spirit is similar. The high-level description of the robots is, however, somewhat simpler. Each robot is a small (90 mm) cubic unit equipped with a pair of arms that can rotate within the xz-vertical-plane around the y-axis and can be extended or retracted to connect and disconnect robots. The robots can change their bonding configurations in the xz-plane using their arms: they connect their arms through a lock-and-key structure, extend them to disconnect their bodies, rotate them by an appropriate angle, and contract them. When the arms are contracted, bonding between two robots is made through the magnetic faces of the cube. A simple alignment correction mechanism is implemented on the faces to avoid the propagation and amplification of alignment errors. The behavior of each robot is determined by its own state and by the states of adjacent robots, that is, robots that have a face in common. Two adjacent robots communicate explicitly. In some configurations, depending on the task to be performed, displacement of one of the robots results from the communication between two adjacent robots.

FIGURE 6.23 Illustration of self-repair. After Yoshida et al. [337]. Reprinted by permission © *IEEE Press.*

What tasks can be carried by such a group of robots? Hosokawa et al. [179, 180] suggest two possible tasks, depicted in Figure 6.24: the formation of a stair-like structure or the formation of a bridge.

The first task was used by Hosokawa et al. [179, 180] to illustrate the capabilities of their robotic system, which needs to be programmed so that robots will

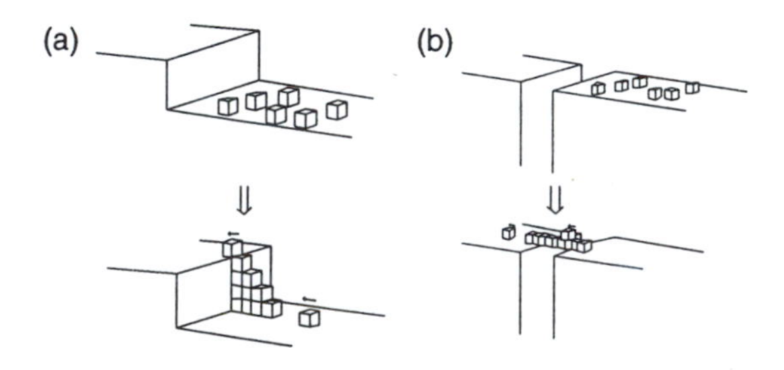

FIGURE 6.24 Examples of applications. After Hosokawa et al. [180]. Reprinted by permission © *IEEE Press.*

reconfigure to form a stair-like structure. The following algorithm is used to move a generic robot A:

1. *To move up.* If there is a robot B on the left of A, no robot on top of A, and no robot on the right of A, then A sends a *Lift-Me-Up* request to B. If there is no robot on top of B, and if there is a robot below B or on the left of B, B accepts the request and A is moved on top of B. Otherwise the request is rejected.
2. *To move left.* If there is a robot B below A, no robot on top of A, and no robot on the left of A, A sends a *Move-Me-Left* request to B. If there is a robot on the left of B, then B accepts the request and A is moved to the left. Otherwise the request is rejected.

These two behaviors are sufficient to generate a stair-like structure, starting from a linear configuration of robots. Figure 6.25 illustrates the formation of a stair-like structure with four robots, starting from an arrangement where robots are in line. In order to get the last robot to climb the obstacle, another rule must be added that permits the transfer of a robot on top of the obstacle. Hosokawa et al. [179] also studied the deconstruction of a stair-like structure.

6.4.2.4 Three-dimensional self-reconfiguration Yoshida et al. [337] recently started experimenting with three-dimensional metamorphic robots. Each unit has six connecting arms, as shown in Fig. 6.26. Each arm can rotate independently. Unit motion is implemented as follows. Consider two connected units X and Y (Figure 6.27(a)). In order for unit Y to move to the location represented in Figure 6.27(b), unit X rotates around the $b - b$ axis after the connection of Y to Z has been released.

The system can be reconfigured from any initial configuration to any final configuration using such elementary moves. Figure 6.28 shows an example of a simulated reconfiguration in 70 steps of 12 units, initially arranged linearly, into a "box." The sequence of moves was found using the same algorithm as for the fracta robots, combined with a form of simulated annealing. Yoshida et al. [337] designed a prototype unit that exhibits the expected reconfiguration properties. This work is extremely promising.

6.4.3 ARCHITECTURAL DESIGN

In the context of architectural design, Coates et al. [75] have used three-dimensional cellular automata to explore new possible types of architecture, at a time when the constraints of structure have almost disappeared, with "form becoming the precursor of function rather than its determinant." In other words, any architectural function can be implemented in virtually any form. According to Coates et al. [75], experimenting with cellular automata permits a "more rigorous analysis of the basic determinants of form, where the global form of an object not only should but actually cannot be predetermined on an aesthetic whim." Coates et al. [75] use cellular automata as a *generative mechanism* to find forms that are totally embedded in the

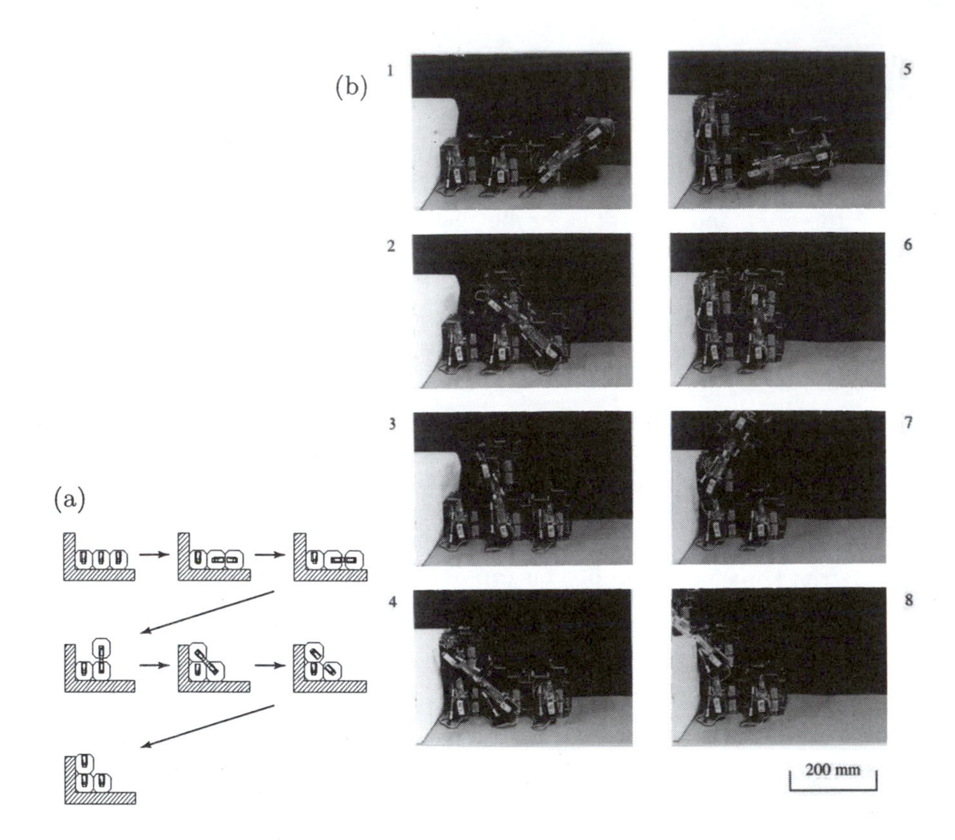

FIGURE 6.25 (a) Schematic representation of the formation of a stair-like structure. (b) Eight successive steps in the formation of a stair-like structure. After Hosokawa et al. [180]. Reprinted by permission © *IEEE Press.*

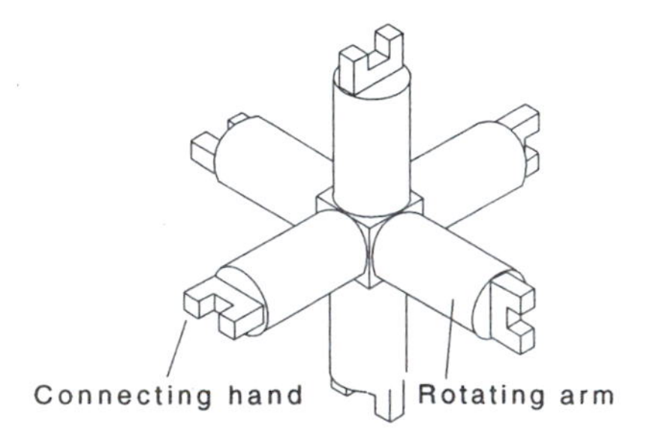

FIGURE 6.26 Schematic representation of a unit. After Yoshida et al. [337]. Reprinted by permission © *IEEE Press.*

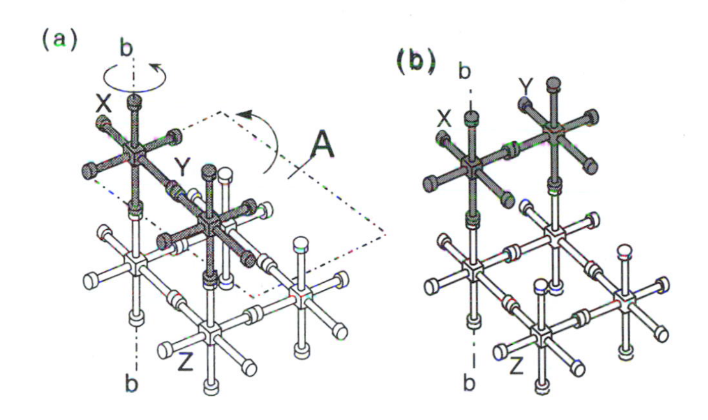

FIGURE 6.27 (a) Position of six units before unit Y moves. (b) Position of the six units after unit Y has moved. After Yoshida et al. [337]. Reprinted by permission © *IEEE Press*.

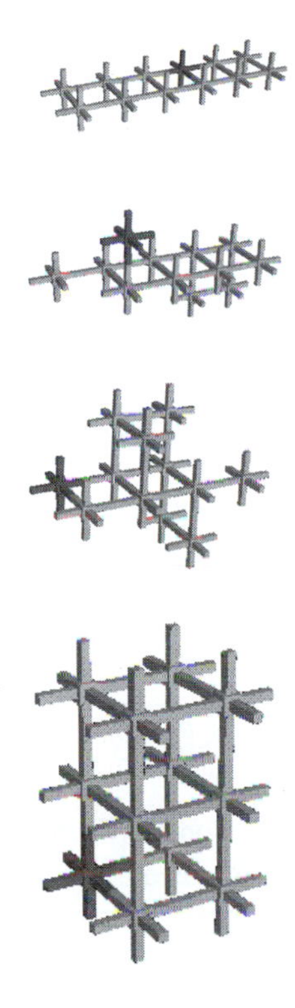

FIGURE 6.28 Illustration of a reconfiguration. After Yoshida et al. [337]. Reprinted by permission © *IEEE Press*.

function to be solved. In that process, the central role of the architect and the role of the "architectural ego" are minimized. The model of collective building in wasps described in this chapter can be viewed as another generative mechanism, in line with a recent trend in architecture that proposes to use self-organizing development in natural systems as a source of new architectural forms. This approach, however, is quite limited if all that is done is a random exploration, with no explicit function and no selection of forms on the basis of their function. In the next section, we see that this limitation can be overcome by resorting to artificial evolution: the fitness function orients the search for new forms, based on aesthetic or functional criteria, or a combination of both.

6.4.4 EVOLVING FUNCTION

In all the examples of sections 6.4.2 and 6.4.3, the algorithm used to reconfigure modules or grow structures was planned in advance, prewired, tuned by hand, or randomly explored. Following the example of section 6.3.4, it is possible and useful to make use of a more systematic way of exploring the space of behaviors of a "metamorphic" robotic system or the space of forms generated by a growth model.

Evolutionary design (ED) [18], the creation of new designs by computers using evolutionary techniques, is a promising approach. The emphasis of ED is on the generation of novelty and not so much on optimization. ED involves two separate aspects: (1) a generative mechanism and (2) a mechanism for artificial evolution. The generative mechanism is extremely important because it must be able to produce enough diversity and novelty. Artificial evolution is equally important because its function is to pick from among the many forms produced by the generative mechanism those that meet certain *functional* or *aesthetic* criteria.

Open-ended artificial evolution requires that the generative mechanism itself be open-ended, which often implies that the designs, or the algorithms that generate them, be represented with variable-length genotypes. For example, in section 6.3.4 building algorithms were represented with variable-length genotypes. Mappings from the representation used in the evolutionary algorithm to the final form, the fitness of which is evaluated, range from very simple (for example, direct one-to-one mapping between genes and pixels of an image) to quite complex (the representation may code for differential equations or cellular automata that generate forms).

Design selection is made on the basis of the design's degree of functionality, which can (often) be evaluated automatically, or on the basis of an observer's goals and desires, which cannot generally be coded. Sometimes both types of fitness evaluation coexist, when aesthetic criteria are used in addition to functional requirements. For example, Rosenman [276] used a combined approach, that is, a coded fitness function based on function and human interactions, to obtain complex arrangements of elements representing house plans, using a design grammar of rules to define how polygons should be constructed out of edge vectors.

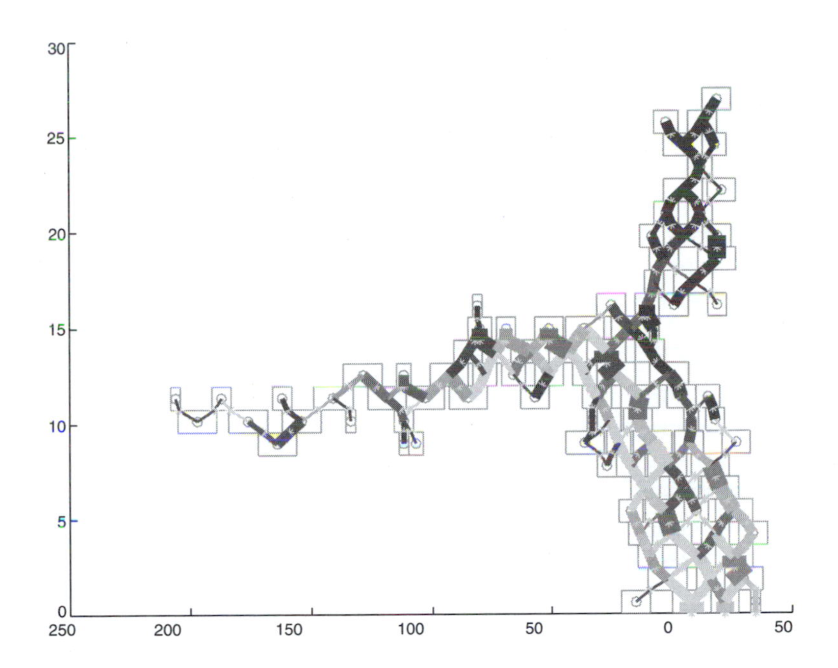

FIGURE 6.29 Representation of a 1mg bridge built with simulated Lego™ blocks obtained by Funes and Pollack [137]. Centers of mass have been marked with circles. Each start is a joint between two centers of them and line thickness is proportional to the capacity of each joint. The bridge is made out of 97 bricks and is 1.67 m long. The x scale is compressed at a variable rate for visualization of the entire structure. After Funes and Pollack [137]. Reprinted by permission © *MIT Press.*

Most examples, however, involve one type of fitness evaluation. Sims [289] or Todd and Latham [313] evolve computer graphics based on interactions with humans. Sims [290] evolves creatures that move in a simulated physical world and compete to get one resource: here, the fitness function, which does not involve any observer, depends on how quickly the creature can move to reach the resource. Broughton et al. [46] use genetic programming (GP) [202] to evolve three-dimensional structures generated by L-systems [224]. One fitness function is the ability of the generated architecture to avoid being hit by a stream of particles moving across the area where the architecture is constructed. Another fitness function is the ability to capture (or be hit by) as many particles as possible. Yet another fitness function, perhaps more useful in the context of architecture, is the volume enclosed by the architecture compared to the volume of material used to grow the architecture itself. They also used symbiotic coevolution of two types of L-systems that produce space and enclosure, respectively. For example, "enclosure"-L-systems must generate forms that enclose as much space as possible generated by "space"-L-systems.

An interesting use of ED is Funes and Pollack's [137] GP-based evolution of buildable Lego™ structures to perform specific tasks (crane arm, rotating crane, scaffold, or bridge). Funes and Pollack [137] have developed a simple physical model of the union of two Lego™ bricks together. Their fitness function is based on such concrete attributes as the length of the structure in one direction, the normalized distance to a target point, or the maximum external weight that the structure can support under the physical constraints of their model. After obtaining a structure that satisfies the physical constraints (basically, as long as there is a way to distribute the weights among the network of bricks such that no joint is stressed beyond its maximum capacity, the structure does not break), the corresponding real Lego™ structure is built to validate the artificial structure. Figure 6.29 shows an example of a structure they obtain when the goal is to generate a long bridge.

Funes and Pollack's [137] model does not rely on self-assembly. But it is without any doubt possible to combine a concrete functional perspective with a self-assembling design algorithm. For example, the self-assembling algorithm leading to the formation of a functional bridge by a homogeneous group of robots such as Hosokawa et al.'s [179, 180] robots or Yoshida et al.'s [337] units could be obtained using ED. The space of multirobot configurations can be explored in the same way as the nest-building algorithm but with function being taken into account explicitly. Evolving self-assembling functional systems is definitely a promising avenue of research.

6.5 POINTS TO REMEMBER

- Stigmergy, that is, indirect interactions among nestmates through modifications of the environment, appears to be an important mechanism underlying nest construction in social insects.
- Two types of stigmergy exist: quantitative or continuous stigmergy, in which the different stimuli that trigger behavior are quantitatively different, and qualitative or discrete stigmergy, in which stimuli can be classified into different classes that differ qualitatively. Quantitative stigmergy results in the emergence of pillars in termites. Qualitative stigmergy operates in the building behavior of the paper wasps *Polistes dominulus*.
- A model of multiagent building in a three-dimensional discrete lattice, motivated by the observation of the building behavior of wasps, produces structured patterns without any direct or explicit communication between the agents. This model, however, produces many random or space-filling patterns.
- When a structured pattern is being generated, the architecture itself governs the dynamics of construction by providing constraints as to where new brick deposits should be made. These constraints permit the construction of compact architec-

tures which require sequences of correlated deposits. Moreover, most structured patterns involve the repetition of a modular subpattern.

- These important properties of structured patterns provide a basis for defining a fitness function to be used with an evolutionary algorithm. A genetic algorithm (GA) based on this fitness function does find a few interesting patterns, some of them making use of existing submodules, but the progression of the GA is impeded by epistasis, that is, interactions among genes. In effect, correlations among brick deposits are necessary to produce structured patterns but such correlations are destroyed by mutation and crossover. The evolutionary approach remains promising if one finds a way of making sure that certain, appropriately selected, brick deposits remain correlated over several generations.
- The discrete-stigmergy model of building can help design self-assembling macroscopic systems, such as self-assembling or self-reconfigurable robotic systems. Combined with evolutionary algorithms, self-assembling models can be explored to generate functional self-assembling systems. In addition to function, aesthetic criteria can be used to guide artificial evolution.

CHAPTER 7
Cooperative Transport by Insects and Robots

7.1 OVERVIEW

Collective robotics is a booming field, and cooperative transport—particularly cooperative box-pushing—has been an important benchmark in testing new types of robotic architecture. Although this task in itself is not especially exciting, it does provide insight into the design of collective problem-solving robotic systems. One of the swarm-based robotic implementations of cooperative transport that seems to work well is one that is closely inspired by cooperative prey retrieval in social insects.

Ants of various species are capable of collectively retrieving large prey that are impossible for a single ant to retrieve. Usually, a single ant finds a prey item and tries to move it alone; when successful, the ant moves the item back to the nest. When unsuccessful, the ant recruits nestmates through direct contact or trail laying. If a group of ants is still unable to move the prey item for a certain time, specialized workers with large mandibles may be recruited in some species to cut the prey into smaller pieces. Although this scenario seems to be fairly well understood in the species where it has been studied, the mechanisms underlying cooperative transport—that is, when and how a group of ants move a large prey item to the nest—remain unclear. No formal description of the biological phenomenon has been developed, and, surprisingly, roboticists went further than biologists in trying to model cooperative transport: perhaps the only convincing model so far is one that has been introduced and studied by roboticists [207] and, although this model was not aimed at describing the behavior of real ants, few adjustments would be required to make it biologically plausible. This chapter first describes empirical research on cooperative transport in ants, and then describes the work of Kube and Zhang [205, 206, 207, 209].

7.2 COOPERATIVE PREY RETRIEVAL IN ANTS

A small prey or food item is easily carried by a single ant. But how can ants "co-operate" to carry a large item? Cooperative prey (or large food item) retrieval and transport has been reported in several species of ants [303, 274, 317]: weaver ants *Oecophylla smaragdina* [173] and *Oecophylla longinoda* [172, 333], army ants *Eciton burchelli* [128], African driver ants *Dorylus* [155, 240], and other species such as *Pheidole crassinoda* [302], *Myrmica rubra* [304], *Formica lugubris* [304], *Lasius neoniger* [316], the desert ants *Aphaenogaster* (ex-*Novomessor*) *cockerelli* and *Aphaenogaster albisetosus* [175, 234], *Pheidologeton diversus* [240], *Pheidole pallidula* [93, 94], *Formica polyctena* [65, 66, 238, 319], *Formica schaufussi* [274, 275, 317] and the ponerine ants *Ectatomma ruidum* [263] and possibly *Paraponera clavata* [41]. This cooperative behavior can be quite impressive. For example, Moffett [240] reports that a group of about 100 ants *Pheidologeton diversus* was able to transport a 10-cm earthworm weighing 1.92 g (more than 5000 times as much as a single 0.3-mg to 0.4-mg minor worker) at 0.41 cm/s on level ground. By comparison, ants engaged in solitary transport of food items on the same trail were carrying burdens weighing at most 5 times their body weight at about 1 cm/s: this means that ants engaged in the cooperative transport of the earthworm were holding at least 10 times more weight than did solitary transporters, with only a modest loss in velocity [240].

We believe that the phenomenon of cooperative transport is much more common in ants than these few studies suggest: to the best of our knowledge, these studies are the only ones that report detailed observations of cooperative prey transport. This phenomenon involves several different aspects:

1. Is there an advantage to group transport as opposed to solitary transport? Is worker behavior in group transport different than in solitary transport?
2. When and how does an ant know that it cannot carry an item alone because it is either too large or too heavy?
3. How are nestmates recruited when help is needed?
4. How do several ants cooperate and coordinate their actions to actually transport the item?
5. How do ants ensure that there is the right number of individuals involved in carrying the item?
6. How does a group of transporting ants handle deadlocks and, more generally, situations where the item to be transported is stuck, either because of antago-

FIGURE 7.1 Weaver ants (*Oecophylla longinoda*) collectively carrying a large prey (a grasshopper).

nistic forces or because of the presence of an obstacle or heterogeneities in the susbtrate?

All these questions, that have been more or less satisfactorily dealt with in the above-mentioned studies, are of enormous interest in view of implementing a decentralized cooperative robotic system to transport objects the locations and sizes of which are unknown. Let us try to answer each of these questions with available data and personal observations.

7.2.1 SOLITARY TRANSPORT VERSUS GROUP TRANSPORT

In *Pheidologeton diversus*, single worker ants usually carry burdens (grasping them between their mandibles, lifting them from the ground and holding them ahead as they walk forward) rather than drag them [240]. By contrast, in cooperative transport, one or both forelegs are placed on the burden to aid in lifting it, mandibles are open and usually lay against the burden without grasping it. The movement patterns of group-transporting ants corresponding to their positions around the perimeter of a burden with reference to the direction of transport are also different than those of ants engaged in solitary transport: workers at the forward margin walk backward, pulling the burden, while those along the trailing margin walk forward, apparently pushing the burden; ants along the sides of the burden shuffle their legs sideways and slant their bodies in the direction of transport [240].

By contrast, Sudd [302, 304] observes that individual *Pheidole crassinoda*, *Myrmica rubra*, and *Myrmica lugubris* ants appear to exhibit the same behavioral patterns in solitary and group transport: in group transport, all three species use the same method as when they work alone, including realignment and repositioning.

This, however, does not exclude cooperative behavior: group transport in these species is particularly interesting because the same individual behavior is functional either in isolation or in group, and may even lead to increasing returns (up to a maximum group size: see section 7.2.2) despite the lack of direct response of individuals to the presence of their nestmates.

In general, whether ants behave similarly or differently when engaged in solitary and group transport, group transport is more efficient than solitary transport for large prey. Ants can dismantle a large food item into small enough pieces to be carried by individual ant workers. Moffett [240] observed that a large piece of cereal, which would have required 498 solitary *Pheidologeton diversus* transporters if broken down into small enough pieces, could be transported collectively by only 14 ants. More generally, he observed that the weight carried by ant increases with group size: the total weight carried by a group of N workers increases as $W \propto N^{2.044}$, which means that the weight carried by each ant increases on average as $N^{1.044}$. Franks [128] made similar observations on *Eciton burchelli*: let W_i be the dry weight of transported items and W_a the total dry weight of the group of transporting ants, the relationship between both is $W_i \propto W_a^{1.377}$, which, assuming that W_i is proportional to N, implies that the dry weight carried by ant increases as $N^{0.377}$. Franks [128] also observed that items were always retrieved at a standard speed, relatively independent of group size: he hypothesized that the increased efficiency of group transport with group size results from the group's ability to overcome the rotational forces necessary to balance a food item. Along the same lines, we already mentioned Moffett's [240] experiment in which he showed that group-transporting ants could carry more than 10 times more weight than did solitary transporters at a speed only divided by 2. Figure 7.2 shows the velocity of transport as a function of the number of *Pheidologeton diversus* carriers: the velocity decreases only for large group sizes. Figure 7.3 shows transport efficiency per ant, measured by the product of burden weight by transport velocity divided by the number of carriers, as a function of the number of carriers: this quantity increases with group size up to a maximum for groups of 8 to 10 ants, and then declines [240].

7.2.2 FROM SOLITARY TO GROUP TRANSPORT

All reports of how the decision is made to switch from solitary to group transport describe variants of the same phenomenon. A single ant first tries to carry the item, and then, if the item resists motion, to drag it (although dragging is rare in *Pheidologeton diversus*). Resistance to transport seems to determine whether the item should be carried or dragged [93, 94, 302, 304]. The ant spends a few seconds testing the resistance of the item to dragging before realigning the orientation of its body without releasing the item: modifying the direction of the applied force may be sufficient to actually move the item. In case realignment is not sufficient, the ant releases the item and finds another position to grasp the item. If several repositioning attempts are unsuccessful, the ant eventually recruits nestmates. Re-

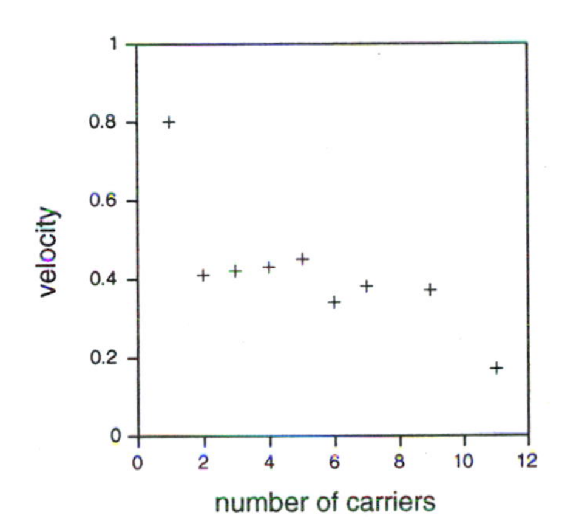

FIGURE 7.2 Velocity of transport as a function of the number of carriers (*Pheidologeton diversus*). Average over n observations for each number of carrier ants: sample size $n = 50$ for one ant, $n = 25$ for two ants, $n = 47$ for three ants, $n = 42$ for four ants, $n = 38$ for five ants, $n = 16$ for six ants, $n = 20$ for seven ants, $n = 12$ for eight to ten ants, and $n = 10$ for eleven ants and more. After Moffett [240]. Reprinted by permission © *National Geographic Research*.

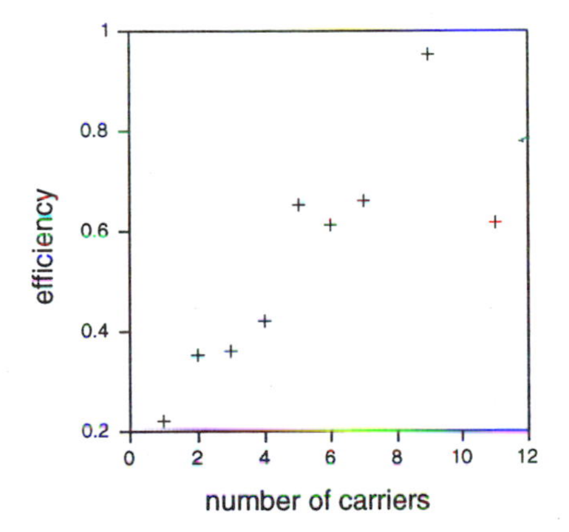

FIGURE 7.3 Transport efficiency (burden weight *times* velocity/number of carriers) as a function of the number of carriers (*Pheidologeton diversus*). Average over n observations for each number of carrier ants; same sample sizes n as in Figure 7.2. After Moffett [240]. Reprinted by permission © *National Geographic Research*.

cruitment per se is examined in the next section. Sudd [304] reports that the time spent attempting to move the item decreases with the item's weight: for example, an ant may spend up to 4 minutes for items less than 100 mg, but only up to 1 minute for items more than 300 mg. Detrain and Deneubourg [93, 94] have shown that in *Pheidole pallidula*, it is indeed resistance to traction, and not directly prey size, that triggers recruitment of nestmates, including majors, to cut the prey: they studied recruitment through individual trail laying for prey of different sizes (fruit flies versus cockroaches), or of the same size but with different levels of retrievability (free fruit flies versus fruit flies covered by a net). A slow recruitment to free fruit flies was observed, in connection to weak individual trail laying; in contrast, strong recruitment and intense individual trail laying were observed when large prey or small but irretrievable prey were offered. It is therefore the ability or inability of an individual or a group that governs recruitment.

7.2.3 RECRUITMENT OF NESTMATES

Hölldobler et al. [172] studied recruitment in the context of cooperative prey retrieval in two *Novomessor* species: *Novomessor albisetosus* and *Novomessor cockerelli*. They showed that recruitment for collective transport falls within two categories: short-range recruitment (SRR) and long-range recruitment (LRR). In SRR, a scout releases a poison gland secretion in the air immediately after discovering a large prey item; nestmates already in the vicinity are attracted from up to 2 m. If SRR does not attract enough nestmates, a scout lays a chemical trail with a poison gland secretion from the prey to the nest: nestmates are stimulated by the pheromone alone (no direct stimulation necessary) to leave the nest and follow the trail toward the prey. Figure 7.4 shows the number of ants (*Novomessor albisetosus*) following an "artificial" poison gland pheromone trail as a function of time. 30-cm-long trails, drawn on a piece of cardboard with the secretion of one gland of a worker, were introduced into the test arena after periods of 1, 2, 3, 4, and 5 minutes elapsed. The number of ants following the trail was counted during the minute after introduction. Means and error bars correspond to 10 trials for each time value (1, 2, 3, 4, and 5 min). The rather short-lived poison gland secretion releases no persistent trail-following behavior but rather an outrush of a group of workers that move in a phalanx-like formation along the trail. Figure 7.5 illustrates a sequence of SRR and LRR: a freshly killed grasshopper was pinned to the ground 6 m from the nest of a *Novomessor cockerelli* colony; soon after the first worker discovered the prey, other workers were attracted (SRR); when the first scouts returned to the nest laying a trail (vertical bars), the number of nestmates at the site increased significantly (LRR).

Hölldobler [173] reports short-range, and more rarely long-range (rectal gland-based), recruitment in *Oecophylla smaragdina* in the context of prey retrieval, during which secretions from the terminal sternal gland and alarm pheromones from the mandibular glands interact. This short-term recruitment attracts nestmates located in the vicinity, which quickly converge toward the intruder or prey item,

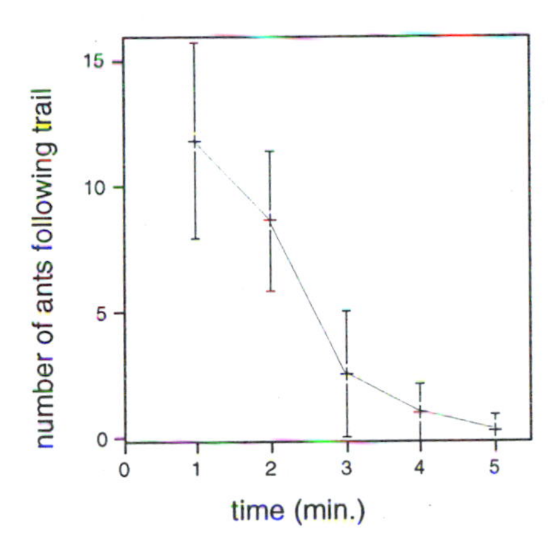

FIGURE 7.4 Number of ants (*Novomessor albisetosus*) following an "artificial" poison gland pheromone trail as a function of time. Average and std are 10 trials. After Hölldobler et al. [175]. Reprinted by permission © *Springer-Verlag*.

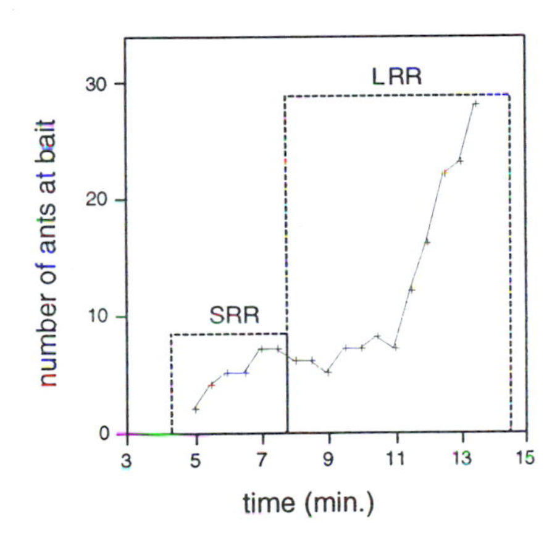

FIGURE 7.5 Short-range and long-range recruitment in *Novomessor cockerelli*. After Hölldobler et al. [175]. Reprinted by permission © *Springer-Verlag*.

which is retrieved into the nest when dead. In a series of experiments with 20 freshly killed cockroaches placed at randomly selected locations in a colony's territory, the prey were discovered within several minutes (average: 8.05 min); ants in the vicinity were attracted by short-range recruitment signals; 5 to 8 ants grasped the prey item and held it on the spot for several minutes (average: 11.6 min) before jointly retrieving it to the nest. This last phase involved 5.3 ants on average. In *Oecophylla longinoda*, even when the prey were pinned to the ground and the ants were unable to retrieve them, long-range recruitment was not used [172]. By contrast, long-range recruitment was observed in *Oecophylla smaragdina* when the cockroaches were pinned to the substrate and several workers had attempted without success to remove the prey: recruiting ants moved back to the nearest leaf nest (although there was only one queen, as is usual in this species, the nest of the considered colony was composed of 19 separate leaf nests, which is also common in this species) where they recruit nestmates which soon moved out of the leaf nest toward the prey. From 25 to 59 could be recruited, whereas between 9 and 19 ants were involved in actually retrieving the prey to the nest once the prey were eventually retrieved. This indicates that the ants do not estimate the size or weight of the prey but rather adapt their group sizes to the difficulty encountered in first moving the prey. Hölldobler [173] reports that the recruited ants were gathering around the prey, seeking to get access, and sometimes grasped nestmates that were already working at the prey, thereby forming a pulling chain, a common behavior in weaver ants (see, for example, Figure 1.2(a)). The prey were usually first transported to the leaf nest from which helpers had been recruited.

7.2.4 COORDINATION IN COLLECTIVE TRANSPORT

Coordination in collective transport seems to occur through the item being transported: a movement of one ant engaged in group transport is likely to modify the stimuli perceived by the other group members, possibly producing, in turn, orientational or positional changes in these ants. This is an example of a mechanism that we have encountered in virtually every chapter of this book: stigmergy [158]. The ubiquity of this concept in swarm intelligence underlines its importance. Here, stigmergy is a promising step toward a robotic implementation, because it suggests that a group of robots can cooperate in group transport without direct communication among robots; moreover, robots do not have to change their behaviors depending on whether or not other robots are engaged in the task of carrying (or dragging, or pulling, or pushing) the item. The coordination mechanism used by ants in cooperative transport is not well understood, and has never really been modeled. The swarm of robots described in section 7.3 is just such a model, which shows that the biology of social insects and swarm-based robotics can both benefit from each other.

7.2.5 NUMBER OF ANTS ENGAGED IN GROUP TRANSPORT

Apparently, the number of ants engaged in transporting an item is an increasing function of the item's weight, which indicates that group size is adapted to the item's characteristics. For example, Figure 7.6 shows the number of *Pheidologeton diversus* carriers as a function of burden weight. Inverting the relationship described in section 7.2.1, we obtain $N \propto W^{0.489}$. The fit is remarkable, suggesting that the adaptation of group size is accurate. Using the same notations as in section 7.2.1, Franks [128] finds that $W_a \propto W_i^{0.726}$ for *Eciton burchelli*. However, as mentioned in section 7.2.3, Hölldobler's [173] observations suggest that the ants adapt group

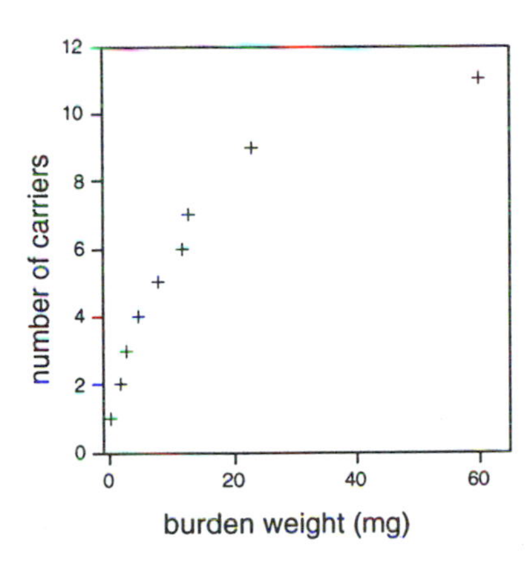

FIGURE 7.6 Number of carriers (*Pheidologeton diversus*) as a function of burden weight. Average over n observations for each number of carrier ants; same sample sizes n as in Figure 7.2. After Moffett [240]. Reprinted by permission © *National Geographic Research*.

size to the difficulty encountered in first moving the prey: decisions rely on how difficult it is to carry the prey, and not simply on weight. A prey item that resists (either actively or passively) stimulates the ant(s) to recruit other ants. Success in carrying a prey item in one direction is followed by another attempt in the same direction. Finally, recruitment ceases as soon as a group of ants can carry the prey in a well-defined direction: in that way, group size is adapted to prey size.

7.2.6 DEADLOCK AND STAGNATION RECOVERY

Sometimes, the item's motion can no longer progress either because forces are applied by ants in opposite directions and cancel one another, or because the group has encountered an obstacle or any significant heterogeneity on the substrate. We have already mentioned that a single ant, who first discovers a food item, tries to transport it alone: the ant first tries to carry it, then to drag it; an unsuccessful ant tries another direction and/or another position and then, if still unsuccessful, gives up the prey temporarily to recruit nestmates. The same phenomenon occurs when ants are engaged in group transport: if, for any reason, the item is stuck, ants exhibit realigning and repositioning behaviors [302, 304]. The frequency of spatial rearrangements, which may result from the ants' response to the reactive forces communicated through the item being transported [304], increases with time, and so does the rate of transport. As is the case for solitary transporters, realignments tend to occur before, and are much frequent than, repositionings: only when realignment is not sufficient do ants try to find other slots around the prey.

Along the same lines, Moffett [240] reports that ants (*Pheidologeton diversus*) gather around food items at the site of their discovery, gnawing on them and pulling

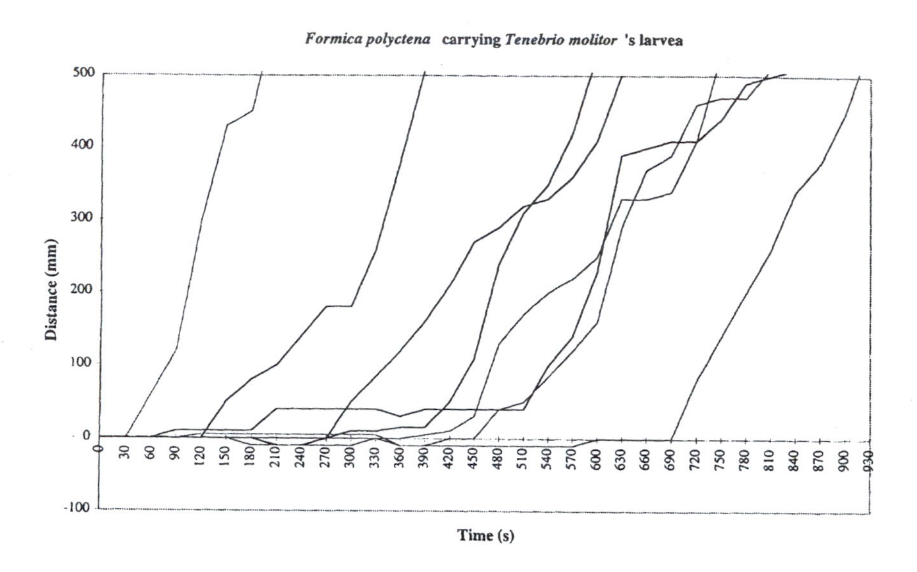

FIGURE 7.7 Distance over which a larva of *Tenebrio molitor* has been transported by *Formica polyctena* ants as a function of time. Eight experiments are shown. Deneubourg [319], reprinted by permission.

them; during the first ten minutes or so, the item is moved about slowly in shifting directions, before ants "sort out" their actions and actual transport can begin. During these ten minutes, a lot of spatial rearrangements take place.

Personal observations of weaver ants *Oecophylla longinoda* confirm the existence of such spatial rearrangements in this species too.

Deneubourg [319] studied cooperative transport of *Tenebrio molitor*'s larvae (a worm) in the ant *Formica polyctena*, and found that after a period of unsuccessful attempts to transport the larvae individually or in group, transport suddenly becomes successful, one possible reason being that the forces applied by the various individuals engaged in cooperative transport become aligned. Figure 7.7 shows the distance over which a larva has been transported as a function of the time elapsed since the larva was discovered. Distance is positive when progress has been made toward the nest and negative otherwise. It can be clearly seen that a "phase transition" occurs at some point (which, however, cannot be predicted: it varies from experiment to experiment), when group transport suddenly becomes successful. After that transition, transport proceeds smoothly until the larva reaches the nest.

7.3 COOPERATIVE TRANSPORT BY A SWARM OF ROBOTS

7.3.1 INTRODUCTION

From the previous section, it seems that the study of group transport in ants may be useful to guide roboticists in the design of noncommunicating, decentralized groups of robots for cooperative transport. Cooperative box-pushing has indeed become one of the benchmark study tasks in swarm-based robotics [10, 57, 106, 113, 206, 207, 209, 250, 277, 280, 296, 297, 322], together with clustering, foraging, and group marching (see Cao et al. [61, 62]; Kube and Zhang [209]).

In particular, Kube and Zhang [206, 207, 209]; Kube [208]) have been the most consistent in their ant-based approach to the problem of cooperative transport (moreover, other approaches, with the exception of Stilwell and Bay's [296, 297] have relied on either global planning or a combination of local and global planning with centralized conflict resolution). They have first introduced a simulation model that they have implemented in a group of five homogeneous robots [206]; they have then improved their basic model to include stagnation recovery [207]; finally, introducing a method (which is briefly described in the next subsection) called "task modeling," where multirobot tasks are described as series of steps, each step possibly consisting of substeps, they have designed a group of 11 homogeneous robots implementing cooperative box-pushing with stagnation recovery [209].

This impressive series of works has demonstrated that a nontrivial task requiring the "coordinated" efforts of several robots could be realized without direct communication among individuals exhibiting identical characteristics. The collective behavior of their system appears to be very similar to that of real ants. Although convergence toward the goal may be slower and more indirect than with more complex robots, this system has the desirable features of being cheap, robust, and flexible. This section is dedicated to a description of this work.

7.3.2 BASIC MODEL AND ITS IMPLEMENTATION

Kube and Zhang [206] have designed a group of five robots that coordinate their actions to carry, or push, an object that one single agent is unable to carry because it is too heavy. Carrying the item requires the coordinated efforts of at least two robots that push in the same direction. This cooperation is implemented without any direct communication among the agents. More precisely, the task to be performed is defined as follows: the objective, that is shared by the whole group, is to localize a lit box within a given arena, and to push this item toward an edge of the arena.

Kube and Zhang [206] first performed simulations of a group of robots with a view to implementing them. In what follows, we mix the description of the simulated robots with that of the real robots, while stressing important differences. The robots are designed following a subsumption architecture [44, 45]; (see also section 4.6.2). A behavior maps sensor inputs to actuators outputs: it realizes a certain stimulus-response relationship. The box-pushing task makes use of:

1. Three sensors—a goal sensor, an obstacle sensor, and a robot sensor. Obstacles are sensed using left and right infrared sensors. The goal (the box) is detected by left and right photocells with adjustable thresholds.
2. One actuator—a steering actuator. Steering the robot is achieved by differentially powering the left and right wheel motors. Only one motor is working at any one time. For example, to turn right, the left motor is turned on with the right motor turned off.
3. Five behaviors:

    (a) A GOAL behavior, that directs the robot toward the box. When active, it turns the robot by a fixed angle toward the goal.
    (b) An AVOID behavior, dedicated to handling collisions. When this behavior becomes active, the robot turns by a fixed angle in the opposite direction of the potential collision.
    (c) A FOLLOW behavior, that allows a robot to follow another one.
    (d) A SLOW behavior, which adjusts motor speed to avoid rear-end collisions.
    (e) A FIND behavior, which is activated during exploration, until the goal has been detected. The FIND behavior is created by adjusting the photocells' thresholds.

In the simulated robots, each behavior receives inputs from sensors it is connected to at each time step, computes a response and sends the corresponding command not directly to the appropriate actuators but to a behavior arbitration module. Each behavior is given a priority. The simplest possible choice for the behavior arbitration module is a fixed, prewired module, the function of which does not vary with time. Even in this case, numerous variants are possible, such as the inclusion of a memory "behavior," which integrates past stimuli and sends a "response proposal" to the arbitration module according to these past stimuli. The subsumption approach usually resorts to a hierarchy of behaviors with fixed priorities (without learning): at any time, actuators are therefore commanded by the current active prioritary behavior. This approach requires that the designer know all possible configurations of the system so as to determine priorities, a task that can rapidly become intractable as the number of atomic behaviors increases. The subsumption-based behavior arbitration module used by Kube and Zhang [206] is sketched in Figure 7.8. The lowest-priority is the default FIND behavior, the output of which can be suppressed if any other behavior is or is becoming active. The next lowest-priority behavior is FOLLOW. When a robot follows another robot, it may get too close to this or another robot, in which case the SLOW behavior is activated, which acts to reduce the robot's speed while active. Finally, the GOAL behavior, which is activated by the photocells, can only be suppressed by the highest-priority behavior AVOID, which is activated when collision with an obstacle is imminent, and remains active as long as obstacle-detecting sensors are active.

Kube and Zhang's [206] also used a method based on Adaptive Logic Networks (ALN) to construct a behavior arbitration module. ALNs are simplified neu-

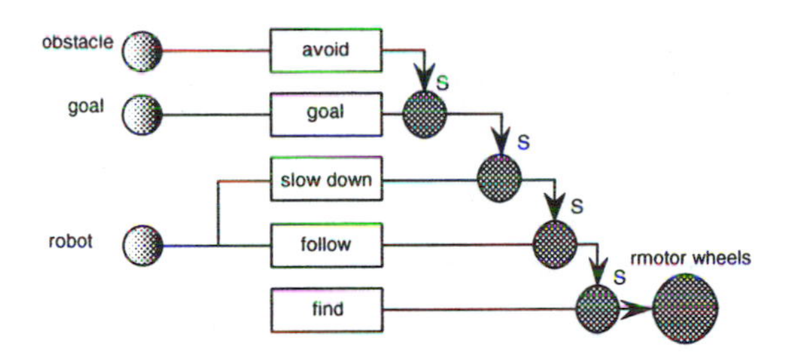

FIGURE 7.8 Schematic representation of the behavior arbitration module. The "obstacle," "goal," and "robot" inputs correspond to sensors that allow the detection of, respectively, an obstacle, the goal or another robot. These inputs are connected to behaviors which are hierarchically organized: a higher-priority behavior can suppress (S) a lower-priority behavior. The system's output is eventually connected to the actuators, namely the two motor wheels. After Kube and Zhang [206]. Reprinted by permission © *MIT Press*.

ral networks designed to synthesize functions using a binary tree of logical AND and OR [4]. Such binary trees are then easily implemented in programmable array logic. During the learning phase, ALNs learn to determine priorities among behaviors on the basis of input vectors that sample input space. Tools to build the appropriate behavior arbitration module are however not restricted to ALNs, and other methods, such as classical artificial neural networks can also be used. The behavior arbitration module obtained with an ALN was simpler to design, but the subsumption-based approach proved more efficient in terms of accomplishing the task.

In order to allow the swarm generated by the FOLLOW behavior to be maintained, parasite signals, such as those resulting from the passage of robots walking in the opposite direction, must be inhibited: a selective attention process can be implemented so as to restrict the swarming robots' attentions to robots that belong to the swarm. This set of behaviors allow a dispersion of the robots around the item to be pushed as well as the implementation of a unidirectional effort. Indeed, the FOLLOW and GOAL behaviors generate a coordinated motion, while the AVOID behavior leads to a dispersion of the robots around the item: as soon as an empty slot appears at a given location around the box, the AVOID behavior is deactivated, which permits the GOAL behavior of another neighboring robot to become prioritary and allows the robot to fill the empty slot. Figure 7.9 shows the logical wiring of the arbitration module in the real robots. Sensor inputs S_1 and S_2 correspond to the left and right infrared sensors, whereas S_3 and S_4 correspond to the left and right photocells; A_1 and A_2 represent the left and right wheel motors.

Kube and Zhang's [206] five robots are able to locate the box, to converge toward it and to push it in a number of directions depending on how many robots are present on each side. The robotic implementation of Kube and Zhang [206] shows

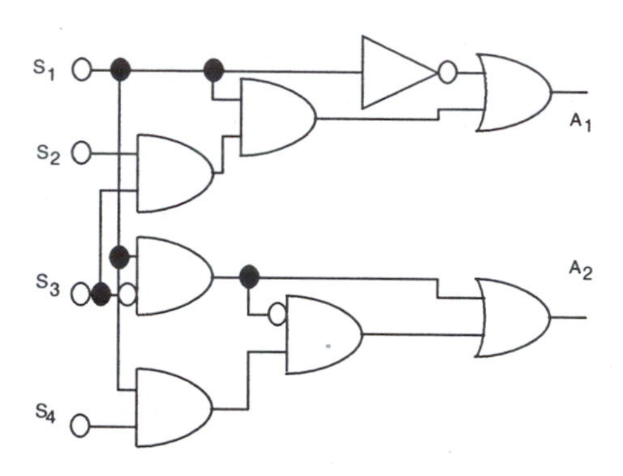

FIGURE 7.9 Combinatorial logic representation of the robots' behavior arbitration module circuit. Sensor inputs S_1 and S_2 correspond to the left and right infrared sensors, whereas S_3 and S_4 correspond to the left and right photocells; A_1 and A_2 represent the left and right wheel motors. After Kube and Zhang [206]. Reprinted by permission © *MIT Press*.

that it is possible for a group of simple robots to perform a task cooperatively without any direct communication (indirect communication occurs through stigmergy), if they have a common goal and behave according to a "noninterference" principle (although the actions of an agent may influence another's subsequent actions, again through stimergy). The group of robots eventually moves the box to the periphery of the arena, in a direction that depends on the initial configuration. Both execution time and average success rate increase with group size up to a point (dependent on the respective sizes of the box, the arena and the robots), where interference takes over. Some limitations associated to this type of approach have also been highlighted, such as the stagnation problem (when there is no improvement in task performance for a long time) or the problem of deadlocks.

7.3.3 STAGNATION RECOVERY: MODEL AND ROBOTIC IMPLEMENTATION

In order to overcome the problem of stagnation, Kube and Zhang [207] have simulated mechanisms inspired by social insects: realignment and repositioning [302, 304]. Each robot exerts a force on the box side at a certain angle, producing a resultant force and torque applied to the center of the box: if the total resultant force is greater than a threshold, the box translates in the plane, while if the total torque exceeds a torque threshold, the box rotates. Stagnation here refers to any situation where a robot is in contact with the box and the box is not moving. Kube and Zhang [207] have implemented several recovery strategies; the various stagnation recovery behaviors each have a counter, which is reset each time the robot moves. If, for example, the box has not been moved during a time greater than the realignment time-out threshold, the corresponding behavior randomly changes the direction of the applied force. When realignment is not sufficient to move the box, the repositioning behavior is activated: the robot repositions randomly on the box.

Repositioning has the potential to induce larger changes in the applied force than realignment. Figure 7.10 shows the percentage of success in a simulated task with four different strategies as a function of group size: (1) no explicit stagnation recovery; (2) realignment only; (3) repositioning only; (4) realignment and repositioning. The task consists of moving the box 200 grid units from its initial position within 2000 simulation steps. It can be seen that stagnation recovery always gives better results; strategy (2) is best when group size is small while strategy (3) is significantly worse when the number of robots is small, but better as the number of robots increases; strategy (4) exhibits intermediate performance, as expected because realignment is activated before repositioning. Figure 7.11 shows the average number of time steps necessary to complete the task in the successful trials as a function of group size for the four different strategies. When the number of robots is smaller than 15, the strategy that involves no stagnation recovery converges more quickly *when it is successful*, while the mixed strategy (4) is slowest when group size is small but becomes both the fastest and the most reliable when group size increases; strategies (2) and (3) exhibit intermediate performance. It appears therefore that the mixed strategy, which combines realignment and repositioning, implements the best combination of reliability and efficiency, providing us with a possible explanation of why such a combination is used in the species of ants described by Sudd [302, 304].

Kube and Zhang [209] have developed a method, called task modeling, where multirobot tasks are described as series of steps, each step possibly consisting of substeps, to design a group of 11 homogeneous robots implementing cooperative box-pushing with stagnation recovery [209]. The actual implementation of the recovery behaviors, here backing off and repositioning, requires the addition of a touch sensor (a forward facing microswitch that provides on/off contact sensing): the robot backs away from objects contacted by its touch sensors, while repositioning moves the robot in a backward arc. Moreover, the task is now different from that of the previous section, since the box has to be pushed toward a goal and not simply to the periphery of the arena, requiring the addition of a goal sensor (a forward facing rotating phototransistor that detects the direction of the goal where the box should be pushed). Bascially, the task of finding a box to be pushed toward a goal can be divided into three subtasks: (1) find the box, (2) move to the box, and (3) push the box toward the goal. Subtask (3) is new with respect to previous section's system: if the box is located between the robot and the goal, then the robot pushes the box; otherwise, the robot steps back and repositions itself.

Although it is beyond the scope of this book to describe Kube and Zhang's [209] work in detail, let us mention that their robotic implementation seems to reproduce important features of real ants (Figure 7.12). Their videotaped experiments, available and worth viewing at http://www.cs.ualberta.ca/ kube/, clearly show the similarities between their swarm of robots and a group of ants engaged in collective transport. Despite the antagonistic forces present in ant group transport [240], the

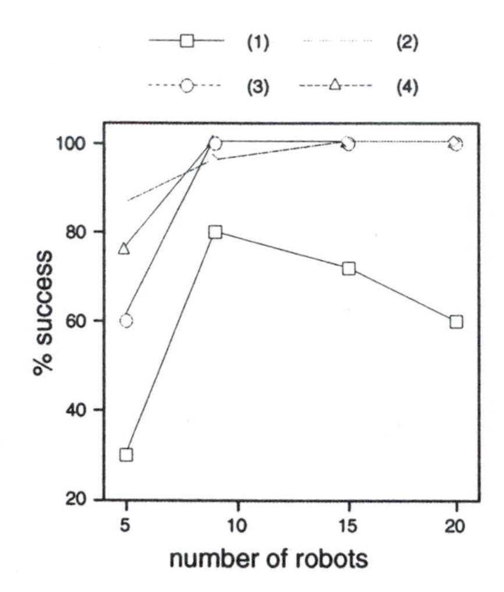

FIGURE 7.10 Percentage of success in a box-pushing task as a function of group size for four different stagnation recovery strategies: (1) no explicit stagnation recovery; (2) realignment only; (3) repositioning only; (4) realignment and repositioning. Percentage over 25 trials for each strategy. After Kube and Zhang [207]. Reprinted by permission © *IEEE*.

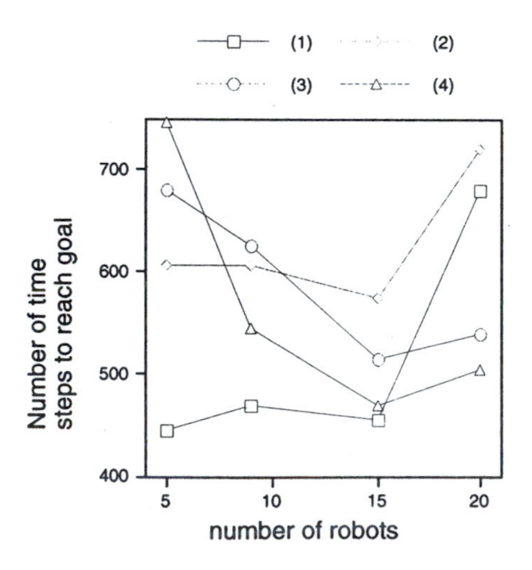

FIGURE 7.11 Average number of time steps necessary to complete the task in the successful trials as a function of group size for the four different stagnation recovery strategies: (1) no explicit stagnation recovery; (2) realignment only; (3) repositioning only; (4) realignment and repositioning. After Kube and Zhang [207]. Reprinted by permission © *IEEE*.

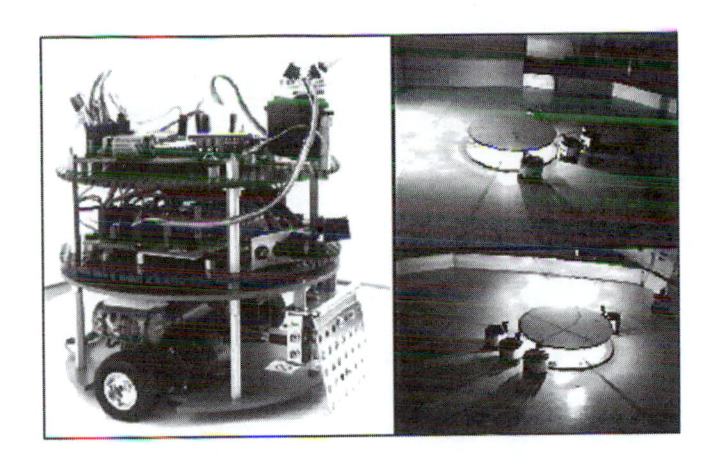

FIGURE 7.12 Picture of a robot (left), and two pictures showing a group of robots pushing a circular lit box toward a lit area [208]. Reprinted by permission.

item is eventually always taken back to the nest (unless it escapes!). In the swarm of robots, the path taken to the goal is certainly not optimal and often varies from experiment to experiment, and progress to the goal is not continuous. "There are even temporary setbacks as the box is moved incorrectly. At times, the robots can lose contact with the box, be blocked by other robots, or be forced to relocate as the box rotates ([209], p. 109). Using a centralized controller, one would likely obtain a better and more consistent performance. But the solution found by the swarm of robots is a feasible solution to the problem given the limited abilities of the individual robots. Let us reemphasize at this point that swarm-based robotics—a booming field indeed—despite being currently limited to solving ridiculously simple problems in toy worlds, must be considered in the perspective of miniaturization and low cost.

7.4 POINTS TO REMEMBER

In several species of ants, workers cooperate to retrieve large prey. Usually, one ant finds a prey item, tries to move it, and, when unsuccessful for some time, recruits nestmates through direct contact or chemical marking. When a group of ants tries to move large prey, the ants change position and alignment until the prey can be moved toward the nest. A robotic implementation of this phenomenon has been described. Although the robotic system may not appear to be very efficient, it is an interesting example of decentralized problem-solving by a group of robots, and it provides the first formalized *model* of cooperative transport in ants.

CHAPTER 8
Epilogue

After seven chapters of swarm-based approaches, where do we stand? First of all, it is clear that social insects and, more generally, natural systems, can bring much insight into the design of algorithms and artificial problem-solving systems. In particular, artificial swarm-intelligent systems are expected to exhibit the features that may have made social insects so successful in the biosphere: flexibility, robustness, decentralized control, and self-organization. The examples that have been described throughout this book provide illustrations of these features, either explicitly or implicitly. The swarm-based approach, therefore, looks promising, in face of a world that continually becomes more complex, dynamic, and overloaded with information than ever.

There remain some issues, however, as to the application of swarm intelligence to solving problems.

1. First, it would be very useful to define methodologies to "program" a swarm or multiagent system so that it performs a given task. There is a similarity here with the problem of training neural networks [167]: how can one tune interaction weights so that the network performs a given task, such as classification, recognition, etc. The fact that (potentially mobile) agents in a swarm can take actions asynchronously and at any spatial location generally makes the problem extremely hard. In order to solve this "inverse" problem and find the appropriate individual algorithm that generates the desired collective pattern, one can either

systematically explore the behaviors of billions of different swarms, or search this huge space of possible swarms with some kind of cost function, assuming a reasonable continuity of the mapping from individual algorithms to collective productions. This latter solution can be based, for example, on artificial evolutionary techniques such as genetic algorithms [152, 171] if individual behavior is adequately coded and if a cost function can be defined.

2. Second, and perhaps even more fundamental than the issue of programming the system, is that of defining it: How complex should individual agents be? Should they be all identical? Should they have the ability to learn? Should they be able to make logical inferences? Should they be purely reactive? How local should their knowledge of the environment be? Should they communicate directly? What information, if any, should they exchange? If they communicate indirectly, what should be the nature of that communication? And so forth. There is certainly no systematic way of answering these questions, which are problem specific. A reasonable approach consists of starting to solve a problem with the simplest possible agents and of making them progressively more complex if needed.
3. Last, swarm-intelligent systems face two major problems common to all adaptive problem-solving systems:

   (a) Their lack of absolute reliability. In particular, it is difficult to predict how such a system will behave when confronted with unexpected events, and the variance in the performance of an adaptive algorithm can be high, which means that the risk is high. This is one of the reasons why, for example, adaptive algorithms are trusted to a limited extent in the context of finance, where decisions are eventually most often made by human beings.
   (b) The lack of standard benchmarks that would allow the evaluation of their performance.

While point 3(a) can be "solved" by extensively exploring the behavior of a given adaptive system (for lack of formal proof of the system's behavior), point 3(b) requires the creation of a set of problems suited to the evaluation of adaptive algorithms in general, and swarm-intelligent systems in particular. A crucial feature of such algorithms is that they often work as poor-to-average heuristics in static problems (except in specific cases) and fully express their capabilities only in the context of dynamically changing problems. Designing benchmarks that take account of time is an interesting line of research, as it raises important questions about the very nature of adaptivity: What should one expect from an adaptive system? What are the natural time scales of a system's behavior, and how should they relate to the time scale of the problem's variations? How can one evaluate the performance of the system? etc. A classification of all possible types of "relevant" dynamics is also required: there are many potential ways of being dynamic, and many ways of being nonstationary, but it *may* be possible to show the existence of large classes of dynamics and nonstationarities that would have similar properties in terms of how difficult it is to solve the corresponding dynamic problem. Although a full body of theory is currently missing, the reward for answering these questions can be huge. One hint at how big the incentive is, is the example of ant-based routing in telecommunications networks: such networks are fundamentally dynamic and nonstationary; work on ant-based routing is only beginning, and not enough network traffic configurations have been tested, but, in all tested situations, it appears that, whatever the dynamics, ant-based routing with agents patrolling the network outperforms all other routing algorithms.

In view of these issues and promises, it is fair to conclude that, although artificial swarm intelligence is on the right track, a lot more work is needed to take advantage of its enormous potential. We hope that the present book will act as a catalyst, by showing the promise of this approach and stimulating researchers to overcome some of the identified issues.

Bibliography

[1] Aarts, E. H. L., and J. H. M. Korst. *Simulated Annealing and Boltzmann Machines.* New York, NY: Wiley & Sons, 1988.

[2] Adamatzky, A., and O. Holland. "Electricity, Chemicals, Ants, and Agents: A Spectrum of Swarm Based Techniques." Presented at ANTS '98—From Ant Colonies to Artificial Ants: First International Workshop on Ant Colony Optimization. IRIDA, Université Libre de Bruxelles, October 15–16, 1998.

[3] Appleby, S., and S. Steward. "Mobile Software Agents for Control in Telecommunications Networks." *British Telecom Technol. J.* **12** (1994): 104–113.

[4] Armstrong, W., A. Dwelly, J. Liang, D. Lin, and S. Reynolds. "Learning and Generalization in Adaptive Logic Networks, In Artificial Neural Networks." In *Proceedings 1991 International Conference on Artificial Neural Networks*, edited by T. Kohonen, K. Makisara, O. Simula, and J. Kangas. New York: Elsevier Science, 1991.

[5] Aron, S., J.-L. Deneubourg, S. Goss, and J. M. Pasteels. "Functional Self-Organisation Illustrated by Inter-Nest Traffic in the Argentine Ant *Iridomyrmex humilis*." In *Biological Motion*, edited by W. Alt and G. Hoffman, 533–547. Berlin: Springer-Verlag, 1990.

[6] Baluja, S., and R. Caruana. "Removing the Genetics From the Standard Genetic Algorithm." In *Proceedings Twelfth International Conference on Machine Learning, ML-95*, edited by A. Prieditis, and S. Russell, 38–46. Palo Alto, CA: Morgan Kaufmann, 1995.

[7] Barker, G., and M. Grimson. "The Physics of Muesli." *New Scientist* **126** (1990): 37–40.

[8] Barto, A. G., R. S. Sutton, and C. W. Anderson. "Neuronlike Adaptive Elements that Can Solve Difficult Learning Control Problems." *IEEE Transactions on Systems, Man, and Cybernetics* **13** (1983): 834–846.

[9] Battiti, R., and G. Tecchiolli. "The Reactive Tabu Search." *ORSA J. Computing* **6** (1994): 126–140.

[10] Bay, J. S. "Design of the Army-Ant Cooperative Lifting Robot." *IEEE Robotics and Automation Mag.* **2** (1995): 36–43.

[11] Beckers, R., J.-L. Deneubourg, and S. Goss. "Trails and U-Turns in the Selection of a Path by the Ant *Lasius niger*." *J. Theor. Biol.* **159** (1992): 397–415.

[12] Beckers, R., O. E. Holland, and J.-L. Deneubourg. "From Local Actions to Global Tasks: Stigmergy and Collective Robotics." In *Artificial Life IV*, edited by R. Brooks and P. Maes, 181–189. Cambridge, MA: MIT Press, 1994.

[13] Beni, G. "The Concept of Cellular Robotic System." In *Proceedings 1988 IEEE Int. Symp. on Intelligent Control*, 57–62. Los Alamitos, CA: IEEE Computer Society Press, 1988.

[14] Beni, G., and J. Wang. "Swarm Intelligence." In *Proceedings Seventh Annual Meeting of the Robotics Society of Japan*, 425–428. Tokyo: RSJ Press, 1989.

[15] Beni, G., and J. Wang. "Theoretical Problems for the Realization of Distributed Robotic Systems." In *Proceedings 1991 IEEE International Conference on Robotic and Automation*, 1914–1920. Los Alamitos, CA: IEEE Computer Society Press, 1991.

[16] Beni, G., and S. Hackwood. "Stationary Waves in Cyclic Swarms." In *Proceedings 1992 IEEE Int. Symp. on Intelligent Control*, 234–242. Los Alamitos, CA: IEEE Computer Society Press, 1992.

[17] Bennett, C. H. "Logical Depth and Physical Complexity." In *The Universal Turing Machine: A Half-Century Survey*, edited by R. Haken. Oxford: Oxford University Press, 1988.

[18] Bentley, P. J., and J. P. Wakefield. "Conceptual Evolutionary Design by Genetic Algorithms." *Eng. Design & Automation J.* **3** (1997): 119–131.

[19] Benzecri, J. P. *L'Analyse des Données. II. L'Analyse des Correspondances.* Paris: Dunod, 1973.

[20] Bersini, H., C. Oury, and M. Dorigo. "Hybridization of Genetic Algorithms." *Technical Report IRIDIA/95-22.* Belgium: Université Libre de Bruxelles, 1995.

[21] Bersini, H., M. Dorigo, S. Langerman, G. Seront, and L. M. Gambardella. "Results of the First International Contest on Evolutionary Optimization." In *Proceedings 1996 IEEE International Conference on Evolutionary Computation*, 611–615. Los Alamitos, CA: IEEE Computer Society Press, 1996.

[22] Bertsekas D. *Dynamic Programming and Optimal Control.* Bellmont, MA: Athena Scientific, 1995.

[23] Bertsekas, D., and R. Gallager. *Data Networks.* Englewood Cliffs, NJ: Prentice Hall, 1992.

[24] Bilchev, G., and I. C. Parmee. "The Ant Colony Metaphor for Searching Continuous Design Spaces." In *Proc. of AISB Workshop on Evolutionary Computing* Lecture Notes in Computer Science 993, edited by T. C. Fogarty, 25–39. Berlin: Springer-Verlag, 1995.

[25] Boender, C. G. E., A. H. G. Rinnooy Kan, L. Strougie, and G. T. Timmer. "A Stochastic Method for Global Optimization." *Math. Prog.* **22** (1982): 125–140.

[26] Bonabeau, E., G. Theraulaz, E. Arpin, and E. Sardet. "The Building Behavior of Lattice Swarms." In *Artificial Life IV*, edited by R. Brooks, and P. Maes, 307–312. Cambridge, MA: MIT Press, 1994.

[27] Bonabeau, E. "Marginally Stable Swarms are Flexible and Efficient." *J. Phys. I France* **6** (1996): 309–320.

[28] Bonabeau, E., and F. Cogne. "Oscillation-Enhanced Adaptability in the Vicinity of a Bifurcation: The Example of Foraging in Ants." In *Proceedings Fourth International Conference on Simulation of Adaptive Behavior: From Animals to Animats 4*, edited by P. Maes, M. Mataric, J.-A. Meyer, J. Pollack, and S. Wilson, 537–544. Cambridge, MA: MIT Press, 1996.

[29] Bonabeau, E., G. Theraulaz, and J.-L. Deneubourg. "Quantitative Study of the Fixed Threshold Model for the Regulation of Division of Labour in Insect Societies." *Proceedings Roy. Soc. London B* **263** (1996): 1565–1569.

[30] Bonabeau, E. "From Classical Models of Morphogenesis to Agent-Based Models of Pattern Formation." *Artificial Life* **3** (1997): 191–209.

[31] Bonabeau, E., G. Theraulaz, J.-L. Deneubourg, S. Aron, and S. Camazine. "Self-Organization in Social Insects." *Trends in Ecol. Evol.* **12** (1997): 188–193.

[32] Bonabeau, E., A. Sobkowski, G. Theraulaz, and J.-L. Deneubourg. "Adaptive Task Allocation Inspired by a Model of Division of Labor in Social Insects." In *Bio-Computation and Emergent Computing*, edited by D. Lundh, B. Olsson, and A. Narayanan, 36–45. Singapore: World Scientific, 1997.

[33] Bonabeau, E., F. Hénaux, S. Guérin, D. Snyers, P. Kuntz, and G. Theraulaz, G. . "Routing in Telecommunications Networks with Smart Ant-Like Agents. In *Proceedings Intelligent Agents for Telecommunications Applications, IATA'98*. Berlin, Springer-Verlag, 1998.

[34] Bonabeau, E., G. Theraulaz, and J.-L. Deneubourg. "Fixed Response Thresholds and the Regulation of Division of Labour in Insect Societies. *Bull. Math. Biol.* **60** (1998): 753–807.

[35] Bonabeau, E., G. Theraulaz, J.-L. Deneubourg, N. R. Franks, O. Rafelsberger, J.-L. Joly, and S. Blanco. "A Model for the Emergence of Pillars, Walls and Royal Chambers in Termite Nests." *Phil. Trans. Roy. Soc. London B* **353** (1998): 1561–1576.

[36] Bonabeau, E., S. Guérin, D. Snyers, P. Kuntz, G. Theraulaz, and F. Cogne. "Complex Three-Dimensional Architectures Grown by Simple Agents: An Exploration with a Genetic Algorithm." *Evol. Comp.* (1998): submitted.

[37] Bounds, D. G. "New Optimization Methods from Physics and Biology." *Nature* **329** (1987): 215–219.

[38] Bowden, N., A. Terfort, J. Carbeck, and G. M. Whitesides. "Self-Assembly of Mesoscale Objects Into Ordered Two-Dimensional Arrays." *Science* **276** (1997): 233–235.

[39] Boyan, J. A., and M. L. Littman. "Packet Routing in Dynamically Changing Networks: A Reinforcement Learning Approach." In *Proceedings Sixth Conference on Neural Information Processing Systems, NIPS-6*, 671–678. San Francisco, CA: Morgan Kaufmann, 1994.

[40] Branke, J., M. Middendorf, and F. Schneider. "Improved Heuristics and a Genetic Algorithm for Finding Short Supersequences." *OR-Spektrum* **20** (1998): 39–46.

[41] Breed, M. D., J. H. Fewell, A. J. Moore, and K. R. Williams. "Graded Recruitment in a Ponerine Ant." *Behav. Ecol. Sociobiol.* **20** (1987): 407–411.

[42] Breed, M. D., G. E. Robinson, and R. E. Page. "Division of Labor During Honey Bee Colony Defense." *Behav. Ecol. Sociobiol.* **27** (1990): 395–401.

[43] Brian, M. V. *Social Insects: Ecology and Behavioural Biology.* Chapman & Hall, 1983.

[44] Brooks, R. A. "A Robust Layered Control System For A Mobile Robot." *J. of Robotics and Automation* **2** (1986): 14–23.

[45] Brooks, R. A. "New Approaches to Robotics." *Science* **253** (1991): 1227–1232.

[46] Broughton, T., A. Tan, and P. S. Coates. "The Use of Genetic Programming in Exploring 3D Design Worlds." In *Proceedings of CAAD Futures*, edited by R. Junge. München: Kluwer Academic Publishers, 1997. (Related paper available at http://www.uel.ac.uk/faculties/arch/latest-draft/chapter12.html.)

[47] Bruckstein, A. M. "Why the Ant Trails Look So Straight and Nice." *The Mathematical Intelligencer* **15:2** (1993): 59–62.

[48] Bruinsma, O. H. "An Analysis of Building Behaviour of the Termite *Macrotermes subhyalinus* (Rambur)." Thesis, The Netherlands: Landbouwhoge School, Wageningen, 1979.

[49] Bullnheimer, B., R. F. Hartl, and C. Strauss. "A New Rank Based Version of the Ant System: A Computational Study." Working paper #1, SFB Adaptive Information Systems and Modelling in Economics and Management Science, Vienna, 1997.

[50] Bullnheimer, B., R. F. Hartl, and C. Strauss. "An Improved Ant System Algorithm for the Vehicle Routing Problem." POM Working Paper No. 10/97, University of Vienna, 1997.

[51] Burkard, R. E., S. Karish, and F. Rendl. "QAPLIB—A Quadratic Assignment Problem Library." *Eur. J. Oper. Res.* **55** (1991): 115–119.

[52] Burton, J. L., and N. R. Franks. "The Foraging Ecology of the Army Ant *Eciton rapax*: An Ergonomic Enigma?" *Ecol. Entomol.* **10** (1985): 131–141.

[53] Butrimenko, A. V. "On the Search for Optimal Routes in Changing Graphs." *Izv. Akad. Nauk SSSR Ser. Tekhn. Kibern.* **6** (1964).

[54] Calabi, P. "Behavioral Flexibility in Hymenoptera: A Re-Examination of the Concept of Caste." In *Advances in Myrmecology*, edited by J. C. Trager, 237–258. Leiden: Brill Press, 1988.

[55] Calderone, N. W., and R. E. Page. "Genotypic Variability in Age Polyethism and Task Specialization in the Honey Bee *Apis mellifera* (Hymenoptera: Apidae)." *Behav. Ecol. Sociobiol.* **22** (1988): 17–25.

[56] Calderone, N. W., and R. E. Page. "Temporal Polyethism and Behavioural Canalization in the Honey Bee, *Apis mellifera*." *Anim. Behav.* **51** (1996): 631–643.

[57] Caloud, P., W. Choi, J.-C. Latombe, C. Le Pape, and M. Yim. "Indoor Automation With Many Mobile Robots." In *Proceedings 1990 IEEE/RSJ International Conference on Intelligent Robots and Systems*, 67-72. Los Alamitos, CA: IEEE Computer Society Press, 1990.

[58] Camazine, S. "Self-Organizing Pattern-Formation on the Combs of Honey Bee Colonies." *Behav. Ecol. Sociobiol.* **28** (1991): 61–76.

[59] Camazine, S., and J. Sneyd. "A Model of Collective Nectar Source Selection by Honey Bees: Self-Organization Through Simple Rules." *J. Theor. Biol.* **149** (1991): 547–571.

[60] Camazine, S., J.-L. Deneubourg, N. R. Franks, J. Sneyd, G. Théraulaz, and E. Bonabeau. *Self-Organized Biological Superstructures.* Princeton, NJ: Princeton University Press, 1998.

[61] Cao, Y. U., A. S. Fukunaga, and A. B. Kahng. "Cooperative Mobilerobotics: Antecedents and Directions." In *Proceedings 1995 IEEE International Conference on Intelligent Robots and Systems*, 226–234. Los Alamitos, CA: IEEE Computer Society Press, 1995.

[62] Cao, Y. U., A. S. Fukunaga, and A. B. Kahng. "Cooperative Mobile Robotics: Antecedents and Directions." *Autonomous Robots* **4** (1997): 7–27.

[63] Ceusters, R. " Etude du degré de Couverture du ciel par la Végétation audessus des nids de Formica polyctena Foerst." *Biol. Ecol. Médit.* **VII (3)** (1980): 187–188.

[64] Ceusters, R. "Simulation du nid Naturel des Fourmis par des nids Artificiels Placés sur un Gradient de Température." *Actes Coll. Insect. Soc.* **3** (1986): 235–241.

[65] Chauvin, R. "Sur le transport collectif des proies par *Formica polyctena*." *Insectes Sociaux* **25** (1968): 193–200.

[66] Chauvin, R. "Les lois de l'ergonomie chez les fourmis au cours du transport d'objets." *C. R. Acad. Sc. Paris D* **273** (1971): 1862–1865.

[67] Chen, K. "A Simple Learning Algorithm for the Traveling Salesman Problem." *Phys. Rev. E* **55** (1997): 7809.

[68] Cherix, D. "Note Préliminaire sur la Structure, la Phénologie et le Régime Alimentaire d'une Super-colonie de *Formica lugubris* Zett." *Insectes Sociaux* **27** (1980): 226–236.

[69] Chirikjian, G. S. "Kinematics of a Metamorphic Robotic System." In *Proceedings 1994 IEEE International Conference on Robotics and Automation*, 449–455. Los Alamitos, CA: IEEE Computer Society Press, 1994.

[70] Chirikjian, G. S., A. Pamecha, and I. Ebert-Uphoff. "Evaluating Efficiency of Self-Rreconfiguration in a Class of Modular Robots." *J. Robotic Systems* **13** (1996): 317–338.

[71] Choi S. P. M., and D.-Y. Yeung. "Predictive Q-Routing: A Memory-Based Reinforcement Learning Approach to Adaptive Traffic Control." *Proceedings 8th Conference on Neural Information Processing Systems, NIPS-8*, 945–910. Cambridge, MA: MIT Press, 1996.

[72] Chrétien, L. "Organisation Spatiale du Matériel Provenant de l'excavation du nid chez Messor Barbarus et des Cadavres d'ouvrières chez *Lasius niger* (Hymenopterae: Formicidae)." Ph.D. dissertation, Université Libre de Bruxelles, 1996.

[73] Clearwater, S. H., and B. A. Huberman. "Thermal Markets for Controlling Building Environments." *Energy Eng.* **91** (1994): 26–56.

[74] Clearwater, S. H. *Market-Based Control: A Paradigm for Distributed Resource Allocation.* Singapore: World Scientific, 1995.

[75] Coates, P. S., N. Healy, C. Lamb, and W. L. Voon. "The Use of Ccellular Automata to Explore Bottom Up Architectonic Rules." Paper presented at Eurographics UK Chapter, 14th Annual Conference, 26–28 March 1996, Imperial College of Science, Technology and Medicine, London, 1996. (Paper available at http://www.uel.ac.uk/faculties/arch/paper-bits/paper.html.)

[76] Colorni, A., M. Dorigo, and V. Maniezzo. "Distributed Optimization by Ant Colonies." In *Proceedings First Europ. Conference on Artificial Life*, edited by F. Varela and P. Bourgine, 134–142. Cambridge, MA: MIT Press, 1991.

[77] Colorni, A., M. Dorigo, and V. Maniezzo. "An Investigation of Some Properties of An Ant Algorithm." In *Proceedings 1992 Parallel Problem Solving from Nature Conference*, edited by R. Männer and B. Manderick, 509–520. Amsterdam: Elsevier, 1992.

[78] Colorni, A., M. Dorigo, V. Maniezzo, and M. Trubian. "Ant System for Job-Shop Scheduling." *JORBEL—Belgian Journal of Operations Research, Statistics and Computer Science* **34** (1994): 39–53.

[79] Cook, S., and C. Rackoff. "Space Lower Bounds for Maze Threadability on Restricted Machines." *SIAM J. Comput.* **9** (1980): 636–652.

[80] Costa, D., A. Hertz, and O. Dubuis. "Embedding of a Sequential Algorithm Within an Evolutionary Algorithm for Coloring Problems in Graphs." *J. of Heuristics* **1** (1995): 105–128.

[81] Costa, D., and A. Hertz. "Ants Can Colour Graphs." *J. Op. Res. Soc.* **48** (1997): 295–305.

[82] Darchen, R. "Les Techniques de la Construction chez Apis mellifica." Ph.D. dissertation, Université de Paris, 1959.

[83] Deneubourg, J.-L. "Application de l'ordre par Fluctuations á la Description de Certaines étapes de la Construction du nid chez les Termites." *Insect. Soc.* **24** (1977): 117–130.

[84] Deneubourg, J.-L., S. Goss, J. M. Pasteels, D. Fresneau, and J.-P. Lachaud. "Self-Organization Mechanisms in Ant Societies (II): Learning in Foraging and Division of Labour." *Experientia Suppl.* **54** (1987): 177–196.

[85] Deneubourg, J.-L., and S. Goss. "Collective Patterns and Decision Making." *Ethol. Ecol. & Evol.* **1** (1989): 295–311.

[86] Deneubourg, J.-L., S. Goss, N. R. Franks, and J. M. Pasteels. "The Blind Leading the Blind: Modelling Chemically Mediated Army Ant Raid Patterns." *J. Insect Behav.* **2** (1989): 719-725.

[87] Deneubourg, J.-L., S. Aron, S. Goss, and J.-M. Pasteels. "The Self-Organizing Exploratory Pattern of the Argentine Ant." *J. Insect Behavior* **3** (1990): 159–168.

[88] Deneubourg, J.-L., S. Goss, N. Franks, A. Sendova-Franks, C. Detrain, and L. Chretien. "The Dynamics of Collective Sorting: Robot-Like Ant and Ant-Like Robot." In *Proceedings First Conference on Simulation of Adaptive Behavior: From Animals to Animats*, edited by J. A. Meyer and S. W. Wilson, 356–365. Cambridge, MA: MIT Press, 1991.

[89] Deneubourg, J.-L., G. Théraulaz, and R. Beckers. "Swarm-Made Architectures." In *Proceedings First European Conference on Artificial Life: Toward a Practice of Autonomous Systems*, edited by F. J. Varela and P. Bourgine, 123–133. Cambridge, MA: MIT Press, 1992.

[90] Detrain, C., J. M. Pasteels, and J.-L. Deneubourg. "Polyéthisme dans le tracé et le suivi de la piste chez *Pheidole pallidula* (Formicidae)." *Actes Coll. Ins. Soc.* **4** (1988): 87–94.

[91] Detrain, C., and J. M. Pasteels. "Caste Differences in Behavioral Thresholds as a Basis for Polyethism During Food Recruitment in the Ant *Pheidole pallidula* (Nyl.) (Hymenoptera: Myrmicinae)." *J. Ins. Behav.* **4** (1991): 157–176.

[92] Detrain, C., and J. M. Pasteels. "Caste Polyethism and Collective Defense in the Ant *Pheidole pallidula*: The Outcome of Quantitative Differences in Recruitment." *Behav. Ecol. Sociobiol.* **29** (1992): 405–412.

[93] Detrain, C., J.-L. Deneubourg. "Origine de la Diversité des Réponses Collectives des Fourmis lors de la récolte de proies." *Actes Coll. Ins. Soc.* **10** (1996): 57–65.

[94] Detrain, C., J.-L. Deneubourg. "Scavenging by *Pheidole pallidula*: A Key for Understanding Decision-Making Systems in Ants." *Anim. Behav.* **53** (1997): 537–547.

[95] Deveza, R., D. Thiel, A. Russell, and A. Mackay-Sim. "Odor Sensing for Robot Guidance." *Int. J. Robotics Res.* **13** (1994): 232–239.

[96] Di Caro, G., and M. Dorigo. "Mobile Agents for Adaptive Routing." In *Proceedings 31st Hawaii International Conference on System Sciences (HICSS-31)*, 74–83. Los Alamitos, CA: IEEE Computer Society Press, 1998.

[97] Di Caro, G., and M. Dorigo. "AntNet: Distributed Stigmergetic Control for Communications Networks." *J. Art. Int. Res.* **9** (1998): 317–365.

[98] Di Caro, G., and M. Dorigo. "Extending AntNet for Best Effort Quality-of-Service Routing." Unpublished presentation at the First Int. Workshop on Ant Colony Optimization: From Ant Colonies to Artificial Ants, ANTS'98, October 15–16, 1998, Brussels, Belgium.

[99] Di Caro, G., and M. Dorigo. "Ant Colonies for Adaptive Routing in Packet-Switched Communications Networks." In *Proceedings of PPSN V—Fifth International Conference on Parallel Problem Solving From Nature*, 673–682. Springer-Verlag, 1998.

[100] Di Caro, G., and M. Dorigo. "An Adaptive Multi-Agent Routing Algorithm Inspired by Ants Behavior." In *Proceedings of PART98—Fifth Annual Australasian Conference on Parallel and Real-Time Systems*, edited by K. A. Hawick and H. A. James, 261–272. Singapore: Springer-Verlag, 1998.

[101] Di Caro, G., and M. Dorigo. "Two Ant Colony Algorithms for Best-Effort Routing in Datagram Networks." In *Proceedings of the 10th IASTED International Conference on Parallel and Distributed Computing and Systems (PDCS'98)*, edited by Y. Pan, S. G. Akl, and K. li, 541–546. Anaheim, CA: IASTED/ACTA Press, 1998.

[102] Di Caro, G., and M. Dorigo. "Adaptive Learning of Routing Tables in Communication Networks." In *Proceedings of the Italian Workshop on Machine Learning (IWML-97)*, Torino, December 9–10, 1997.

[103] Di Caro, G., and M. Dorigo. "AntNet: A Mobile Agents Approach to Adaptive Routing." Technical Report IRIDIA/97-12, Université Libre de Bruxelles, Belgium, 1997.

[104] Di Caro, G., and M. Dorigo. "A Study of Distributed Stigmergetic Control for Packet-Switched Communitcations Networks." Technical Report IRIDIA/97-20, Université Libre de Bruxelles, Belgium, 1997.

[105] Di Caro, G., and M. Dorigo. "Distributed Reinforcement Agents for Adaptive Routing in Communication Networks." Third European Workshop on Reinforcement Learning (EWRL-3), Rennes, France, October 13–14, 1997.

[106] Donald, B., J. Jennings, and D. Rus. "Analyzing Teams of Cooperating Mobile Robots." In *Proceedings 1994 IEEE International Conference on Robotics and Automation*, 1896–1903. Los Alamitos, CA: IEEE Computer Society Press, 1994.

[107] Dorigo, M., V. Maniezzo, and A. Colorni. "Positive Feedback as a Search Strategy." Tech. Rep. No. 91-016, Politecnico di Milano, Italy, 1991.

[108] Dorigo, M. "Ottimizzazione, Apprendimento Automatico, ed Algoritmi Basati su Metafora Naturale." Ph.D. Dissertation, Politecnico di Milano, Italy, 1992.

[109] Dorigo, M., V. Maniezzo, and A. Colorni. "The Ant System: Optimization by a Colony of Cooperating Agents." *IEEE Trans. Syst. Man Cybern. B* **26** (1996): 29–41.

[110] Dorigo, M., and L. M. Gambardella. "A Study of Some Properties of Ant-Q." In *Proceedings Fourth International Conference on Parallel Problem Solving From Nature, PPSN IV*, 656–665. Berlin: Springer-Verlag, 1996.

[111] Dorigo, M., and L. M. Gambardella. "Ant Colony System: A Cooperative Learning Approach to the Traveling Salesman Problem." *IEEE Trans. Evol. Comp.* **1** (1997): 53–66.

[112] Dorigo, M., and L. M. Gambardella. "Ant Colonies for the Traveling Salesman Problem." *BioSystems* **43** (1997): 73–81.

[113] Doty, K. L., and R. E. van Aken. "Swarm Robot Materials Handling Paradigm for a Manufacturing Workcell." In *Proceedings 1993 IEEE International Conference on Robotics and Automation*, 778–782. Los Alamitos, CA: IEEE Computer Society Press, 1993.

[114] Downing, H. A., and R. L. Jeanne. "Nest Construction by the Paperwasp Polistes: A Test of Stigmergy Theory." *Anim. Behav.* **36** (1988): 1729–1739.

[115] Downing, H. A., and R. L. Jeanne. "The Regulation of Complex Building Behavior in the Paperwasp *Polistes fuscatus*." *Anim. Behav.* **39** (1990): 105–124.

[116] Dukas, R., and P. K. Visscher. "Lifetime Learning by Foraging Honey Bees." *Anim. Behav.* **48** (1994): 1007–1012.

[117] Durbin, R., and D. Willshaw. "An Analogue Approach to the Travelling Salesman Problem Using an Elastic Net Method." *Nature* **326** (1987): 689–691.

[118] Eggenberger, P. "Evolving Morphologies of Simulated 3D Organisms Based on Differential Gene Expression." In *Proceedings Fourth European Conference on Artificial Life*, edited by P. Husbands and I. Harvey, 206–213. Cambridge, MA: MIT Press, 1997.

[119] Eilon, S., C. D. T. Watson-Gandy, and N. Christofides. "Distribution Management: Mathematical Modeling and Practical Analysis." *Oper. Resh. Quart.* **20** (1969): 37–53.

[120] Escudero, L. F. "An Inexact Algorithm for the Sequential Ordering Problem." *Eur. J. Op. Res.* **37** (1988): 232–253.

[121] Farmer, J. D., N. H. Packard, and A. S. Perelson. "The Immune System, Adaptation, and Machine Learning." *Physica D* **22** (1986): 187–204.

[122] Feynman, R. P. *Surely You're Joking, Mr. Feynman!* Toronto: Bantam Books, 1985.

[123] Fleurent, C., and J. Ferland. "Genetic Hybrids for the Quadratic Assignment Problem." *DIMACS Series in Mathematics and Theoretical Computer Science* **16** (1994): 173–187.

[124] Fleurent, C., and J. Ferland. "Genetic and Hybrid Algorithms for Graph Coloring." *Ann. Oper. Res.* **63** (1996): 437–461.

[125] Fogel, D. B. "Applying Eevolutionary Programming to Selected Traveling Salesman Problems." *Cybern Syst.: Int. J.* **24** (1993): 27–36.

[126] Forrest, S. "Genetic Algorithms: Principles of Natural Selection Applied to Computation." *Science* **261** (1993): 872–878.

[127] Foulser, D. E., M. Li, and Q. Yang. "Theory and Agorithms for Plan Merging." *Artificial Intelligence* **57** (1992): 143–181.

[128] Franks, N. R. "Teams in Social Insects: Group Retrieval of Prey by Army Ants (*Eciton burchelli*, Hymenoptera: Formicidae)." *Behav. Ecol. Sociobiol.* **18** (1986): 425–429.

[129] Franks, N. R., N. Gomez, S. Goss, and J.-L. Deneubourg. "The Blind Leading the Blind in Army Ant Raid Patterns: Testing A Model of Self-Organization (Hymenoptera: Formicidae)." *J. Insect Behav.* **4** (1991): 583–607.

[130] Franks, N. R., and A. B. Sendova-Franks. "Brood Sorting by Ants: Distributing the Workload Over The Work Surface." *Behav. Ecol. Sociobiol.* **30** (1992): 109–123.

[131] Franks, N. R., A. Wilby, V. W. Silverman, and C. Tofts. "Self-Organizing Nest Construction in Ants: Sophisticated Building by Blind Buldozing." *Anim. Behav.* **44** (1992): 357–375.

[132] Franks, N. R., and J.-L. Deneubourg. "Self-Organizing Nest Construction in Ants: Individual Worker Behaviour and the Nest's Dynamics." *Anim. Behav.* **54** (1997): 779–796.

[133] Freisleben, B., and P. Merz. "New Genetic Local Search Operators for the Traveling Salesman Problem." In *Proceedings 4th International Conference Parallel Problem Solving from Nature, PPSN IV*, edited by H.-M. Voigt, W. Ebeling, I. Rechenberg, and H.-S. Schwefel, 890–899. Berlin: Springer-Verlag, 1996.

[134] Freisleben, B., and P. Merz. "Genetic Local Search Algorithm for Solving Symmetric and Asymmetric Traveling Salesman Problems." In *Proceedings IEEE Int. Conf. Evolutionary Computation*, 616–621. IEEE-EC96, 1996. Operators for the Traveling Salesman Problem."

[135] Frumhoff, P. C., and J. Baker. "A Genetic Component to Division of Labour Within Hhoney Bee Colonies." *Nature* **333** (1988): 358–361.

[136] Fukuda, T., S. Nakaggawa, Y. Kawauchi, and M. Buss. "Structure Decision Method for Self-Organizing Robots Based on Cell Structure—CEBOT." In *Proceedings IEEE International Conference on Robotics and Automation*, 695–700. Los Alamitos, CA: IEEE Computer Society Press, 1989.

[137] Funes, P., and J. Pollack. "Computer Evolution of Buildable Objects." In *Proceedings Fourth European Conference on Artificial Life*, edited by P. Husbands and I. Harvey, 358–367. Cambridge, MA: MIT Press, 1997.

[138] Gambardella, L. M., and M. Dorigo. "Ant-Q: A Reinforcement Learning Approach to the Traveling Salesman Pproblem." In *Proceedings Twelfth International Conference on Machine Learning, ML-95*, 252–260. Palo Alto, CA: Morgan Kaufmann, 1995.

[139] Gambardella, L. M., and M. Dorigo. "HAS-SOP: Hybrid Ant System for the Sequential Ordering Problem." Technical Report IDSIA 11-97, IDSIA, Lugano, Switzerland, 1997. To appear in *INFORMS J. Comp.*, 2000.

[140] Gambardella, L. M., E. D. Taillard, and M. Dorigo. "Ant Colonies for the QAP." Technical Report IDSIA 4-97, IDSIA, Lugano, Switzerland, 1997. Published in *J. Oper. Resh. Soc.* **50(2)** (1999): 167–176.

[141] Garbers, J., H. J. Promel, and A. Steger. "Finding Clusters in VLSI Circuits." In *Proceedings 1990 IEEE International Conference on Computer-Aided Design*, 520–523. Los Alamitos, CA: IEEE Computer Society Press, 1990.

[142] Garey, M. R., D. S. Johnson, and L. Stockmeyer. "Some Simplified NP-Complete Graph Problems." *Theor. Comp. Sci.* **1** (1976): 237–267.

[143] Garey, M. R., D. S. Johnson, and R. Sethi. "The Complexity of Flowshop and Jobshop Scheduling." *Math. Oper. Res.* **1** (1976): 117–129.

[144] Garey M. R., R. L. Graham, and D. S. Johnson. "The Complexity of Computing Steiner Minimal Trees." *SIAM J. Appl. Math.* **32** (1977): 835–859.

[145] Garey, M. R., and D. S. Johnson. *Computers and Intractability: A Guide to the Theory of NP-Completeness.* San Francisco, CA: W. H. Freeman, 1979.

[146] Garey, M. R., and D. S. Johnson. "Crossing Number is NP-Complete." *SIAM J. Alg. Disc. Meth.* **4** (1983): 312–316.

[147] Gaussier, P., and S. Zrehen. "Avoiding the World Model Trap: An Acting Robot Does Not Need to be Smart." *Robotics and Computer-Integrated Manufacturing* **11** (1994): 279–286.

[148] Gaussier, P., and S. Zrehen. "A Constructivist Approach for Autonomous Agents." In *Artificial Life and Virtual Reality*, edited by N. Magnenat Thalmann and D. Thalmann, 97–113. New York, NY: Wiley & Sons, 1994.

[149] Gilmore, P. "Optimal and Suboptimal Algorithms for the Quadratic Assignment Problem." *J. SIAM* **10** (1962): 305–313.

[150] Glover, F. "Tabu Search. Part I." *ORSA J. Computing* **1** (1989): 190–206.

[151] Glover, F. "Tabu Search. Part II." *ORSA J. Computing* **2** (1990): 4–32.

[152] Goldberg, D. E. *Genetic Algorithms in Search, Optimization and Machine Learning.* Reading, MA: Addison-Wesley, 1989.

[153] Gordon, D. M "The Organization of Work in Social Insect Colonies." *Nature* **380** (1996): 121–124.

[154] Goss, S., S. Aron, J.-L. Deneubourg, and J. M. Pasteels. "Self-Organized Shortcuts in the Argentine Ant." *Naturwissenchaften* **76** (1989): 579–581.

[155] Gotwald, W. H. "Death on the March." *Rotunda* **Fall/Winter** (1984/1985): 37–41.

[156] Graham, R. L., E. L. Lawler, J. K. Lenstra, and A. H. G. Rinnooy Kan. "Optimization and Approximation in Deterministic Sequencing and Scheduling: A Survey." *Ann. Disc. Math.* **5** (1979): 287–326.

[157] Grassé, P.-P. "La Reconstruction du nid et les Coordinations Inter-Individuelles chez *Bellicositermes Natalensis* et *Cubitermes sp.* La théorie de la Stigmergie: Essai d'interprétation du Comportement des Termites Constructeurs." *Insect. Soc.* **6** (1959): 41–80.

[158] Grassé, P.-P. "Termitologia, Tome II." *Fondation des Sociétés. Construction.* Paris: Masson, 1984.

[159] Guérin, S. "Optimisation Multi-Agents en Environnement Dynamique: Application au Routage dans les Réseaux de Télécommunications". DEA Dissertation, University of Rennes I, France, 1997.
[160] Gutowitz, H. "Complexity-Seeking Ants." (1993): Unpublished report.
[161] Hackwood, S., and G. Beni. "Self-Organizing Sensors by Deterministic Annealing." In *Proceedings 1991 IEEE/RSJ International Conference on Intelligent Robot and Systems, IROS'91*, 1177–1183. Los Alamitos, CA: IEEE Computer Society Press, 1991.
[162] Hackwood, S., and G. Beni. "Self-Organization of Sensors for Swarm Intelligence." In *Proceedings IEEE 1992 International Conference on Robotics and Automation*, 819–829. Los Alamitos, CA: IEEE Computer Society Press, 1992.
[163] Hagen, L., and A. B. Kahng. "New Spectral Methods for Ratio Cut Partitioning and Clustering." *IEEE Trans. Computer-Aided Design* **11** (1992): 1074–1085.
[164] Haken, H. *Synergetics*. Berlin: Springer-Verlag, 1983.
[165] Hansell, M. H. *Animal Architecture and Building Behavior*. London: Longman, 1984.
[166] Heinrich, B. "The Regulation of Temperature in the Honey Bee Swarm." *Sci. Am.* **244** (1981): 146–160.
[167] Hertz, J., A. Krogh, and R. G. Palmer. *Introduction to the Theory of Neural Computation*. Santa Fe Institute Studies in the Sciences of Complexity, Lect. Notes Vol. I Redwood City, CA: Addison-Wesley, 1991.
[168] Hertz, A., E. Taillard, and D. de Werra. "A Tutorial on Tabu Search." (1997): Unpublished.
[169] Heusse, M., S. Guérin, D. Snyers, and P. Kuntz. "Adaptive Agent-Driven Routing and Load Balancing in Communication Networks." Technical Report RR-98001-IASC, ENST Bretagne, Brest, France, 1998.
[170] Holden, C. "On the Scent of a Data Trail." *Science* **278** (1997): 1407.
[171] Holland, J. H. *Adaptation in Natural and Artificial Systems*. Ann Arbor, MI: The University of Michigan Press, 1975.
[172] Hölldobler, B., and E. O. Wilson. "The Multiple Recruitment Systems of the African Weaver ant *Oecophylla longinoda* (Latreille)." *Behav. Ecol. Sociobiol.* **3** (1978): 19–60.
[173] Hölldobler, B. "Territorial Behavior in the Green Tree Ant (*Oecophylla smaragdina*)." *Biotropica* **15** (1983): 241–250.
[174] Hölldobler, B., and E. O. Wilson. *The Ants*. Cambridge, MA: Harvard University Press, 1990.
[175] Hölldobler, B., R. C. Stanton, and H. Markl. "Recruitment and Food-Retrieving Behavior in *Novomessor* (Formicidae: Hymenoptera). I. Chemical Signals." *Behav. Ecol. Sociobiol.* **4** (1978): 163–181.
[176] Hosokawa, K., I. Shimoyama, and H. Miura. "Dynamics of Self-Assembling Systems. Analogy with Chemical Kinetics." *Artificial Life* **1** (1995): 413–427.

[177] Hosokawa, K., I. Shimoyama, and H. Miura. "Two-Dimensional Micro-Self-Assembly Using the Surface Tension of Water." *Sensors and Actuators A* **57** (1996): 117–125.

[178] Hosokawa, K., I. Shimoyama, and H. Miura. "Self-Assembling Microstructures." In *Artificial Life V*, edited by C. G. Langton and K. Shimohara, 362–369. Cambridge, MA: MIT Press, 1997.

[179] Hosokawa, K., T. Tsujimori, T. Fujii, H. Kaetsu, H. Asama, Y. Kuroda, and I. Endo. "Self-Organizing Collective Robots with Morphogenesis in a Vertical Plane." In *Proceedings 1998 IEEE International Conference Robotics and Automation*, 2858–2863. Los Alamitos, CA: IEEE Computer Society Press, 1998.

[180] Hosokawa, K., T. Tsujimori, T. Fujii, H. Kaetsu, H. Asama, Y. Kuroda, and I. Endo. "Mechanisms for Self-Organizing Robots Which Reconfigure in a Vertical Plane." In *Proceedings Distributed Autonomous Robotic Systems 3, DARS'98*, edited by T. Lueth, R. Dillmann, P. Dario, and H. Worn, 111–118. Berlin: Springer-Verlag, 1998.

[181] Huang, Z. Y., and G. E. Robinson. "Honey Bee Colony Integration: Worker-Worker Interactions Mediate Hormonally Regulated Plasticity." *Proceedings Nat. Acad. Sci. USA* **89** (1992): 11726–11729.

[182] Huberman, B. A., ed. *The Ecology of Computation.* Amsterdam, North-Holland: Elsevier, 1988.

[183] Huberman, B. A. "The Performance of Cooperative Processes." *Physica D* **42** (1990): 38–47.

[184] Huberman, B. A., and T. Hogg. "Distributed Computation as an Economic System." *J. Econom. Perspect.* **9** (1995): 141–152.

[185] Huberman, B. A., R. M. Lukose, and T. Hogg. "An Economics Approach to Hard Computational Problems." *Science* **275** (1997): 51–54.

[186] Jaisson, P., D. Fresneau, and J.-P. Lachaud. "Individual Traits of Social Behaviour in Ants." In *Interindividual Behavioral Variability in Social Insects*, edited by R. L. Jeanne, 1–51. Boulder, CO: Westview Press, 1988.

[187] Jeanne, R. L. "The Adaptativeness of Social Wasp Nest Architecture." *Quart. Rev. Biol.* **50** (1975): 267–287.

[188] Jeanne, R. L. "The Evolution of the Organization of Work in Social Insects." *Monit. Zool. Ital.* **20** (1986): 119–133.

[189] Jeanne, R. L. "Regulation of Nest Construction Behaviour in *Polybia occidentalis*." *Anim. Behav.* **52** (1996): 473–488.

[190] Johnson, R. A. "Learning, Memory, and Foraging Efficiency in Two Species of Desert Seed-Harvester Ants." *Ecology* **72** (1991): 1408–1419.

[191] Johnson, D. S., and L. A. McGeoch. "The Travelling Salesman Problem: A Case Study in Local Optimization." In *Local Search in Combinatorial Optimization*, edited by E. H. L. Aarts and J. K. Lenstra. New York, NY: Wiley & Sons, 1997.

[192] Kamada, T. *Visualizing Abstract Objects and Relations.* Singapore: World Scientific, 1989.

[193] Kamada, T., and S. Kawai. "An Algorithm for Drawing General Undirected Graphs." *Inf. Proceedings Lett.* **31** (1989): 7–15.

[194] Karsai, I., and Z. Penzes. "Comb Building in Social Wasps: Self-Organization and Stigmergic Script." *J. Theor. Biol.* **161** (1993): 505–525.

[195] Kelly, F. P. "Modelling Communication Networks, Present and Future." *Phil. Trans. R. Soc. London A* **354** (1995): 437–463.

[196] Kephart, J. O., T. Hogg, and B. A. Huberman. "Dynamics of Computational Ecosystems." *Phys. Rev. A* **40** (1989): 404–421.

[197] Khanna A., and J. Zinky. "The Revised ARPANET Routing Metric." In *ACM SIGCOMM Computer Communication Review* **19(4)** (1989): 45–56.

[198] Kim, E., Y. Xia, and G. M. Whitesides. "Polymer Microstructures Formed by Moulding in Capillaries." *Nature* **376** (1995): 581–584.

[199] Kirkpatrick, S., C. Gelatt, and M. Vecchi. "Optimization by Simulated Annealing." *Science* **220** (1983): 671–680.

[200] Kokaji, S., S. Murata, H. Kurokawa, and K. Tomita. "Clock Synchronization Algorithm for a Distributed Autonomous System." *J. Rob. Mechatronics* **8** (1996): 317–338.

[201] Koopmans, T. C., and M. J. Beckman. "Assignment Problems and the Location of Economic Activities." *Econometrica* **25** (1957): 53–76.

[202] Koza, J. R. *Genetic Programming. On the Programming of Computers by Means of Natural Selection.* Cambridge, MA: MIT Press, 1992.

[203] Krieger, M. Personal communication.

[204] Krink, T., and F. Vollrath. "Analysing Spider Web-Building Behaviour with Rule-Based Simulations, and Genetic Algorithms." *J. Theor. Biol.* **185** (1997): 321–331.

[205] Kube, C. R., and H. Zhang. "Collective Robotic Intelligence." In *Proceedings Second International Conference on Simulation of Adaptive Behavior: From Animals to Animats 2*, 460–468. Cambridge, MA: MIT Press, 1992.

[206] Kube, C. R., and H. Zhang. "Collective Robotics: From Social Insects to Robots." *Adaptive Behavior* **2** (1994): 189–218.

[207] Kube, R. C., and H. Zhang. "Stagnation Recovery Behaviors for Collective Robotics." In *Proceedings 1994 IEEE/RSJ/GI International Conference on Intelligent Robots and Systems*, 1883–1890. Los Alamitos, CA: IEEE Computer Society Press, 1995.

[208] Kube, C. R. "Collective Robotics: From Local Perception to Global Action." Ph.D. Thesis, University of Alberta, 1997.

[209] Kube, C. R., H. Zhang. "Task Modelling in Collective Robotics." *Auton. Robots* **4** (1997): 53–72.

[210] Kuntz, P., and P. Layzell. "A New Stochastic Approach to Find Clusters in Vertex Set of Large Graphs with Applications Partitioning in VLSI Technology." *Tech. Rep. LIASC*, Ecole Nationale Supérieure des Télécommunications de Bretagne, 1995.

[211] Kuntz, P., P. Layzell, and D. Snyers. "A Colony of Ant-Like Agents for Partitioning in VLSI Technology." In *Proceedings Fourth European Conference*

on Artificial Life, edited by P. Husbands and I. Harvey, 417–424. Cambridge, MA: MIT Press, 1997.

[212] Kurose, J. F., and R. Simha. "A Microeconomic Approach to Optimal Resource Allocation in Distributed Computer Systems." *IEEE Trans. Computers* **38** (1989): 705–717.

[213] Lawler, E. "The Quadratic Assignment Problem." *Management Science* **9** (1963): 586–599.

[214] Lawler, E. L., J. K. Lenstra, A. H. G. Rinnooy-Kan, and D. B. Shmoys, eds. *The Travelling Salesman Problem.* New York, NY: Wiley & Sons, 1985.

[215] Lefebvre, J. *Introduction aux Analyses Statistiques Multi-Dimensionnelles.* Paris: Masson, 1980

[216] Leighton, F. T. "A Graph Coloring Algorithm for Large Scheduling Problems." *Journal of Research of the National Bureau of Standards* **84** (1979): 489–505.

[217] Lenoir, A. "Le Comportement Alimentaire et la Division du Travail chez la Fourmi *Lasius niger*." *Bull. Biol. France & Belgique* **113** (1979): 79–314.

[218] Li Y., P. M. Pardalos, and M. G. C. Resende. "A Greedy Randomized Adaptive Search Procedure for the Quadratic Assignment Problem." *DIMACS Series in Discrete Mathematics and Theoretical Computer Science* **16** (1994): 237–261.

[219] Liggett, R. S. "The Quadratic Assignment Problem: An Experimental Evaluation of Solution Strategies." *Management Science* **27** (1981): 442–458.

[220] Lin, S. "Computer Solutions of the Traveling Salesman Problem." *Bell Syst. Journal* **44** (1965): 2245–2269.

[221] Lin, S., and B. W. Kernighan. "An Effective Heuristic Algorithm for the Traveling Salesman Problem." *Oper. Res.* **21** (1973): 498–516.

[222] Lin, F.-T., C.-Y. Kao, and C.-C. Hsu. "Applying the Genetic Approach to Simulated Annealing in Solving Some NP-Hard Problems." *IEEE Trans. Syst. Man Cybern.* **23** (1993): 1752–1767.

[223] Lindauer, M. "Ein Beitrag zur Frage der Arbeitsteilung im Bienenstaat." *Z. Vgl. Physiol.* **34** (1952): 299–345.

[224] Lindenmayer, A., and P. Prusinkiewicz. *The Algorithmic Beauty of Plants.* Berlin: Springer-Verlag, 1988.

[225] Lumer, E., and B. Faieta. "Diversity and Adaptation in Populations of Clustering Ants." In *Proceedings Third International Conference on Simulation of Adaptive Behavior: From Animals to Animats 3*, 499–508. Cambridge, MA: MIT Press, 1994.

[226] Lumer, E., and B. Faieta. "Exploratory Database Analysis via Self-Organization." (1995): Unpublished manuscript.

[227] Lüscher, M. "Der Lufterneuerung im Nest der Termite *Macrotermes natalensis* (Hav.)." *Insectes Sociaux* **3** (1956): 273–276.

[228] Lüscher, M. "Air-Conditioned Termite Nests." *Sci. Am.* **205** (1961): 138–145.

[229] Ma, Q., P. Steenkiste, and H. Zhang. "Routing High-Bandwidth Traffic in Max-Min Fair Share Networks." *ACM SIGCOMM Computer Communication Review* **26** (1996): 206–217.

[230] Maeterlinck, M. *The Life of the White Ant.* London: George Allen and Unwin, 1927.

[231] Maniezzo, V. "Exact and Approximate Nondeterministic Tree-Search Procedures for the Quadratic Assignment Problem." Research Report CSR 98-1, *Corso di Laurea in Scienze dell'Informazione*, Universitá di Bologna, Sede di Cesena, Italy, 1998.

[232] Maniezzo, V., and A. Colorni. "The Ant System Applied to the Quadratic Assignment Problem." *IEEE Trans. Knowledge and Data Engineering* **11(5)** (1998): 769–778.

[233] Maniezzo, V., A. Colorni, and M. Dorigo. "The Ant System applied to the Quadratic Assignment Problem." Tech. Rep. IRIDIA/94-28, Université Libre de Bruxelles, Belgium, 1994.

[234] Markl, H. and B. Hölldobler. "Recruitment and Food-Behavior in *Novomessor* (Formicidae, Hymenoptera). II. Vibration Signals." *Behav. Ecol. Sociobiol.* **4** (1978): 183-216.

[235] Martinoli, A., A. J. Ijspeert, and F. Mondada. "Understanding Collective Aggregation Mechanisms: From Probabilistic Modelling to Experiments with Real Robots." *Robotics and Autonomous Systems* **29** (1999): 51–63.

[236] Martinoli, A., E. Franzi, and O. Matthey. "Towards a Reliable Set-Up For Bio-Inspired Collective Experiments With Real Robots." In *Proceedings Fifth Int. Symp. on Experimental Robotics, ISER 97*, edited by A. Casals, and A. T. de Almeida, 597–608. Lecture Notes in Control and Information Sciences. Berlin: Springer-Verlag, 1997.

[237] Melhuish, C., O. Holland, and S. Hoddell. "Collective Sorting and Segregation in Robots with Minimal Sensing." (1998): Unpublished preprint.

[238] Meyer, J.-A. "Sur les lois Régissant l'Accomplissement d'une Tâche Collective Complexe Chez *Formica polyctena*." *Compte Rendus de l'Academie des Sciences Paris D* **270** (1970): 2111–2114.

[239] Michel, R., and M. Middendorf. "An Island Model Based Ant System with Lookahead for the Shortest Supersequence Problem." In *Proc.of PPSNV—Fifth Int. Conf. on Parallel Problem Solving from Nature*, 692–701. Springer-Verlag, 1998.

[240] Moffett, M. W. "Cooperative Food Transport By an Asiatic Ant." *National Geog. Res.* **4** (1988): 386–394.

[241] Mondada, F., E. Franzi, and P. Ienne. "Mobile Robot Miniaturization: A Tool for Investigation in Control Algorithms." *Proceedings of the Third International Symposium on Experimental Robotics ISER-93* (1993): 501–513.

[242] Morley, R. "Painting Trucks at General Motors: The Effectiveness of a Complexity-Based Approach." In *Embracing Complexity: Exploring the Application of Complex Adaptive Systems to Business*, 53–58. Cambridge, MA: The Ernst & Young Center for Business Innovation, 1996.

[243] Morley, R., and G. Ekberg. "Cases in Chaos: Complexity-Based Approaches to Manufacturing." In *Embracing Complexity: A Colloquium on the Application of Complex Adaptive Systems to Business*, 97–702. Cambridge, MA: The Ernst & Young Center for Business Innovation, 1998.

[244] Moy J. "Link-State Routing." In *Routing in Communications Networks*, edited by M. E. Steenstrup, 137–157. Englewood Cliffs, NJ: Prentice-Hall, 1995.

[245] Murata, S., H. Kurokawa, and S. Kokaji. "Self-Assembling Machines." In *Proceedings 1994 IEEE International Conference on Robotics and Automation*, 441–448. Los Alamitos, CA: IEEE Computer Society Press, 1994.

[246] Murciano, A., J. del R. Millan, and J. Zamora. "Specialization in Multi-Agent Systems Through Learning." *Biol. Cybern.* **76** (1997): 375–382.

[247] Murray, J. D. *Mathematical Biology.* New York, NY: Springer-Verlag, 1989.

[248] Nicolis, G., and I. Prigogine. *Self-Organization in Non-Equilibrium Systems.* New York, NY: Wiley & Sons, 1977.

[249] Nijhout, H. F., and S. M. Paulsen. "Developmental Models, and Polygenic Characters." *Am. Nat.* **149** (1997): 394–405.

[250] Noreils, F. R. "An Architecture for Cooperative and Autonomous Mobile Robots." In *Proceedings 1992 IEEE International Conference on Robotics and Automation*, 2703–2710. Los Alamitos, CA: IEEE Computer Society Press,

[251] Oprisan, S. A., V. Holban, and B. Moldoveanu. "Functional Self-Organization Performing Wide-Sense Stochastic Processes." *Phys. Lett. A* **216** (1996): 303–306.

[252] Oster, G. F. "Modeling Social Insect Populations. I. Ergonomics of Foraging and Population Growth in Bumblebees." *Am. Nat.* **110** (1976): 215–245.

[253] Oster, G., and E. O. Wilson. *Caste and Ecology in the Social Insects.* Princeton, NJ: Princeton University Press, 1978.

[254] Paessens, H. "The Savings Algorithm for the Vehicle Routing Problem." *European Journal of Operational Research* **34** (1988): 336–344.

[255] Page, R. E., and G. E. Robinson. "The Genetics of Division of Labour in Honey Bee Colonies." *Adv. Ins. Physiol.* **23** (1991): 117–169.

[256] Page, R. E. "The Evolution of Insect Societies." *Endeavour* **21** (1997): 114–120.

[257] Pamecha, A., C.-J. Chiang, D. Stein, and G. S. Chirikjian. "Design and Implementation of Metamorphic Robots." In *Proceedings 1996 ASME Design Engineering Technical Conference and Computers and Engineering Conference*, 1–10. New York, NY: ASME Press, 1996.

[258] Pamecha, A., I. Ebert-Uphoff, and G. S. Chirikjian. "Useful Metrics for Modular Robot Motion Planning." *IEEE Trans. Robot. Autom.* **13** (1997): 531–545.

[259] Pasteels, J. M., J.-L. Deneubourg, and S. Goss. "Self-Organization Mechanisms in Ant Societies (I): Trail Recruitment to Newly Discovered Food Sources." *Experientia Suppl.* **54** (1987): 155–175.

[260] Penrose, L. S. "Self-Reproducing Machines." *Sci. Am.* **200** (1959): 105–114.

[261] Plowright, R. C., and C. M. S. Plowright. "Elitism in Social Insects: A Positive Feedback Model." In *Interindividual Behavioral Variability in Social Insects*, edited by R. L. Jeanne, 419–431. Boulder, CO: Westview Press, 1988.

[262] Potvin, J.-Y. "The Traveling Salesman Problem: A Neural Network Perspective." *ORSA J. Comp.* **5** (1993): 328–347.

[263] Pratt, S. C. "Recruitment and Other Communication Behavior in the Ponerine Ant *Ectatomma ruidum*. *Ethology* **81** (1989): 313–331.

[264] Prescott, T. J., and C. Ibbotson. "A Robot Trace Maker: Modeling the Fossil Evidence of Early Investebrate Behavior." *Artificial Life* **3(4)** (1997): 289–306.

[265] Rabin, M. O. "The Choice Coordination Problem." *Acta Informatica* **17** (1982): 121–134.

[266] Räihä, K.-J., and E. Ukkonen. "The Shortest Common Supersequence Problem over Binary Alphabet is NP-Complete." *Theoret. Comp. Sci.* **16** (1981): 187–198.

[267] Reinelt, G. *The Traveling Salesman: Computational Solutions for TSP Applications.* Berlin: Springer-Verlag, 1994.

[268] Resnick, M. *Turtles, Termites, and Traffic Jams.* Cambridge, MA: MIT Press, 1994.

[269] Rettenmeyer, C. W. "Behavioral Studies of Army Ants." *Univ. Kans. Sci. Bull.* **44** (1963): 281–465.

[270] Robinson, G. E. "Modulation of Alarm Pheromone Perception in the Honey Bee: Evidence for Division of Labour Based on Hormonally Regulated Response Thresholds." *J. Comp. Physiol. A* **160** (1987): 613–619.

[271] Robinson, G. E., and R. E. Page. "Genetic Determination of Guarding and Undertaking in Honey Bee Colonies." *Nature* **333** (1988): 356–358.

[272] Robinson, G. E. "Regulation of Vision of Labor in Insect Societies." *Annu. Rev. Entomol.* **37** (1992): 637–665.

[273] Robinson, G. E., R. E. Page, and Z.-Y. Huang. "Temporal Polyethism in Social Insects is a Developmental Process." *Anim. Behav.* **48** (1994): 467–469.

[274] Robson, S. K., and J. F. A. Traniello."Resource Assesment, Recruitment Behavior, and Organization of Cooperative Prey Retrieval in the Ant *Formica schaufussi* (Hymenoptera: Formicidae)." *J. Insect Behav.* **11** (1998): 1–22.

[275] Robson, S. K., and J. F. A. Traniello. "Key individuals and the organisation of labor in ants." In *Information Processing in the Social Insects*, edited by C. Detrain, J.-L. Deneubourg, and J. M. Pasteels. Basil & Stuttgart, Germany: Birkhauser, in press.

[276] Rosenman, M. A. "A Growth Model for Form Generation Using a Hierarchical Evolutionary Approach" *Microcomp. in Civ. Engineer.* **11** (1996): 161–172.

[277] Rus, D., B. Donald, and J. Jennings. "Moving Furniture with Teams of Autonomous Robots." In *Proceedings IEEE/RSJ IROS*, 235–242. Los Alamitos, CA: IEEE Computer Society Press, 1995.

[278] Russell, R. A. "Mobile Robot Guidance Using a Short-Liived Heat Trail." *Robotica* **11** (1993): 427–431.

[279] Sahni, S., and T. Gonzales. "P-Complete Approximation Problems." *J. ACM* **23** (1976): 555–565.

[280] Sasaki, J., J. Ota, E. Yoshida, D. Kurabayashi, and T. Arai. "Cooperative Grasping of a Large Object by Multiple Mobile Robots." In *Proceedings 1995 IEEE International Conference on Robotics and Automation*, 1205–1210. Los Alamitos, CA: IEEE Computer Society Press, 1995.

[281] Schatz, B. "Modalités de la Recherche et de la Récolte Alimentaire chez la Fourmi *Ectatomma ruidum* Roger: Flexibilités Individuelle et Collective." Ph.D. dissertation, Université Paul Sabatier, Toulouse, France, 1997.

[282] Schneirla, T. C. "Army Ants." In *Army Ants: A Study in Social Organization*, edited by H. R. Topoff. San Francisco, CA: W. H. Freeman, 1971.

[283] Schoonderwoerd, R., O. Holland, J. Bruten, and L. Rothkrantz. "Ant-Based Load Balancing in Telecommunications Networks." *Adapt. Behav.* **5** (1996): 169–207.

[284] Scriabin, M., and R. C. Vergin. "Comparison of Computer Algorithms and Visual-Based Methods for Plant Layout." *Management Science* **22** (1975): 172–181.

[285] Seeley, T. D. "Adaptive Significance of the Age Polyethism Schedule in Honey Bee Colonies." *Behav. Ecol. Sociobiol.* **11** (1982): 287–293.

[286] Seeley, T. D. "The Tremble Dance of the Honey Bee: Message and Meanings." *Behav. Ecol. Sociobiol.* **31** (1992): 375–383.

[287] Seeley, T. D., S. Camazine, and J. Sneyd. "Collective Decision-Making in Honey Bees: How Colonies Choose Among Hectar Sources." *Behav. Ecol. Sociobiol.* **28** (1991): 277–290.

[288] Sendova-Franks, A. B., and N. R. Franks. "Social Resilience in Individual Worker Ants and Its Role in Division of Labour." *Proceedings Roy. Soc. London B* **256** (1994): 305–309.

[289] Sims, K. "Artificial Evolution for Computer Graphics." *Comp. Graph.* **25** (1991): 319–328.

[290] Sims, K. "Evolving 3D Morphology and Behavior by Ccompetition." In *Artificial Life IV*, edited by R. Brooks and P. Maes, 28–39. Cambridge, MA: MIT Press, 1994.

[291] Steenstrup, M. *Routing in Communications Networks.* Englewood Cliffs, NJ: Prentice Hall, 1995.

[292] Stefanyuk, V. L. "On Mutual Assistance in the Collective of Radiostations." *Information Transmission Problems* **7** (1971): 103–107.

[293] Steinberg, M. S. "On The Mechanism of Tissue Reconstruction by Dissociated Cells. III. Free Energy Relationships and the Reorganization of Fused, Heteronomic Tissue Fragments." *Proceedings Nat. Acad. Sci. USA* **48** (1962): 1769–1776.

[294] Steinberg, M. S. "Reconstruction of Tissues by Dissociated Cells." *Science* **141** (1963): 401–408.

[295] Steinberg, M. S., and T. J. Poole. "Strategies for Specifying Form and Pattern: Adhesion-Guided Multicellular Assembly." *Phil. Trans. R. Soc. Lond. B* **295** (1981): 451–460.

[296] Stilwell, D. J., and J. S. Bay. "Toward the Development of a Material Transport System Using Swarms of Ant-Like Robots." In *Proceedings 1993 IEEE International Conference on Robotics and Automation*, 766–771. Los Alamitos, CA: IEEE Computer Society Press, 1993.

[297] Stilwell, D. J., and J. S. Bay. "Optimal Control for Cooperating Mobile Robots Bearing a Common Load." In *Proceedings 1994 IEEE International Conference on Robotics and Automation.* Los Alamitos, CA: IEEE Computer Society Press, 1994.

[298] Stützle, T., and H. Hoos. "The MAX-MIN Ant System and Local Search for the Traveling Salesman Problem." In *Proceedings IEEE International Conference on Evolutionary Computation, ICEC'97*, 309–314. Los Alamitos, CA: IEEE Computer Society Press, 1997.

[299] Stützle, T., and H. Hoos. "Improvements on the Ant System: Introducing MAX-MIN Ant System." In *Proceedings International Conference on Artificial Neural Networks and Genetic Algorithms.* Vienna: Springer-Verlag, 1997.

[300] Stützle, T., and H. Hoos. "MAX-MIN Ant System and Local Search for Combinatorial Optimization Problems." In *Proceedings Second International Conference on Metaheuristics, MIC'97.* Dordrecht: Kluwer Academic, 1998.

[301] Subramanian, D.,P. Druschel, and J. Chen. "Ants and Reinforcement Learning: A Case Study in Routing in Dynamic Networks." In *Proceedings 1997 International Joint Conference on Artificial Intelligence, IJCAI-97*, 832–838. Palo Alto, CA: Morgan Kaufmann, 1997.

[302] Sudd, J. H. "The Transport of Prey by an Ant *Pheidole crassinoda.*" *Behaviour* **16** (1960): 295–308.

[303] Sudd, J. H. "How Insects Work in Groups." *Discovery* **24** (1963): 15–19.

[304] Sudd, J. H. "The Transport of Prey by Ant." *Behaviour* **25** (1965): 234–271.

[305] Sugawara, K., and M. Sano. "Cooperative Acceleration of Task Performance: Foraging Behavior of Interacting Multi-Robot Systems." *Physica D* **100** (1997): 343–354.

[306] Taillard, E. "Robust Taboo Search for the Quadratic Assignment Problem." *Parallel Computing* **17** (1991): 443–455.

[307] Tamassia, R., G. Battista, and C. Battini. "Automatic Graph Drawing and Readability of Diagrams." *IEEE Trans. Syst. Man Cybern.* **18** (1988): 61–79.

[308] Théraulaz G., S. Goss, J. Gervet, and J.-L. Deneubourg. "Task Differentiation in Polistes Wasp Colonies: A Model for Self-Organizing Groups of Robots." In *Proceedings First International Conference on Simulation of Adaptive Behavior: From Animals to Animats*, edited by J.-A. Meyer and S. W. Wilson, 346–355. Cambridge, MA: MIT Press, 1991.

[309] Théraulaz, G., J. Gervet, and S. Semenoff-Tian-Chansky. "Social Regulation of Foraging Activities in *Polistes dominulus* Christ: A Systemic Approach to Behavioural Organization." *Behaviour* **116** (1992): 292–320.

[310] Théraulaz, G., and E. Bonabeau. "Coordination in Distributed Building." *Science* **269** (1995): 686–688.

[311] Théraulaz, G., and E. Bonabeau. "Modelling the Collective Building of Complex Architectures in Social Insects with Lattice Swarms." *J. Theor. Biol.* **177** (1995): 381–400.

[312] Théraulaz, G., E. Bonabeau, and J.-L. Deneubourg. "Threshold Reinforcement and The Regulation of Division of Labour in Insect Societies." *Proceedings Roy. Soc. London B* **265** (1998): 327–335.

[313] Todd, S., and W. Latham. *Evolutionary Art and Computers.* New York, NY: Academic Press, 1992.

[314] Tofts, C., and N. R. Franks. "Doing the Right Thing: Ants, Honey Bees and Naked Mole-Rats." *Trends Ecol. Evol.* **7** (1992): 346–349.

[315] Tomita, K., S. Murata, E. Yoshida, H. Kurokawa, and S. Kokaji. "Reconfiguration Method for a Distributed Mechanical System." In *Distributed Autonomous Robotic Systems 2 (DARS)*, edited by H. Asama, T. Fukuda, T.,Arai, and I. Endo. Tokyo: Springer-Verlag, 1996.

[316] Traniello, J. F. A. "Social Organization and Foraging Success in *Lasius neoniger* (Hymenoptera: Formicidae): Behavioral and Ecological Aspects of Recruitment Communication." *Oecologia* **59** (1983): 94–100.

[317] Traniello, J. F. A., and S. N. Beshers. "Maximization of Foraging Efficiency and Resource Defense By Group Retrieval in the Ant *Formica schaufussi.*" *Behav. Ecol. Sociobiol.* **29** (1991): 283–289.

[318] Tsetlin, M. L. *Automaton Theory and Modelling of Biological Systems.* New York, NY: Academic Press, 1973.

[319] Van Damme, T., and J.-L. Deneubourg. "Cooperative Transport in *Formica polyctena.*" Submitted.

[320] Van der Blom, J. "Individual Differentiation in Behaviour of Honey Bee Workers (*Apis mellifera* L.)." *Insect. Soc.* **40** (1993): 345–361.

[321] Waldspurger, C. A., T. Hogg, B. A. Huberman, and J. O. Kephart. "Spawn: A Distributed Computational Economy." *IEEE Trans. Softw. Engineer.* **18** (1992): 103–117.

[322] Wang, Z.-D., E. Nakano, and T. Matsukawa. "Cooperating Multiple Behavior-Based Robots for Object Manipulation." In *IEEE/RSJ 1994 International Conference on Intelligent Robot and Systems, IROS'94*, 1524–1531. Los Alamitos, CA: IEEE Computer Society Press, 1994.

[323] Watkins, C. J. C. H. "Learning with Delayed Rewards." Ph.D. dissertation, Psychology Department, Univ. of Cambridge, UK, 1989.

[324] Wenzel, J. W. "Evolution of Nest Architecture." In *Social Biology of Wasps*, edited by K. G. Ross and R. W. Matthews, 480–521. Ithaca, NY: Cornell University Press, 1991.

[325] White, T., B. Pagurek, and F. Oppacher. "Connection Management Using Adaptive Mobile Agents." In *Proc. Int. Conf. on Parallel Distributed Processing Techniques and Applications (PDPTA'98)*, 802–809. CSREA Press, 1998.

[326] Whitesides, G. M., J. P. Mathias, and C. T. Seto. "Molecular Self-Assembly and Nanochemistry: A Chemical Strategy for The Synthesis of Nanostructures." *Science* **254** (1991): 1312–1319.

[327] Whitley, D., T. Starkweather, and D. Fuquay. "Scheduling Problems and Travelling Salesman: The Genetic Edge Recombination Operator." In *Proceedings Third International Conference on Genetic Algorithms*, 133–140. Palo Alto, CA: Morgan Kaufmann, 1989.

[328] Wilson, E. O. *The Insect Societies.* Cambridge, MA: Harvard University Press, 1971.

[329] Wilson, E. O. *Sociobiology.* Cambridge, MA: Harvard University Press, 1975.

[330] Wilson, E. O. "The Relation Between Caste Ratios and Division of Labour in the Ant Genus Pheidole (Hymenoptera: Formicidae)." *Behav. Ecol. Sociobiol.* **16** (1984): 89–98.

[331] Withers, G. S., S. E. Fahrbach, and G. E. Robinson. "Selective Neuroanatomical Plasticity and Division of Labour in the Honey Bee." *Nature* **364** (1993): 238–240.

[332] Wodrich, M. "Ant Colony Optimization." B.Sc. Thesis, Department of Electrical and Electronic Engineering, University of Cape Town, South Africa, 1996.

[333] Wojtusiak, J., E. J. Godzinska, and A. Dejean. "Capture and Retrieval of Very Large Prey By Workers of the African Weaver Ant *Oecophylla longinoda.*" *Tropical Zool.* **8** (1995): 309–318.

[334] Yim, M. "A Reconfigurable Robot with Many Modes of Locomotion." In *Proceedings 1993 JSME International Conference on Advanced Mechatronics*, 283–288. Tokyo: JSME Press, 1993.

[335] Yim, M. "New Locomotion Gaits." In *Proceedings 1994 IEEE International Conference on Robotics and Automation*, 1508–1524. Los Alamitos, CA: IEEE Computer Society Press,

[336] Yoshida, E., S. Murata, K. Tomita, H. Kurokawa, and S. Kokaji. "Distributed Formation Control for a Modular Mechanical System." In *Proceedings IEEE/RSJ International Conference on Intelligent Robot and Systems, IROS'97.* Los Alamitos, CA: IEEE Computer Society Press, 1997.

[337] Yoshida, E., S. Murata, K. Tomita, H. Kurokawa, and S. Kokaji. "Experiment of Self-Repairing Modular Machine." *Proceedings Distributed Autonomous Robotic Systems 3, DARS'98*, edited by T. Lueth, R. Dillmann, P. Dario, and H. Worn, 119–128. Berlin: Springer-Verlag, 1998.

Index